AF904657

NEW ACHIEVEMENTS IN GEOSCIENCE

Edited by **Hwee-San Lim**

New Achievements in Geoscience
Edited by Hwee-San Lim

Published by InTech
Janeza Trdine 9, 51000 Rijeka, Croatia

Copyright © 2012 InTech
All chapters are distributed under the Creative Commons Attribution 3.0 license, which allows users to download, copy and build upon published articles even for commercial purposes, as long as the author and publisher are properly credited, which ensures maximum dissemination and a wider impact of our publications. After this work has been published by InTech, authors have the right to republish it, in whole or part, in any publication of which they are the author, and to make other personal use of the work. Any republication, referencing or personal use of the work must explicitly identify the original source.

As for readers, this license allows users to download, copy and build upon published chapters even for commercial purposes, as long as the author and publisher are properly credited, which ensures maximum dissemination and a wider impact of our publications.

Notice
Statements and opinions expressed in the chapters are these of the individual contributors and not necessarily those of the editors or publisher. No responsibility is accepted for the accuracy of information contained in the published chapters. The publisher assumes no responsibility for any damage or injury to persons or property arising out of the use of any materials, instructions, methods or ideas contained in the book.

Publishing Process Manager Ana Skalamera

Technical Editor Teodora Smiljanic

Cover Designer InTech Design Team

Additional hard copies can be obtained from orders@intechopen.com

New Achievements in Geoscience, Edited by Hwee-San Lim
p. cm.
ISBN 978-953-51-0263-2

We are IntechOpen,
the world's leading publisher of
Open Access books
Built by scientists, for scientists

4,100+
Open access books available

116,000+
International authors and editors

120M+
Downloads

151
Countries delivered to

Our authors are among the
Top 1%
most cited scientists

12.2%
Contributors from top 500 universities

CLARIVATE ANALYTICS
BOOK CITATION INDEX
INDEXED

WEB OF SCIENCE™

Selection of our books indexed in the Book Citation Index
in Web of Science™ Core Collection (BKCI)

Interested in publishing with us?
Contact book.department@intechopen.com

Numbers displayed above are based on latest data collected.
For more information visit www.intechopen.com

For EU product safety concerns:
IN TECH d.o.o., Prolaz Marije Krucifikse Kozulić 3, 51000 Rijeka, Croatia,
info@intechopen.com or visit our website at intechopen.com.

Meet the editor

Lim Hwee San is a lecturer at the school of physics, University Sains Malaysia, teaching courses in meteorology, geophysics and remote sensing. He obtained his B.Sc. from USM in Geophysics in 2001 and M.Sc. from USM in remote sensing in 2003. He obtained his Ph.D in environmental remote sensing from USM in 2006. His research interests are generally in the field of optical remote sensing - both passive and active - and digital image processing, particularly as it applies to spectral image data. In both cases, the primary applications are water quality monitoring, air quality monitoring, land cover/change detection, land surface properties and digital images classification. He has published more than 200 articles in these areas. He is a Member of the International Society for Optical Engineering, US.

Contents

Preface XI

Chapter 1 **Seismic Reflection Contribution to the Study of the Jerid Complexe Terminal Aquifer (Tunisia) 1**
Rihab Guellala, Mohamed Hédi Inoubli, Lahmaidi Moumni and Taher Zouaghi

Chapter 2 **New Techniques for Potential Field Data Interpretation 21**
Khalid S. Essa

Chapter 3 **Geophysics in Near-Surface Investigations 45**
Jadwiga A. Jarzyna, Jerzy Dec, Jerzy Karczewski, Sławomir Porzucek, Sylwia Tomecka-Suchoń, Anna Wojas and Jerzy Ziętek

Chapter 4 **Magnetotelluric Tensor Decomposition: Insights from Linear Algebra and Mohr Diagrams 81**
F.E.M. (Ted) Lilley

Chapter 5 **Heat and SO_2 Emission Rates at Active Volcanoes – The Case Study of Masaya, Nicaragua 107**
Letizia Spampinato and Giuseppe Salerno

Chapter 6 **What Caused the Ice Ages? 131**
Willy Woelfli and Walter Baltensperger

Chapter 7 **Origin of HED Meteorites from the Spalling of Mercury – Implications for the Formation and Composition of the Inner Planets 153**
Anne M. Hofmeister and Robert E. Criss

Chapter 8 **The Eruptions of Sarychev Peak Volcano, Kurile Arc: Particularities of Activity and Influence on the Environment 179**
Alexander Rybin, Nadezhda Razjigaeva, Artem Degterev, Kirill Ganzey and Marina Chibisova

Chapter 9 **Asymmetric Continuum Theories – Fracture Processes in Seismology and Extreme Fluid Dynamics** **199**
Teisseyre Roman

Preface

Geoscience is the scientific field of science covering all related disciplines dealing with Earth and its systems. The scientific field includes geology, physical geography, geophysics, geodesy, soil science, oceanography, hydrology, glaciology and atmospheric sciences. New Achievements in Geoscience is a comprehensive, up-to-date resource for academic researchers in geophysics, environmental science, earth science, natural resource managements and their related support fields. This book attempts to highlight issues dealing with geophysical and earth sciences. It describes the research carried out by world-class scientists in the fields of geoscience.

The content of the book includes selected chapters covering seismic interpretation, potential field data interpretation and several chapters on earth science. It also includs an introductory chapter on new techniques for potential field data interpretation. This book also includes a few chapters on volcanic activity research. The knowledge and skills in regards of Howardite, Eucrite, and Diogenite meteorites (HED) originating in deep spalling of Mercury - Implications for composition and differentiation of the inner planets also described fully in this book. Finally, I'm sure that this book will provide precious opportunities for all the readers or researchers a new stage for earth science research for the benefit of the entire population. The direct contributions from the authors are sincerely appreciated by the editors, as well as the effort by Ana Skalamera for her coordinating to this book.

Hwee-San Lim
School of Physics,
University Sains Malaysia,
Malaysia

Seismic Reflection Contribution to the Study of the Jerid Complexe Terminal Aquifer (Tunisia)

Rihab Guellala[1*], Mohamed Hédi Inoubli[2], Lahmaidi Moumni[3] and Taher Zouaghi[1]
[1]*Laboratoire de Géoressources, CERTE, Pôle Technologique de Borj Cédria,*
[2]*Département des Sciences de la Terre, FST, Université Tunis El Manar,*
[3]*Arrondissement des Ressources en Eaux de Tozeur,*
Tunisia

1. Introduction

The North African Sahara is characterized by the immense aquifer system of the «Complexe Terminal» (Fig.1) covering 655.000 Km^2 of the Algerian- Tunisian-Libyan domain (UNESCO, 1972c; OSS, 2003). The aquifer thickness is on average 340 m and its reserves are estimated of 11.000 10^9 m^3 (Ould Baba Sy, 2005).

According to Kilan (1931) the term « Continental Terminal » concerns the sandy and clayey continental Formations dated Miocene-Pliocene. In 1966, Bel & Demargne highlighted a vertical communication between the aquifer contained in these Formations and the Eocene, Senonian and Turonian aquifers. Consequently, the Continetal Terminal is redefined as a multi-layered aquifer which extends from the Late Cretaceous to the Miocene –Pliocene. Frequently, this hydrogeological system is designated by the term «Complexe Terminal» proposed by Bel & Cuche in 1969.

The «Complexe Terminal» aquifer has been exploited since the XIX th century (Jus, 1890). The drilled wells provided much information about the aquifer. They encouraged the launching of various hydrogeological studies (Cornet, 1964 ; Ricolvi,1970; UNESCO, 1972a ; Mekrazi, 1975 ; Ben Salah & Lessi,1978 ; Levassor, 1978; Ben Baccar, 1982 ; Castany , 1982; PNUD, 1983 ; ARMINES & ENIT, 1984 ; Pizzi & Sartori, 1984 ; Besbès & Zammouri, 1985 ; Mamou, 1990; Zammouri, 1990 ; BRL, 1998; Swezey, 1999 ; Guendouz et al., 2003 ; OSS, 2003 ; Chalbaoui, 2005 ; Ould Baba Sy, 2005 ; Kamel et al., 2006; Guellala, 2010; Guellala et al., 2011).

The Jerid area (Fig.2), located in the Southwestern Tunisia is an arid region where the pluviometry doesn't exceed 200 mm/year. The strong needs in water supply for domestic needs and irrigation render necessary the exploitation of the underground water reserves.

The «Complexe terminal» appears as a potential resource able to provide interesting flows. However, former geological and hydrogeological studies in this region were not sufficient to propose zones and strategies for the exploitation of this resource. Tectonic and sedimentary phenomena and their impact on the aquifer functioning had not been elucidated.

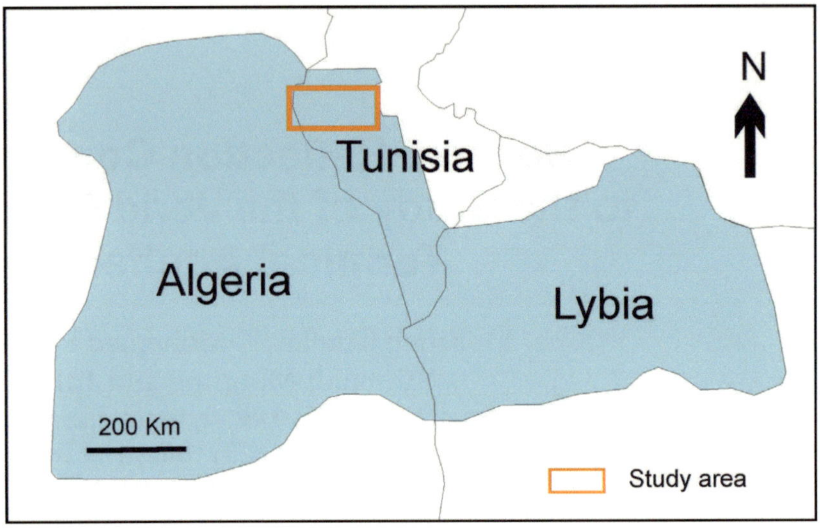

Fig. 1. Extent of the Complexe Terminal aquifer (OSS, 2003, modified).

The aim of this study is therefore to precise the deposits structures in the Jerid area in view to guarantee a good knowledge of the «Complexe Terminal» geometry and an accurate estimation of the relations between the different hydrogeological units.

Usually, the aquifers prospection is the privileged application of the electrical method (Gasmi, 2002 ; Koussoubé et al., 2003; Zouhri et al., 2004; Gouasmia et al., 2006; Asfahani, 2007; Guellala et al., 2005; 2009a; 2009b; 2010; Tizro et al., 2010). In this study, the important depth of the «Complexe Terminal » aquifer (>500 m) incites the use of the seismic method associated to the deep wells data (Jaffal et al., 2002; Zouhri et al. ,2003 ;Larroque & Dupuy 2004; Sumanovac, 2006; Saidane et al., 2008 ; Guellala et al, 2008 ; 2011; Lachaal et al., 2011).

2. Geological context

Part of the Maghreb, Tunisia is characterized by two different geological domains: the folded and faulted Atlas in the north and the stable saharan platform in the south (Caire,1971; Aissaoui, 1984; Addoum, 1995; Jallouli & Mickus, 2000; Bouaziz et al., 2002; Gabtni et al., 2005; Frizon de Lamotte, 2006 ; Missenard, 2006 ; Rigane & Gourmelen, 20011).

The Jerid area occupies an intermediate position between these domains. The anticlinal structures of Draa Jerid and Sidi Bouhlel, situated between El Gharsa Chott and Jerid Chott (Fig.2), constitute the western extent of the Chotts fold belt (Fakraoui, 1990; Zouaghi et al., 2011), which corresponds to the most southern structures of the Atlassic domain (Abdeljaoued,1983; Rabia ,1984 ; Zargouni, 1985; Abbes & Zargouni, 1986; Fakraoui, 1990; Ben Ayed, 1993; Boukadi, 1994; Zouari, 1995; Bouaziz, 1995; Bédir, 1995; Hlaiem, 1999; Zouaghi et al., 2005; Lazzez et al.,2008).

The «Complexe Terminal» Formations ranging from Late Cretaceous to Miocene -Pliocene in age are largely outcropped in the Jerid area (Fig.3).They are characterized by different facies indicating the combined action of continental and marine domains.

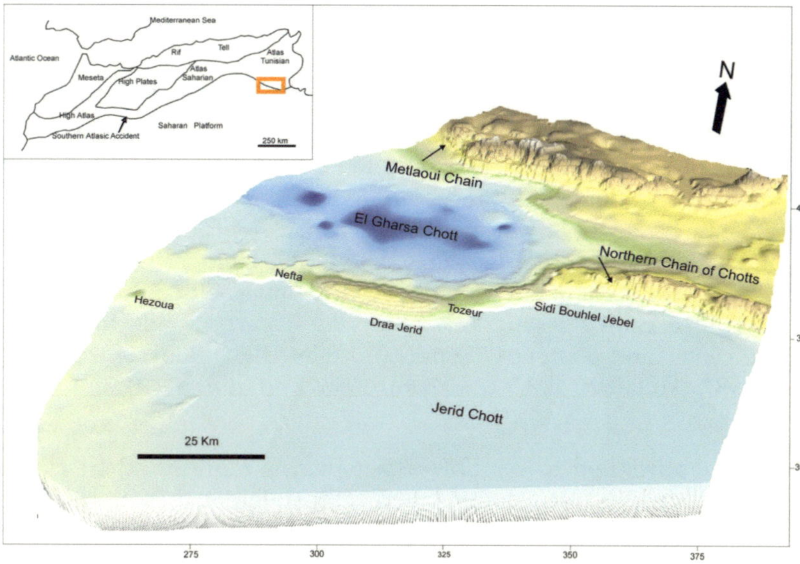

Fig. 2. Study area setting.

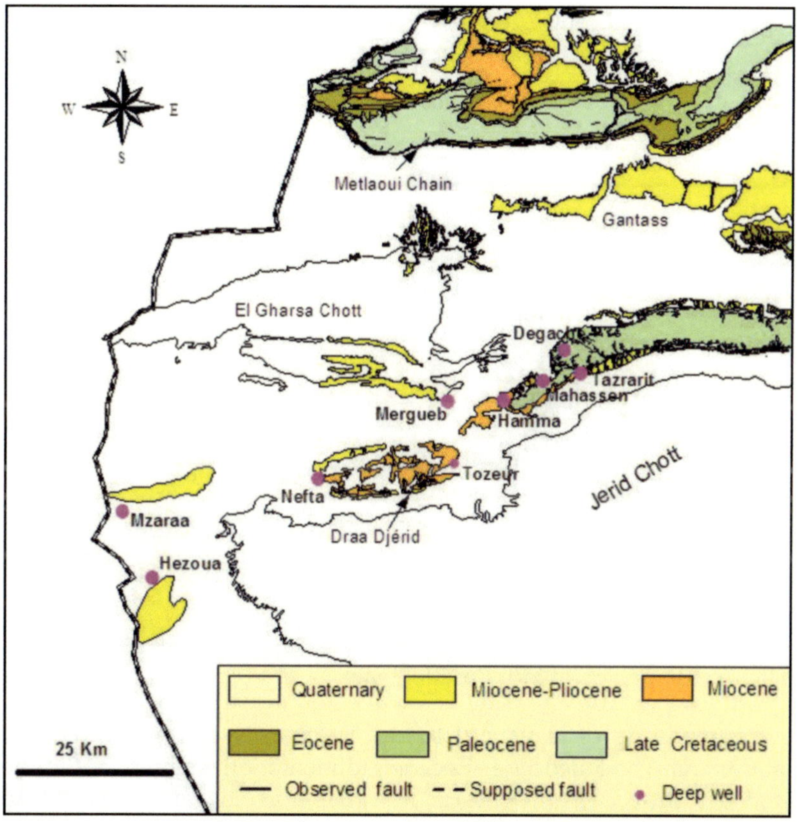

Fig. 3. Geological map of the Jerid area (Fakraoui & Mahjoub, 1995; Mahjoub, 1995; Regaya et al., 2001).

The Late Cretaceous is represented by the lithostratigraphic Formations: Zebbag, Aleg and Berda.

The Zebbag Formation dated Late Albian -Cenomanian -Turonian is recognized from deep well data (Fig.3). It is subdivided in three members. The lower member (Late Albian - Cenomanian), thick on average 150 m, is constituted by dolomites and dolomitic limestones. The Middle member (Cenomanian) is composed of marls, clays, gypsums and thin dolomitic limestones beds. Its power exceeds 600 m in Mergueb and Tazrarit. It varies between 300 and 400 m in the other localities. The upper member (Turonian) is formed of fractured limestones and dolomitic limestones. This member with a thickness ranging from 130 to 300 m is an excellent stratigraphic marker. It corresponds to the «Gattar bar» described in the center and southern Tunisia (Burollet, 1956; Fournié, 1978; Boltenhagen, 1985; M'Rabet, 1987; Abdallah, 1987; Chaabani, 1995; Zouari et al., 1990; Negra, 1994; Ben Youssef, 1999).

The Aleg Formation attributed to the Lower Senonian is represented by clays and marls with limestone and gypsum intercalations. It reveals different thickness: 400 to 470 m in Mergueb, Hezoua and Nefta, 310 to 350 in Degache, Tazrarit and Tozeur and 220 to 280 m in Mahassen, Mzaraa and Hamma. The Aleg Formation constitutes the Sidi Bouhlel anticline core (Fakraoui, 1990).

The Berda Formation dated Late Senonian is marked by friable limestones intercalated by marly beds. It is thick of 120 to 380 m with remarkable thinning towards Sidi Bouhlel Jebel. Mahassen well and Degache well implanted on this structure are drilled on the Berda Formation outcrops (Fig.4).

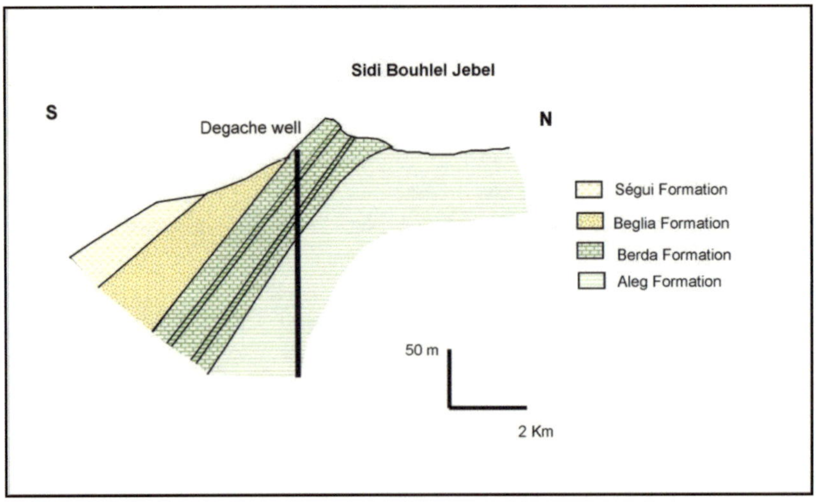

Fig. 4. Geological cross-section at Sidi Bouhlel Jebel.

The Paleocene is made up of clays and marls. The Eocene is represented by limestones, phosphates and marls. The Paleocene -Eocene sedimentation is absent at Sidi Bouhlel Jebel. It is recognized in Hezoua, Mzaraa, Nefta, Mergueb and Tozeur and its thickness doesn't reach 300 m.

The Paleocene -Eocene exposures characterize Metlaoui chain, northern border of the Jerid.

The Miocene –Pliocene largely outcropped in the Jerid area is characterized by continental sedimentation represented by Beglia Formation (Tayech-Mannai & Otera, 2005; Tayech-Mannai , 2006;2009; Swezy,2009) and Segui Formation .

The Beglia Formation (Miocene), thick on average 100 m, is made up of fine to coarse sands with thin clayey intercalations. At Sidi Bouhlel Jebel, this Formation is underlain by the Senonian deposits (Fig.4).

The Segui Formation (Miocene-Pliocene) is essentially clayey. It is enriched in sands and this thickness decreases towards the eastern part of the Jerid. Its power attains 500m in Mzaraa well implanted in the western part (Fig.3).

3. Hydrogeological context

On the basis of the preceding descriptions, the fractured limestones and dolomitic limestones of the Upper member of Zebbag Formation (Turonian), the friable limestones of Berda Formation (Upper Senonian) and the sands of Beglia Formation (Miocene) are the Jerid «Complexe Terminal» reservoirs (Fig.5).The clayey and marly deposits of the Lower and Middle members of Zebbag Formation, Aleg Formation, El Haria Formation and Segui Formation are aquicludes (Fig.5).

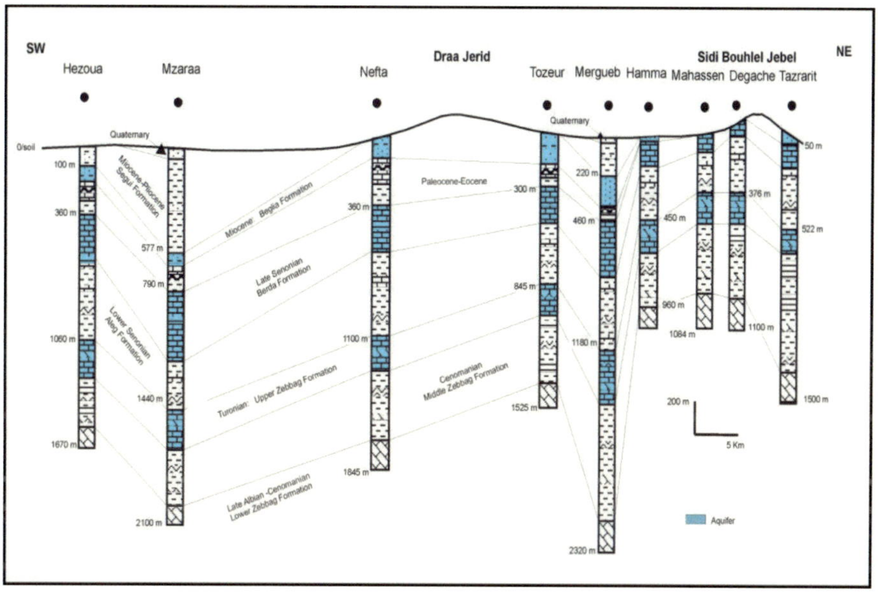

Fig. 5. Deep wells correlation.

The deep wells correlation (Fig.5) shows notable variation of the Complexe Terminal aquifer depth. It is characterized by raised and subsided zones (Fig.5). The Beglia Formation, the most superficial reservoir, is encountered at 570 m in Mzaraa well, at 100 m in Hezoua well and at 220 m in Mergueb well. It outcrops in many localities along Draa Jerid and Sidi Bouhlel structures. The Berda Formation reservoir is reached at 790 m in Mzaraa and about 350 m in Hezoua, Nefta and Tozeur. The Upper Zebbag Formation, the deepest reservoir, is encountered at 370 m in Mahassen and Degache. Its depth exceeds 1000 m in Hezoua, Mzaraa, Nefta and Mergueb.

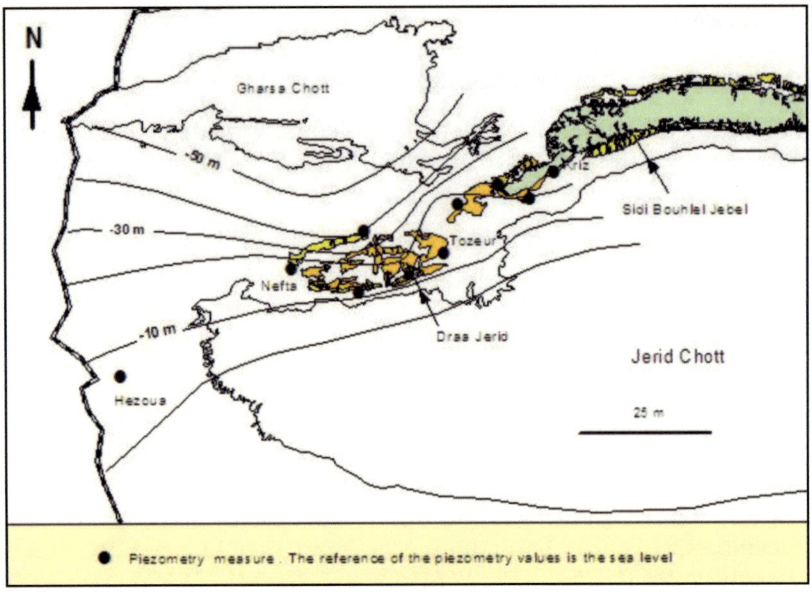

Fig. 6. Piezometric map of the Beglia aquifer (2008).

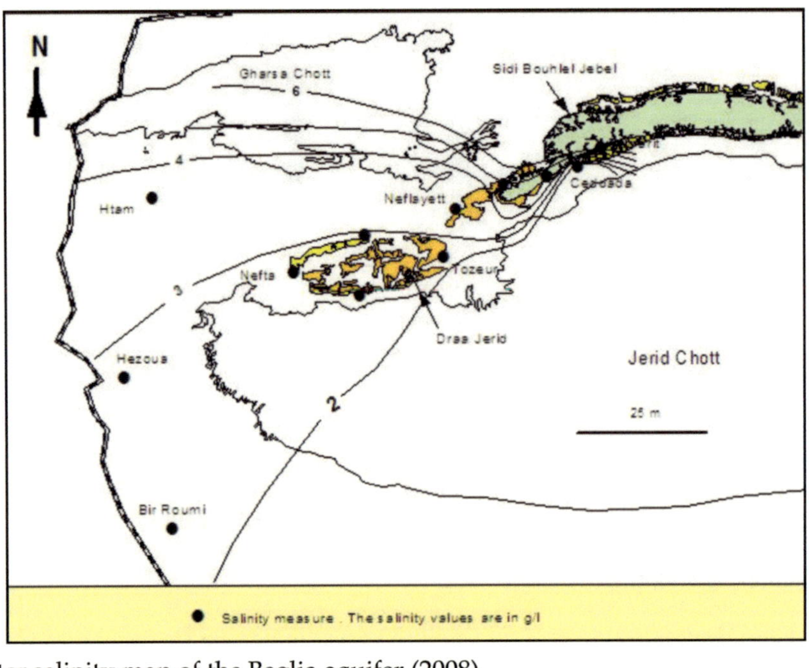

Fig. 7. Water salinity map of the Beglia aquifer (2008).

In the Jerid area, the catchment of the «Complexe Terminal» groundwater is restricted to the Beglia Formation. Piezometric measures (General Direction of Water Resources, 2008), contoured in the form of map (Fig.6) show a piezometry decrease towards the nord. It is -7/0 sea at Hezoua. It drops to -50 m/0 sea in the southern part of Gharsa Chott. Therefore, south to north is the main groundwater flow direction.

It is interesting to note the packed piezometric contours at the north flank of the Draa Jerid structure; the piezometry falls abruptly to -40 m/0 sea.

The water salinity measured for eleven groundwater samples collected from Beglia aquifer ranges from 2, 2 g/l to 5.9 g/l. The salinity map (Fig.7) reveals a clear increase of values from south to north which coincides with the groundwater flow direction. Additionally, this map exposes high values in the eastern part of the Jerid area, at Sidi Bouhlel Jebel: 3.7 g/l in Hamma well, 4.5 g/l in Kriz well and 5.9 g/l in Tazrarit well. Nearby, Ceddada well expresses a value of 2g/l. This sudden salinity change may indicate an obstructed lateral communication between the aquifers.

4. Data and methodology

The present study is based on seismic reflection sections associated to deep wells (Fig.8). These data are provided by the "General Direction of Water Resources" in Tunisia.

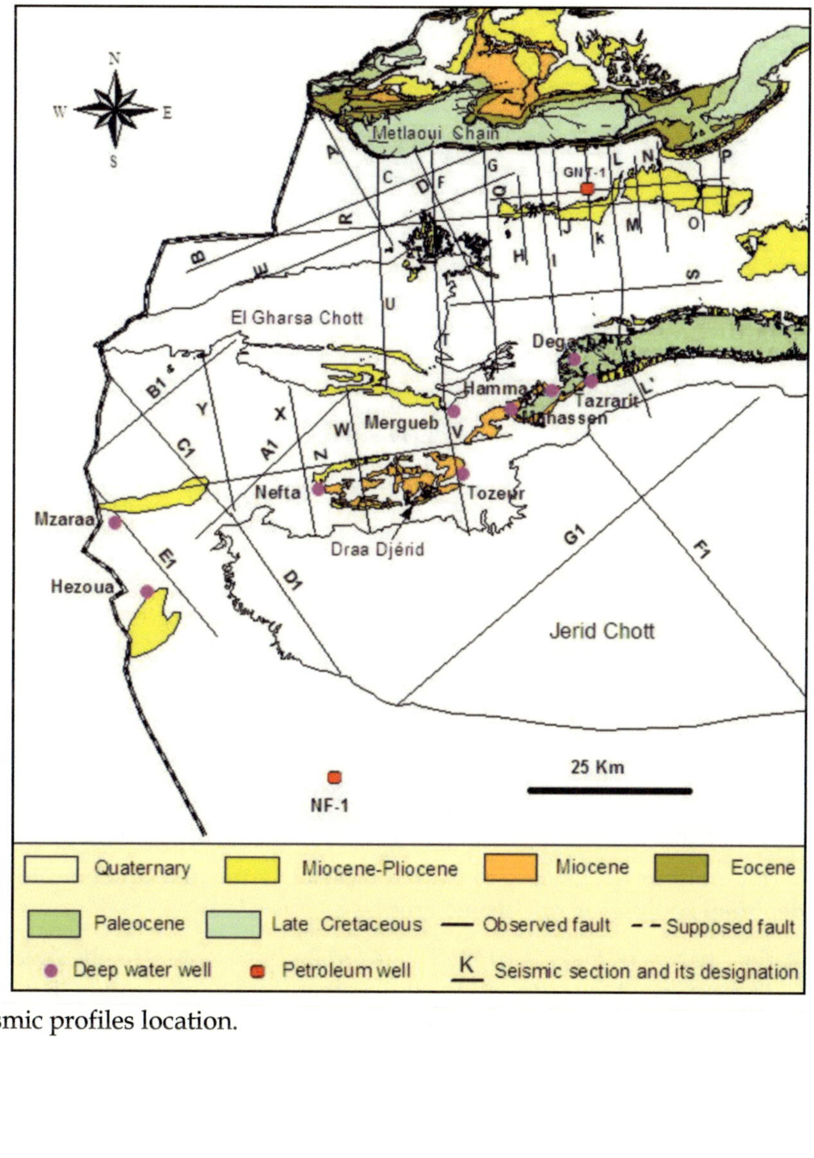

Fig. 8. Seismic profiles location.

The seismic reflection (Lavergne, 1986) is a method of exploration geophysics which estimates the properties of the Earth's subsurface from reflected seismic waves (Cagniard , 1962 ; Telford et al., 1976). It requires a controlled seismic source of energy, such as seismic vibrator (Vibroseis) or dynamite explosion.

Seismic waves will be reflected when they encounter a boundary between two different materials with different acoustic impedances. They are detected using seismometers ; On land, the typical seismometer is the geophone. In water, hydrophone is used.

The recorded signals are plotted on a seismic section after significant amounts of processing (Gardner, 1985; Inoubli, 1993; Henry, 1997; Cox, 1999; Inoubli & Mechler, 1999; Robein, 1999; Mari et al., 2001; Upadhyay, 2004): demultiplexing, filtring, deconvolution, velocity analysis, stacking, migration…

A seismic section resembles a geological cross-section, but the vertical axis is in time, rather than depth. It still needs to be interpreted.

In this study, 31 seismic profiles (Fig.8) covering the Jerid area are interpreted .Realized by oil industry (Tab.1) during four seismic surveys using different parameters of acquisition and processing (Tab.1), these profiles show variable quality.

Survey	**MT**	**MTB**	**GSB**	**MET**
Date	1973	1974	1982	1989
Operator	MOBIL	MOBIL	AGIP	SCHELL
Seismic source	Weight dropping	Dynamite	Dynamite	Vibroseis
Shot interval (m)	150	125	150	25
Distance between geophones (m)	150	125	75	25
Bandpass-filters (Hz)	3/8 -30/40	5/10 -32/42	8/12 - 50/57 7/10 - 40/46 6/8 - 32/37	8-75 8-45
Display for interpretation	Stack	Stack	Stack	Migrated

Table 1. Seismic reflection data.

The Miocene sands (Beglia Formation), the Upper Senonian (Berda Formation) and the Turonian (Upper Zebbag Formation) limestones which are the main reservoirs of the «Complexe Terminal» groundwater in the Jerid area constitute good seismic markers. The abrupt lithological change between these deposits and the clays and marly Formations originates a strong acoustic impedance contrast.

Seismic calibration (Fig.9) was performed using GNT-1 and to a lesser extent NF-1 petroleum wells. Seismic horizons and facies have been tied using the relation time-depth.

The different reflectors corresponding to the «Complexe Terminal » aquifers and aquicludes have been identified and picked on seismic profiles all over the area.

The interpolation between the interpreted seismic sections leads to the isochron maps construction. These maps display the tectonic structures which affect the « Jerid Complexe Terminal » aquifer. Their analysis specifies the reservoirs geometry.

The geoseismic cross sections, integrating wells data and seismic interpretation, clarify the relations between the different hydrogeological units. They allow the comprehension of the Jerid« Complexe Terminal» aquifer functioning.

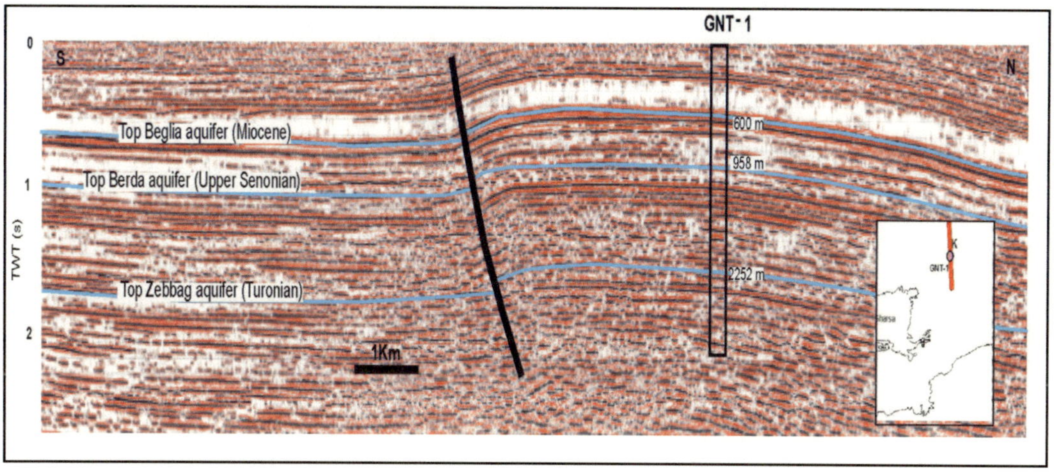

Fig. 9. Seismic profile «**K**» calibrated with the «**GNT-1**» petroleum well.

5. Results and discussions

The isochron map (Fig.10) of the botton of the Beglia Formation (Miocene) reflects the main deformations characterizing the «Complexe Terminal» multilayered aquifer in the Jerid area. In the center, it reveals narrow anticlinals, overfolded towards the south (Fig.10). These structures situated between 200 and 300 ms extend from Sidi Bouhlel Jebel in the east to Hezoua region in the west. They allow the elongation of the Chotts fold belt, most southern Atlassic structures, until the Tunisian-Algerian boundary.

Major reverse faults oriented E-W to NE-SW, mark the limit between anticlinal structures and two large syncline basins highlighted along the northwestern border of Jerid Chott and in the south of Gharsa Chott (Fig.11), where the depth of the Beglia aquifer bottom reach respectively 800 and 600 ms.

The northern basin named «Jerid basin» widens and deepens towards the West (Fig.10). It is limited by a raised structure (400 to 500 ms) characterizing the southern part of Gharsa Chott. Towards the north, this structure evolves to a depression deep of 1600 ms.

Geodynamic studies (Ben Ayed , 1980 ;Ben Ayed & Viguier, 1981; Zargouni et al. ,1985; Fakraoui ,1990; Zouari, 1995; Bouaziz, 1995; Zitouni et al., 1997; Boutib & Zargouni,1998; Zouaghi et al., 2011) describe three main directions: NW - SE, E-W and NE-SW for the accidents which guided the formation of the Atlas folds attributed to the Tortonian compression. These accidents are reactivated in reverse faults during the post-Villafranchian compressive episode, responsible of the Chotts chain actual structure (Fakraoui, 1990).

In this study, the majority of tectonic accidents are exposed by N-S and NW-SE seismic sections indicating the predominance of E-W and NE-SW directions.

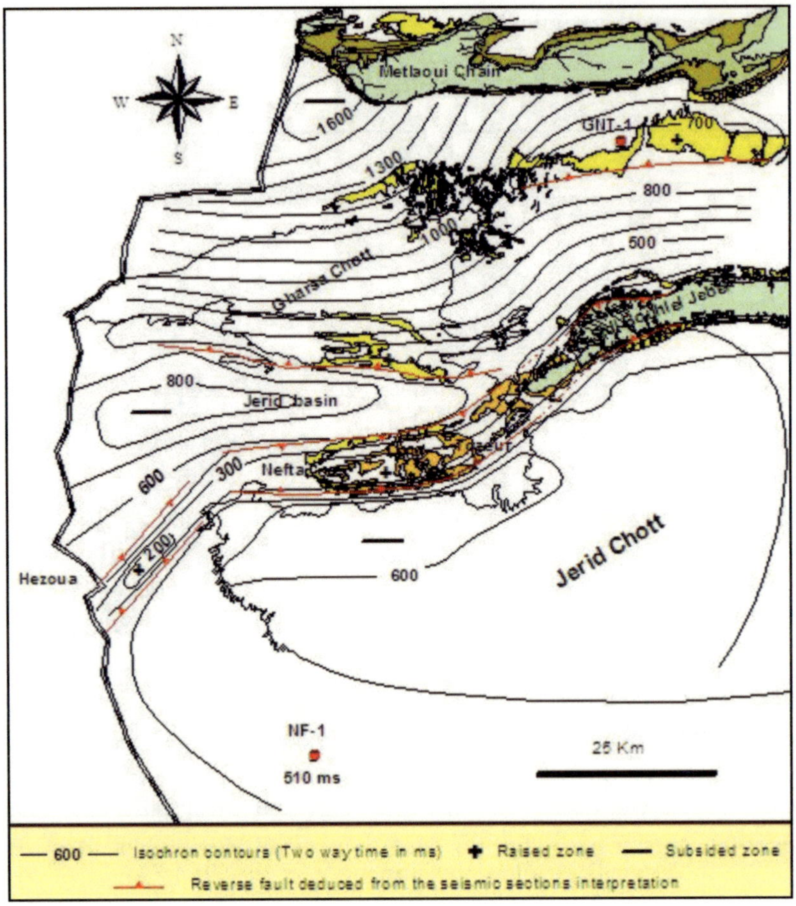

Fig. 10. Isochron map of the Beglia aquifer bottom.

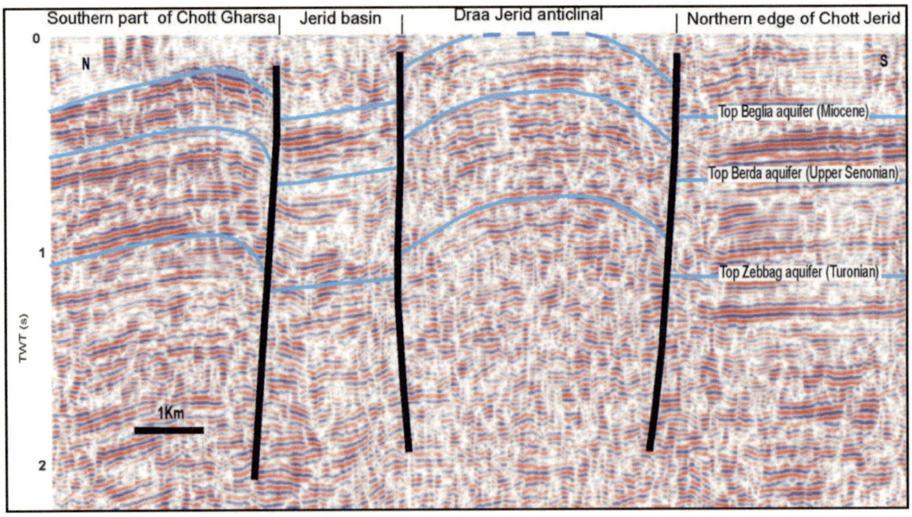

Fig. 11. Seismic section «**V**» interpretation.

Therefore, the resulting structural map of the Jerid area (Fig.10) reveals that in addition to the tectonic traits described on the geological map, exist in subsurface more important structures which must be taken in consideration for the hydrogeological system characterization.

The obtained results highlighted the tectonics influence on the« Complexe Terminal» aquifer geometry; the Tortonian folding, the reverse faulting during the post-Villafranchian compressive phase compartmentalized the aquifer in the form of tilted blocks.

The geosismic cross sections show the variability of this structure implication.

In Mzaraa-Hezoua sector, western part of the Jerid area , the tectonic deformations affecting the «Complexe Terminal» series control the aquifers depth without influencing the groundwater circulation. In fact, the geoseismic cross section corresponding to the profile **E1** (Fig.12) reveals that the reservoirs formations in Hezoua anticlinal are in communication with their equivalents in Jerid basin, at the north and in Chott Jerid, at the south. The Beglia aquifer expresses similar chemical (chemical composition) and isotopic (^{14}C, ^{2}H, ^{18}O tenor) characteristics in Hezoua and Mzaraa regions (Kamel et al., 2005) reflecting this lateral communication.

The geoseismic cross section corresponding to the profile **V** (Fig.13) describes the relations between the different hydrogeological units in Tozeur -Mergueb sector, central part of the Jerid area. It is controlled by Tozeur and Mergueb deep wells and Neflayett and Jhim wells which exploit the Beglia aquifer.

The geoseismic cross section shows an important variation of the aquifers depth between the exposed geological structures. In Draa Jerid anticlinal, the Beglia Formation outcrops

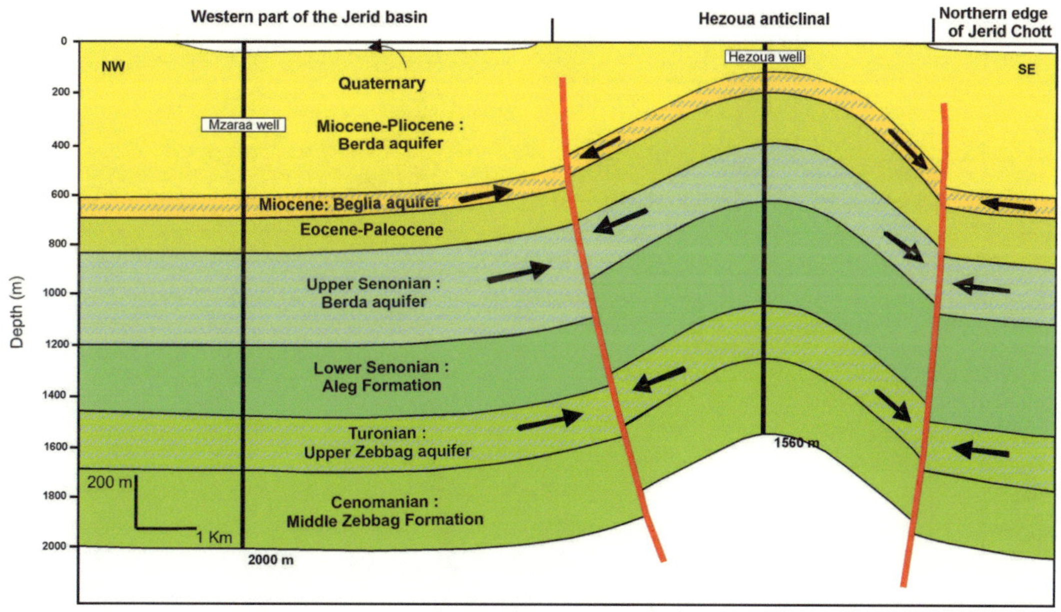

Fig. 12. Geoseismic cross section corresponding to the profile **E1.**

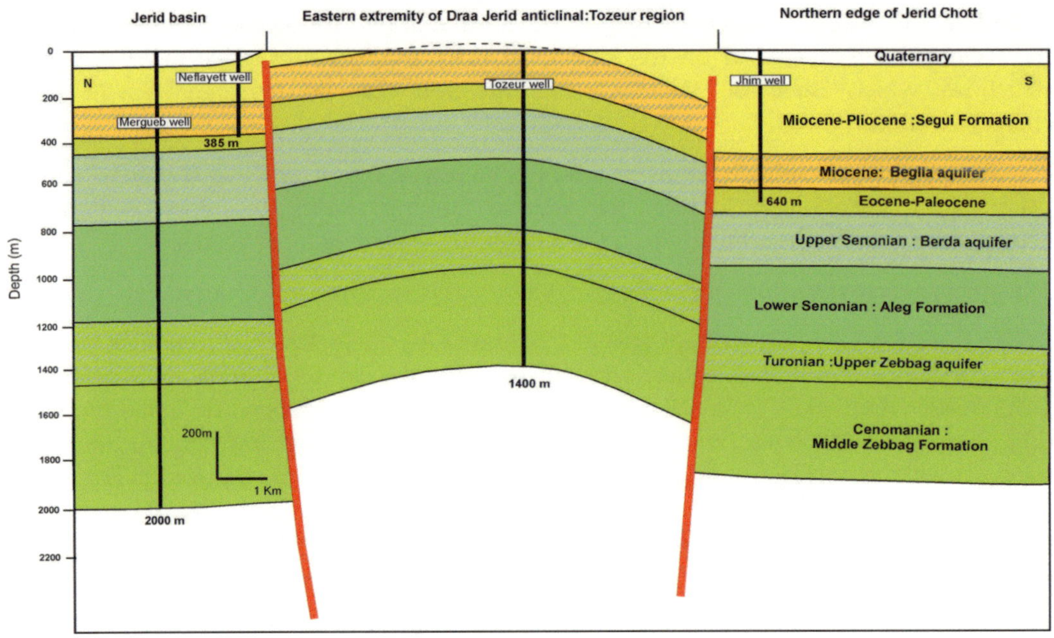

Fig. 13. Geoseismic cross section corresponding to the profile **V.**

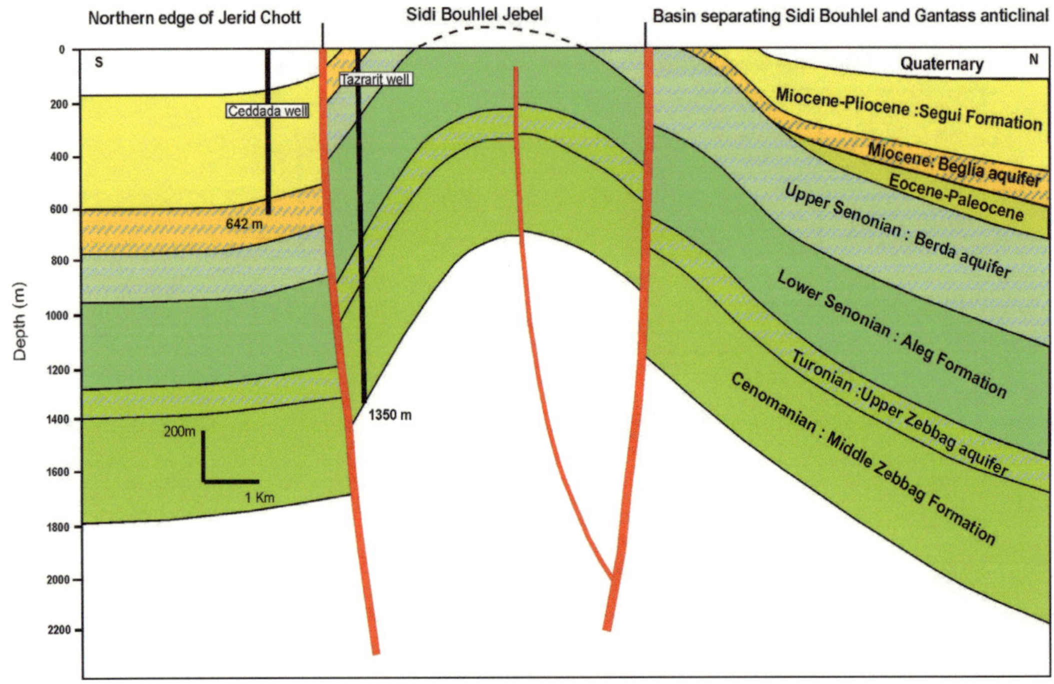

Fig. 14. Geoseismic cross section corresponding to the profiles **L** and **L'.**

The Berda Formation is encountered at 300 m and the Upper Zebbag Formation is recognized at 845 m. In Jerid basin, these reservoirs are respectively reached at 220 m, 460 m and 1180 m.

Additionnaly, this cross section reveals an obstructed communication between the aquifers Formations. In Draa Jerid anticlinal, the Beglia sandy reservoir is isolated. It is wedged between the Segui clays of the Jerid basin and those deposed in the northern edge of the Jerid Chott.

In the same structure (Draa Jerid), the fractured limestones of Berda aquifer are in contact with their equivalents of the Jerid basin and with the Beglia aquifer situated in the Jerid Chott. The Upper Zebbag aquifer collides with the clays and marls of Aleg Formation in the north and in the south.

In the Eastern part of Jerid area, the seismic profiles **L** and **L'** interpretation and the data of Tazrarit and Ceddada wells allow the « Complexe Terminal» aquifer characterization. The established geoseismic cross section (Fig.14) provides informations about the geometry of the aquifer and the groundwater flow.

At Sidi Bouhlel Jebel, the Beglia reservoir exists only in the southern flank , where it outcrops. At the northern edge of Jerid Chott, this reservoir is attained at 600 m. In the basin separating Sidi Bouhlel and Gantass anticlinals, the Beglia Formation depth and thickness decrease towards the south.

Equally, in this sector, the tectonic deformations affecting the «Complexe Terminal» series influence the groundwater circulation, the Beglia aquifer of Sidi Bouhlel anticlinal is opposite the Segui clays deposed in the synclinal basin at the south. This obstructed hydraulic communication between the folded structures explains the significant difference between the salinity of Beglia aquifer at Ceddada well: 2g/l and Tazrarit well: 5.9 g/l (Fig.7).

The Berda aquifer is exposed at Sidi Bouhlel Jebel. In the north it is in contact with its equivalent which shows an important thickness. In the south it is against the clayey Segui Formation.

A vertical communication between Beglia and Berda aquifers is noticed at the northern edge of Jerid Chott and at Sidi Bouhlel anticlinal. At Gantass-Sidi Bouhlel basin, they are separated by the Eocene-Paleocene deposits, pinched out nearby Sidi Bouhlel Jebel.

At Sidi Bouhlel anticlinal are located the most raised Upper Zebbag aquifer of the Jerid area (300 m). This aquifer is blocked between the clays and marls of Aleg Formation.

6. Conclusion

This study based on seismic reflection sections and wells data display the tectonic deformations which affect the multilayered «Complexe Terminal» aquifer in the Jerid area (Fig.15).

The Tortonian folding, the reverse faulting during the post-Villafranchian compressive phase compartmentalized the aquifer in raised and subsided blocks. The geoseismic cross sections reveal that this structure has variable implications; except Hezoua –Mzaraa sector where the reservoirs are in communication, it influences the depth of permeable formations and the circulation of groundwater

These results should be useful for choosing the best sites for the «Complexe Terminal» aquifer exploitation in the Jerid area.

Additionally, the present study shows the interest of the seismic reflection method for the hydrogeological systems comprehension when the well data are limited for a precise characterization. Such prospection appeared particularly suitable in this study in view of the great depth of the aquifer and the importance of the tectonic structures which are not easily detectable by the simple well correlations.

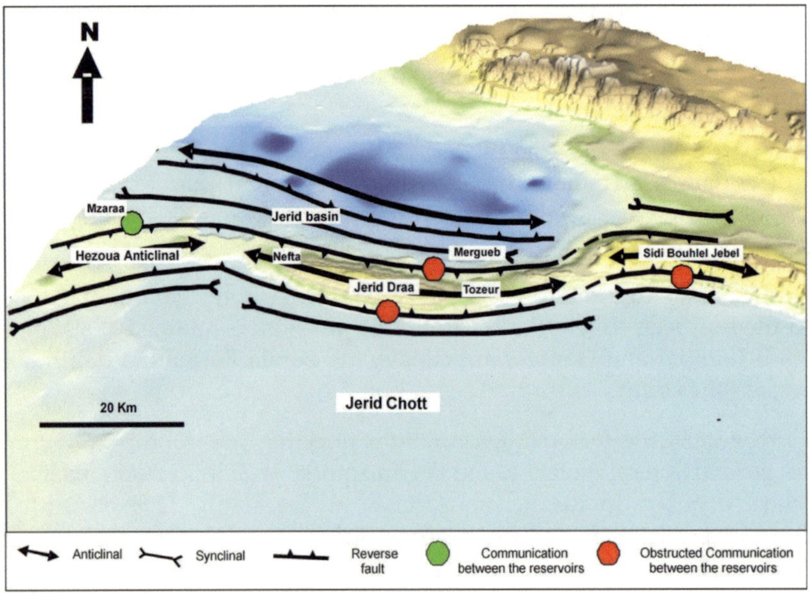

Fig. 15. Geometry of the Jerid «Complexe Terminal» aquifer and communication between compartments.

7. References

Abdallah, H. (1987). Le Crétacé supérieur de la chaine nord des Chotts (Sud tunisien). Biostratigraphie, Sédimentation, Diagenèse. *Thèse Doct., Univ. Bourgogne,* 255p.

Abdeljaoued, S. (1983). Etude sédimentologique et structurale de la partie orientale de la chaine Nord des Chotts. *Thèse Doct., Univ. Tunis II,* 184p.

Addoum, B. (1995). L'Atlas saharien sud-oriental : cinématique des plis- chevauchements et reconstruction du bassin du sud-est constantinois confins algéro-tunsiens). *Thèse Doct., Univ Paris Sud. Orsay.* France, 200 p.

Aissaoui, D. (1984). Les structures liées à l'accident Sud-atlasique entre Biskra et Jebel Manndra, Algérie, évolution géométrique et cinématique. *Thèse Doct., Univ. Louis Pasteur,* 205p.

Asfahani, J. (2007). Geoelectrical investigation for characterizing the hydrogeological conditions in semi-arid region in Khanasser valley, Syria. *Journal of Arid Environments,* 68, pp31-52.

ARMINES & ENIT (1984). Modèle mathématique du Complexe Terminal Nefzaoua -Djerid. *Ministère Agriculture Tunis.* 87p. Tunisie.

Bédir, M. (1995). Mécanismes géodynamiques des bassins associés aux couloirs de coulissements de la marge atlasique de la Tunisie. Seismo-stratigraphie, seismotectonique et implications pétrolières. *Thèse Doct. es Sciences, Univ. Tunis II,* 412 p. Tunisie

Bel, F. & Demargne , F . (1966) .Etude géologique du Continental Terminal .*DEC, ANRH, Alger,* 22p. Algérie.

Bel, F. & Cuche, D. (1969). Mise au point des connaissances sur la nappe du Complexe Terminal. *Projet ERESS ; Ouargla,* 20p. Algérie.

Ben Ayed, N. (1980). Le rôle des décrochements E-W dans l'évolution structurale de l'Atlas tunisien. . *C.R. Acad. Sci,* pp.29-32. France.

Ben Ayed, N. & Viguier, C. (1981). Interprétation structurale de la Tunisie atlasique. *C.R. Acad. Sci,* pp 1445-1448.France.

Ben Ayed, N. (1993). Evolution tectonique de l'avant-pays de la chaine alpine de Tunisie du début du Mésozoïque à l'Actuel. *Annales des Mines et de la Géologie de Tunisie,* n°32. 286 p. Tunisie.

Ben Baccar, B. (1982). Contribution à l'étude hydrogéologique de l' aquifère multicouche de Gabès Sud. *Thèse Doct., Univ. Paris-Sud,* 243p.

Ben Youssef, M. (1999). Stratigraphie génétique du Crétacé de Tunisie : micropaléontologie, stratigraphie séquentielle et géodynamique des bassins de la marge Sud et péritéthysienne. *Thèse Doct. es-Sciences, Univ. Tunis II,* 402 p. Tunisie.

Ben Salah, DH. & Lessi, J. (1978). Construction d'un modèle multicouche de la nappe de la Nefzaoua du Complexe Terminal. Description et résultats des simulations. *Rapport inédit. Direction Générale des Ressources en Eau.* Tunisie.

Besbès, M. & Zammouri, M. (1985). Modèle mathématique du Complexe Terminal. Nefzaoua-Djérid. Note sur les conclusions du rapport ARMINES-ENIT. *Rapport inédit. Direction Générale des Ressources en Eau.* Tunisie.

Boltenhagen C. (1985). Les séquences de sédimentation du Crétacé moyen en Tunisie centrale. *Actes du 1er Cong. Nat. Sci. Terre.* Tunis, Sepembre , 1985

Bouaziz, S. (1995). Etude de la tectonique cassante dans la plate-forme et l'Atlas Sahariens (Tunisie méridionale): Evolution des paléochamps de contraintes et implications géodynamiques. *Thèse Doct. es-Sciences, Univ. Tunis II,* Tunisie, 484 p. Tunisie.

Bouaziz, S. ;Barrier, E. ; Soussi, M. ; Turki, M.M. ; Zouari, H. (2002). Tectonic evolution of the northern African margin in Tunisia from paleostress data and sedimentary record. *Tectonophysics,* 375, pp. 227-253.

Boukadi, N. (1994). Structuration de l'Atlas de Tunisie : signification géométrique et cinématique des noeuds et des zones d'interférences structurales au contact de grands couloirs tectoniques. *Thèse Doct. es-Sciences, Univ. Tunis II,* 249 p.

Boutib, L. & Zargouni, F. (1998). Disposition et géométrie des plis de l'Atlas centro-méridional de Tunisie. Découpage et cisaillement en lanières tectoniques. *C. R. Acad. Sci., 326,* pp. 261-265.

BRL ingénierie (1998) Etude du Plan directeur général de développement des régions sahariennes- Modélisation du Complexe Terminal. *Rapport, ANRH,* Alger, Algérie.

Burollet, P. F. (1956). Contribution à l'étude stratigraphique de la Tunisie centrale. *Annales des Mines et de la Géologie de Tunisie 18,* 350 p.

Caire, A. (1971). Chaînes alpines de la Méditerranée centrale (Algérie et Tunisie septentrionale, Sicile, Calabre et Apennin méridional). *Unesco, Tectonique de l'Afrique, Science de la Terre 6,* pp. 61-90.

Cagniard, L. (1962). *Reflection and Refraction of progressive seismic waves. McGraw.Hill (New York). PUB.ID 101-196-361.*

Castany, G. (1982) . Bassin sédimentaire du Sahara septentrional (Algérie-Tunisie). Aquifers du Continental Intercalaire et du Complexe Terminal. Bull. BRGM, Paris, 3, pp 127-147

Chaabani, F. (1995). Dynamique de la partie orientale du bassin de Gafsa au Crétacé et au Paléocène. Etude minéralogique et géochimique de la série phosphatée éocène, Tunisie méridionale. *Thèse Doct. es-Sciences, Univ. Tunis II*, 428p.

Chalbaoui, M. (2005). Première approche pour l'étude des bassins hydrogéologiques profonds du Sud-Ouest tunisien. *C. R. Acad. Sci.*, 337, pp. 1484-1491

Coque, Ft. (1962). La Tunisie Presaharienne. *Armand Colin, Paris*, 476p.

Cornet, A. (1964). Introduction à l'hydrogéologie saharienne. *Rev. Géog. Phys. et Géol. Dyn.* , pp. 5-72.

Cox, M. (1999). *Static corrections for seismic reflection surveys. Society of Exploration Geophysicists. Digital Library.* USA

Fakraoui, M. (1990). Etude stratigraphique et structurale des chaînes des Chotts (Tunisie méridionale) évolution géométrique et cinématique liée l'accident sud-atlasique. *Thèse Doct., Univ. Tunis II,* 243p.

Fakraoui, M. & Mahjoub, K. (1995). Notice de la carte géologique de la Tunisie 1/ 100.000. Feuille de Hamma Jérid. *Serv. Géol.*, Tunisie.

Fournié, D. (1978). Nomenclature lithostratigraphique des séries du Crétacé supérieur au Tertiaire en Tunisie. *Bull. Cent. Rech. Exploration-production, Elf-Aquitaine, Pau, 2,* pp.97-148.

Frizon de Lamotte, D.; Michard, A. & Saddiqi, O. (2006). Quelques développements récents sur la géodynamique du Maghreb. *C. R. Geoscience,* Vol.336, pp.1-10.

Gabtni, H. ; Jallouli, C.; Mickus, K. & Zouari, H. (2005). Geophysical constraints on the location and nature of the North saharan flexure in southern Tunisia. *Pure andApplied Geophysics.* 162, pp. 2051-2069.

Gardner, G.H.F. (1985). *Migration of seismic data. Society of Exploration Geophysicists. USA.*

Gasmi, M. (2002). Apports de la géophysique à la reconnaissance et la gestion des ressources naturelles. Application en Tunisie. *Thèse Doct. es-Sciences, Univ. Tunis II,* 471p.

General Direction of Water Resources. (2008). Annuaire piézométrique des nappes profondes de la Tunisie ,2008.

Gouasmia, M.; Gasmi, M.; Mhamdi , A. ; Bouri, S. & Ben Dhia, H. (2006). Prospection géoélectrique pour l'étude de l'aquifère thermal des calcaires récifaux Hmeima-Boujaber (Centre-Ouest de la Tunisie). *C.R. Acad. Sci* . pp 1219-1227.

Guellala, R.; Inoubli, M. H.; Alouani, R.; Manaa, M. & Amri, F. (2005). Caractérisation des réservoirs aquifères de la Haute Vallée de la Mejerda (Tunisie). *Jour. Afric. Geosci. Rev.* 12,pp. 189-202.

Guellala, R.; Ben Marzoug, H.; Inoubli, M.H. & Moumni, L. (2008). Identification des structures profondes en Djérid (Tunisie) par sismique réflexion. Implications hydrogéologiques. *22nd Colloquium of African Geology,* Hammamet, Tunisia, Novembre,2008

Guellala, R.; Inoubli, M. H. & Amri, F. (2009a). Nouveaux éléments sur la structure de l'aquifère superficiel de Ghardimaou (Tunisie): contribution de la géophysique électrique. *Hydrol. Sci. J.,* 54, pp. 974-983.

Guellala, R. ; Inoubli, M.H. & Amri, F. (2009b). Apport de la méthode électrique à l'étude hydrogéologique de la plaine de Bousalem (Nord-Ouest de la Tunisie). Quatrième Congrès Maghrébin de Géophysique Appliquée. Hammamet, Tunisia, Mars,2009.

Guellala, R. (2010). Etude géologique et hydrogéologique des séries crétacées inférieures du Jérid (Sud-Ouest de la Tunisie). Apports des méthodes géophysiques. *Thèse Doct., Univ. Tunis II,* 165p

Guellala, R. ; Inoubli, M.H. & Amri, F. (2010). Hydrogeological study of Oued Bouhertma zone. Geoelectrical prospecting contribution. *Proceeding of Tunisia-Japan Symposium. Regional Developement and water Resource*. Tunis: November 28 to December 1st, 2010.

Guellala, R.; Ben Marzoug, H.; Inoubli, M.H. & Moumni, L. (2011). Apports de la Sismique Réflexion à l'étude de l'aquifère du Continental Intercalaire du Jérid (Tunisie). *Hydrol. Sci. J.*,5, pp. 1040-1052.

Guendouz, A.; Moullaa, S.; Edmunds, W.M. ; Zouari, K.; Shands, P. & Mamou, A. (2003). Hydrogeochemical and isotopic evolution of water in the complex terminal aquifer in Algerian Sahara. *Hydogeology Journal* , 11,pp. 483-495.

Henry, G. (1997). *La sismique réflexion, principes et développements ISBN: 2710807254. TECHNIP*, France.

Hlaiem, A. (1999). Halokinesis and structural evolution of the major features in eastern and southern Tunisian Atlas. *Tectonophysics*. pp 79-95.

Inoubli, M.H. (1993) Conception d'une méthodologie d'interprétation en géophysique : Apport du traitement et de l'inversion de données sismiques réelles. *Thèse es Sci. Univ. Tunis II,* 328 p.

Inoubli, M.H & Mechler P. (1999) Expression géologique et apport du champ de vitesse de sommation dans l'amélioration de la résolution sismique. *Notes Serv. Géol. Tunisie*. 54p

Jaffal, M.; Kchikach, A.; Lefort, J.P. & Hanich, L. (2002). Contribution à l'étude d'une partie du bassin d'Essaouira (Maroc) par sismique réflexion. *C.R. Acad. Sci* . pp 229-234.

Jallouli, C. & Mickus, K. (2000). Regional gravity analysis of the crustal structure of Tunisia. *Journal of African Earth Sciences 30,* pp. 63-78.

Jus, H. (1890). Résumé graphique des sondages exécutés dans la province de Constantine de 1er Juin 1856 au 1er Janvier 1890. *Rapport. Constantine. Algérie*

Kamel, S.; Dassi, L.; Zouari , K. & Abidi B. (2005). Geochemical and isotopic investigation of the aquifer system in the Djérid-Nefzaoua basin, southern Tunisia, *Env. Geol.*,49, pp. 159-170

Kamel, S. ; Dassi, L. & Zouari , K (2006). Approche hydrogéologique et hydrochimique des échanges hydrodynamiques entre aquifères profond et superficiel du bassin du Djérid, Tunisie, *Hydrol. Sci. J.*, 51, pp. 713-730.

Kilan, C. (1931). Les principaux complexes continentaux du Sahara. *C.R. Somm. Soc. Géol.Fr.*, pp109-111.

Koussoubé, Y. ; Nakolendoussé, S. ; Bazié, P. ; Savadogo, A.N. (2003). Typologie des courbes des sondages électriques verticaux pour la reconnaissance des formations superficielles et leur incidence en hydrogéologie de socle cristallin du Burkina. *Sud Sciences et Technologies.* , 10, pp. 26-32.

Lachaal, F.; Bedir, M.; Tarhouni,J.; Gacha, A.B.; Leduc, C. (2011). Characterizing a complex aquifer system using geophysics, hydrodynamics and geochemistry: a new distribution of Miocene aquifers in the Zeramdine and Mahdia-Jebeniana blocks (east-central Tunisia). *Journal of African Earth Sciences,* 60, pp. 222-236

Larroque, F. & Dupuy, A. (2004). Apports de la méthode sismique réflexion haute résolution à l'identification des structures profondes des formations tertiaires en Médoc (Gironde, France): implications hydrogéologiques. *C.R. Acad. Sci.* pp 1111-1120.

Lavergne, M. (1986). *Méthodes sismiques. Editions TECHNIP et l'Institut Francais de Petrole. ISBN 2.7108.0514.6.*

Lazzez, M.; Zouaghi, T. & Ben Youssef, M. (2008) . Austrian phase on the northern African margin inferred from sequence stratigraphy and sedimentary records in southern Tunisia (Chotts and Djeffara areas). C.R. Acad. Sci. *340*, pp. 543-552.

Levassor, A. (1978). Simulation et gestion des systèmes aquifères. Application aux nappes du « Complexe Terminal du bas-Sahara algérien. *Thèse Doct., Univ. Paris.*

Mahjoub, K. (1995). Notice de la carte géologique de la Tunisie 1/ 100.000. Feuille de Tozeur. *Serv. Géol.*, Tunisie.

Mamou, A. (1990). Caractéristiques, Evolutions et Gestion des Ressources en Eau du Sud tunisien. *Thèse Doct. es-Sciences, Univ. Paris Sud,* 426p

Mannai-Tayech, B. & Otera, O. (2005). Un nouveau gisement miocène à ichthyofaune au Sud de la chaine des Chotts (Tunisie méridionale). *C.R. Palevol. 4 pp.405-412.*

Mannai-Tayech, B. (2006). Les series silicoclastiques miocènes du Nord-Est au Sud-Ouest de la Tunisie: une mise au point. *Geobios 39,* pp. 71-84.

Mannai-Tayech, B. (2009). The lithostratigraphy of Miocene series from Tunisia, revisited. *Journal of African Earth Sciences 54,* pp. 53-61.

Mekrazi, A. (1975). Contribution à l'étude géologique et hydrogéologique de la région de Gabès Nord. *Thèse Doct., Univ. Bordeaux I.*

Mari, J.L. ; Glangeaud, F. C.; Coppens, F. (2001). *Traitement du signal pour géologues et géophysiciens : Techniques de base. Volume 2. Publications de l'Institut Francais de Pétrole. ISBN. 2710807858.*France.

Missenard, Y. (2006). Le relief des Atlas marocains : contribution des processus asthénosphériques et du raccourcissement crustal, aspects chronologiques. *Thèse Doct., Univ. Cergy Pontoise, France.* 236p.

M'Rabet, A. (1987). Stratigraphie, sédimentation et diagenèse carbonatée des séries du Crétacé inférieur de Tunisie centrale. *Annales des Mines et de la Géologie de Tunisie 30,* 412 p.

Negra, M. E. H. (1994). Les dépôts de plate-forme à bassin du Crétacé supérieur en Tunisie centro-septentrionale (Formation Abiod et faciès associés), stratigraphie, sédimentation, diagenèse et intérêt pétrolier. *Thèse Doct. es-Sciences, Univ. Tunis II,* 649 p.

OSS. (2003). Système Aquifère du Sahara Septentrional. Volume 2 : Hydrogéologie. *Projet SASS. Rapport interne. Direction Génerale des Ressources en Eau , Tunis.* 275p. Tunisie

Ould Baba Sy, M. (2005). Recharge et paleorecharge du système aquifère du Sahara Septentrional. *Thèse Doct., Univ. Tunis II,* 261p

Pizzi, G. & Sartori, L. (1984). Interconnected groundwater systems simulation (IGROSS) - Description of the system and a case history application. *J. Hydrol.*, 75, pp 255-285.

PNUD. (1983). Actualisation de l'Etude des Ressources en Eau du Sahara Septentrional. Projet RAB/80/011 Rapport final. *Rapport interne. Direction Génerale des Ressources en Eau , Tunis.* 490p. Tunisie

Rabia, M.C. (1984). Etude géologique de la region des Chotts (Sud tunisien) par Télédetection spatial, détection de la radioactivité naturelle et analyse hydrogéochimique. *Thèse Doct., Univ. Bordeaux I,* 196p

Regaya, K.; Ben Mamou, A.; Ben Youssef, M. & Ghanmi, M. (2001). Notice de la carte géologique de la Tunisie 1/ 100.000. Feuille de Metlaoui. *Serv. Géol.*, Tunisie.

Rigane, A.& Gourmelen, C. (2011). *Tunisian transtensive basins in tethyan geodynamic context and their post-Tortonian inversion New Frontiers in Tectonic Research - At the Midst of Plate Convergence. ISBN: 978-953-307-594-5. Intech Vienna, Austria.*

Ricolvi, M. (1970). Programme d'exploitation et de surveillance de la nappe de Complexe Terminal dans la région du Djérid. *Rapport interne. Direction Génerale des Ressources en Eau , Tunis.* 21p. Tunisie

Robein, E. (1999). Vitesse et techniques d'imagerie en sismique réflexion : principes et méthodes. *Doc. Lavoisier.* France.

Saidane, H.; Bédir,M.& Zargouni, F. (2008). Le Néogène continental du bassin de Gafsa (Tunisie): Sismo-stratigraphie, structuration et potentialités aquifers. . *Jour. Afric. Geosci. Rev.* 15,pp. 51–68.

Sumanovac, F. (2006). Mapping of thin sandy aquifers by using high resolution reflection seismics and 2-D electrical tomography. *Journal of Applied Geophysics,* 58, pp.144-157

Swezey, C. (1999). The lifespan of the Complexe Terminal Aquifer. Algerian-Tunisian Sahara. *Journal of African Earth Sciences,* 3, pp. 751-756

Swezey, C. (2009). Cenozoic stratigraphy of the sahara , northern Africa. *Journal of African Earth Sciences 53,* pp. 89-121.

Telford, W. M.; Gedrart, L.P.; Sherriff, R.E. & Key, D.A. (1976) Prospection géophysique - Tome 1: Prospection sismique. *Cambridge Univ.*

Tizro, A.T.; Voudouris, K.; Salehzade,M.; Mashayekhi, H. (2010) Hydrogeological framework and estimation of aquifer hydraulic parameters using geoelectrical data: a case study from West Iran. *Hydrogeology Journal* ,18, pp. 917–929

UNESCO. (1972a). ERESS: Etude des ressources en eau du Sahara septentrional. Rapport final . *Rapport interne. Direction Génerale des Ressources en Eau , Tunis.* 78p. Tunisie

UNESCO. (1972c). ERESS: Etude des ressources en eau du Sahara septentrional. Nappe du Complexe Terminal. *Rapport interne. Direction Génerale des Ressources en Eau , Tunis.* 59p. Tunisie

Upadhyay, S.K. (2004). *Seismic reflection processing : with special reference to anisotropy. Springer. Amazon France.*

Zammouri, M. (1990). Contribution à une révision des modèles hydrogéologiques du Sud tunisien. *Thèse Doct., Univ. Tunis II,* 90p.

Zargouni, F. (1985) Tectonique de l'Atlas méridional de Tunisie, évolution géométrique et cinématique des structures en zone de cisaillement. *Thèse Doct. es-Sciences, Univ. Louis Pasteur Strasbourg,* 292 p.

Zargouni, F.; Rabia, M. CH. & Abbès, CH. (1985) Rôle des couloirs de cisaillement de Gafsa et de Négrine -Tozeur dans la structuration du faisceau des plis des Chotts, éléments de l'accident Sud -Atlasique. *C.R. Acad. Sci.,* 301, 831-834.

Zitouni, L. ; Bédir, M. ; Boukadi, N. ; Ibouh, H. ; Tlig, S. & Bobier, C. (1997). Géométrie, chronologie et cinématique des accidents N 120-140. Exemple: accident de Majoura-Mech (données de surface et de subsurface). *Jour. Afric. Geosci. Rev. 15,* pp. 373-380.

Zouaghi, T.; Bédir, M. & Inoubli, M. H. (2005). 2D Seismic interpretation of strike-slip faulting, salt tectonics, and Cretaceous unconformities, Atlas Mountains, central Tunisia. *Journal of African Earth Sciences 43,* pp. 464-486.

Zouaghi, T.; Guellala, R.; Lazzez, M.; Bédir, M.; Ben Youssef, M.; Inoubli, M.H. & Zargouni, F. (2011). *The Chotts Fold Belt of Southern Tunisia, North African Margin: Structural Pattern, Evolution, and Regional Geodynamic Implications. New Frontiers in Tectonic Research - At the Midst of Plate Convergence. ISBN: 978-953-307-594-5. Intech* Vienna, Austria.

Zouari, H. ; Turki, M. M. & Delteil J. (1990). Nouvelles données sur l'évolution tectonique de la chaîne de Gafsa. *Bulletin de la Société Géologique de France 8*, pp. 621-628.

Zouari, H. (1995). Evolution géodynamique de l'Atlas centro-méridional de la Tunisie. Stratigraphie, analyses géométrique, cinématique et tectono-sédimentaire. *Thèse Doct. es-Sciences, Univ. Tunis II*, 278 p.

Zouhri, L.; Gorini, C.; Lamouroux, C.; Vachard, D.& Dakki M. (2003). Interprétation hydrogéologique de l'aquifère des bassins Sud-rifains (Maroc): apport de la sismique réflexion. *C.R. Acad. Sci* . pp 319-326.

Zouhri, L.; Gorini, C.; Mania, J; Deffontaines, B. & Zerouali, A. (2004). Spatial distribution of resistivity in the hydrogeological systems, and identification of the catchment area in the Rharb basin, Morocco: *Hydrological Science Journal*, 49, 387–398.

2

New Techniques for Potential Field Data Interpretation

Khalid S. Essa
Cairo University/ Faculty of Science/ Geophysics Department, Giza, Egypt

1. Introduction

Many of the geological structures in mineral and petroleum exploration can be classified into four categories: spheres, cylinders, dikes and geological contacts. These four simple geometric forms are convenient approximations to common geological structures often encountered in the interpretation of magnetic and gravity data. Two quantitative categories are usually adapted for the interpretation of magnetic and gravity anomalies in order to determine basically the depth, the shape of the buried structure and the other model parameters. First, two parameters are considered the most important problem in exploration geophysics. The first category includes 2D and 3D continuous modelling and inversion methods (Mohan et al., 1982; Nabighian, 1984; Chai & Hinze, 1988; Martin-Atienza & Garcia-Abdeslem 1999; Zhang et al., 2001; Keating & Pilkington, 2004; Numes et al., 2008; Martins et al., 2010; Silva et al., 2010). The solutions of these methods require magnetic susceptibility and/or remanent magnetization for magnetic interpretation and density information for gravity interpretation a as part of the input, along with the same depth information obtained from geological and/or other geophysical data. Thus, the resulting model can vary widely depending on these factors because the magnetic and gravity inverse problem are ill-posed and are, therefore unstable and nonunique (Zhdanov, 2002; Tarantola, 2005). Standard techniques for obtaining a stable solution of an ill-posed inverse problem include those based on regularization methods (Tikhonov & Arsenin, 1977). The second category is fixed simple geometry methods, in which the spheres, cylinders, dikes, and geological contacts models estimate the depth and the shape of the buried structures from residuals and/or observed data. The models may be shifted from geological reality, but they are usually sufficient to determine whether the form and the magnitude of both calculated magnetic and gravity effects are close enough to those observed to make the geological interpretation reasonable. Several methods have been developed for the second category to interpret magnetic and gravity data using a fixed simple geometry (Nettleton, 1976; Stanley, 1977; Atchuta Rao & Ram Babu, 1980; Prakasa Rao & Subrahmanyan, 1988; Abdelrahman & El-Araby, 1993; Zhang et al., 2000; Salem et al., 2004; Abdelrahman & Essa, 2005; Essa, 2007; Essa, 2011). The drawbacks of these methods are that they are highly subjective, need a priori information about the shape (shape factor) of the anomalous body and use few characteristic points when inverting for the remaining parameters. New algorithms for magnetic and gravity inversion have been developed that successively determines the depth

(z) and the other associated model parameters of the buried structure. The inverse problem of the depth estimation from the observed magnetic and gravity data has been transformed into a nonlinear equation of the form f(z) = 0. This equation is then solved for depth by minimizing an objective functional in the least-squares sense. Using the estimated depth, the other model parameters are computed from the measured magnetic and gravity data, respectively. The procedures are applied to synthetic data with and without noise for magnetic and gravity data. These procedures are also tested to complicated regional and interference from neighbouring magnetic rocks. Finally, these techniques are also successfully applied to real data sets for mineral and petroleum exploration, and it is found that the estimated depths and the associated model parameters are in good agreement with the actual values.

2. Formulation of the problems

2.1 A least-squares minimization approach to depth determination from residual magnetic anomaly

Following Gay (1963), Prakasa Rao et al. (1986), and Prakasa Rao & Subrahmanyan (1988), the magnetic anomaly expression produced by most geologic structures with center located at x_i= 0 can be represented by the following equation:

$$T(x_i,z,\theta)=K\frac{\left(az^{2r}+bx_i^2\right)(\sin\theta)^m(\cos\theta)^n+cx_i z^p(\sin\theta)^n(\cos\theta)^m}{\left(x_i^2+z^2\right)^q},i=1,2,3,...L \quad (1)$$

The geometries are shown in Figure 1. In equation (1), z is the depth, K is the amplitude coefficient (effective magnetization intensity), θ is the index parameter (effective magnetization inclination), x_i is the horizontal coordinate position, and q is the shape factor. Values for a, b, c, m, n, r, p, and q, are given in Table 1.

At the origin (x_i =0), equation (1) gives the following relationship:

$$K=\frac{T(0)z^{2q-2r}}{a(\sin\theta)^m(\cos\theta)^n}, \quad (2)$$

where T(0) is the anomaly value at the origin (Fig. 2).

Using equation (2), equation (1) can be rewritten as

$$T(X_i,Z,\theta)=\frac{T(0)z^{2q-2r}}{a}\left[\frac{(az^{2r}+bx_i^2)+cx_iz^p(\tan\theta)^{n-m}}{(x_i^2+z^2)^q}\right], \quad (3)$$

For all shapes (function of q), equation (3) gives the following relationship at x_i=N

$$(\tan\theta)^{n-m}=\frac{aT(N)(N^2+z^2)^q-T(0)z^{2q-2r}(az^{2r}+bN^2)}{cNz^pT(0)z^{2q-2r}}, \quad (4)$$

where T (N) is the anomaly value at x_i = N (Fig. 2).

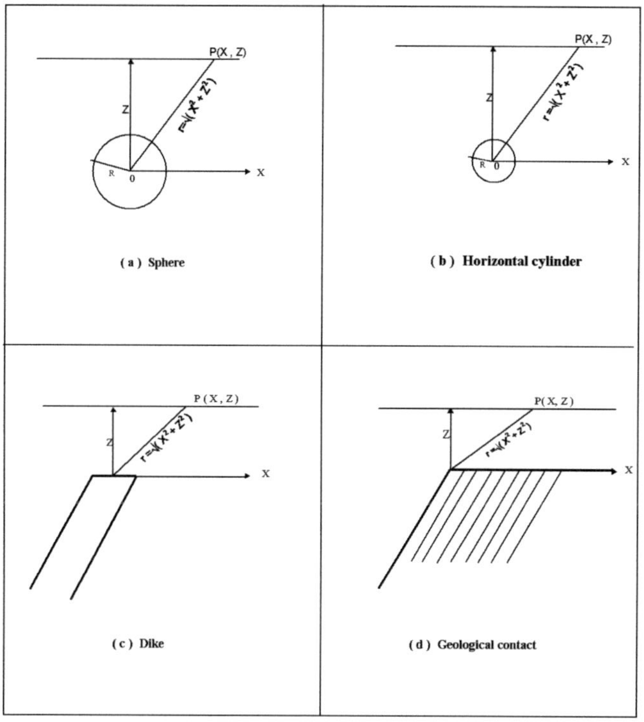

Fig. 1. Various simple geometrical structures diagram.

Model	Magnetization	*a*	*b*	*c*	*m*	*n*	*p*	*r*	*q*
Sphere	Vertical	2	−1	−3	1	0	1	1	2.5
Sphere	Horizontal	−1	2	−3	0	1	1	1	2.5
Horizontal cylinder Dike (F.H.D.)	Total, vertical, horizontal	1	−1	2	0	1	1	1	2
Geologic contact (S.H.D.) Dike	Total, vertical, horizontal	1	0	−1	0	1	0	0.5	1

Table 1. Definition of a, b, c, m, n, p, r, and q values were shown in equation (1). F.H.D. and S.H.D. are the first and the second horizontal derivatives of the magnetic anomaly, respectively.

Substituting equation (4) into equation (3), we obtain the following nonlinear equation in z

$$T(x_i,z)=\frac{NT(0)z^{2q-2r}(az^{2r}+bx_i^2)+ax_iT(N)(N^2+z^2)^q-x_iT(0)z^{2q-2r}(az^{2r}+bN^2)}{aN(x_i^2+z^2)^q}, \quad (5)$$

The unknown depth z in equation (5) can be obtained by minimizing

$$\psi(z)=\sum_i^N\left[L(x_i)-T(x_i,z)\right]^2, \quad (6)$$

where L (x_i) denotes the observed magnetic anomaly at x_i.

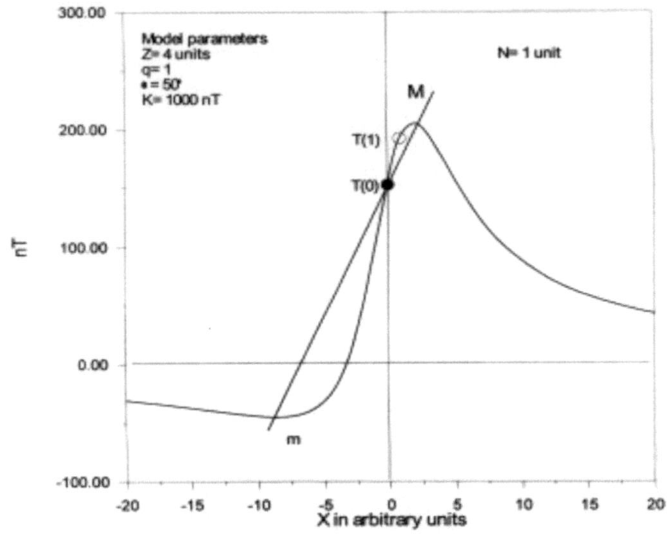

Fig. 2. A typical magnetic anomaly profile over a thin dike. The anomaly value at the origin T(0) and the anomaly value T(N) where N is taken to be 1arbitrary unit in this case, the position of the maximum value (M), and the minimum value (m) are illustrated.

Minimization of ψ(z) in the least - squares sense, i.e., (d/dz) ψ(z)=0 leads to the following equation:

$$f(z) = \sum_{i}^{N} [L(x_i) - T(x_i, z)] T^*(x_i, z) = 0, \tag{7}$$

where

$$T^*(x_i, z) = (d/dz) T(x_i, z)$$

Equation (7) can be solved for z using the standard methods for solving nonlinear equations. Here, it is solved by an iteration method (Press et al., 1986). The iteration form of equation (7) is given as

$$z_f = f(z_j) \ , \tag{8}$$

where z_j is the initial depth and z_f is the revised depth. z_f will be used as z_j for next iteration. The iteration stops when $|z_f - z_j| < e$, where e is a small predetermined real number close to zero. The source body depth is determined by solving one nonlinear equation in z. Any initial guess for z works well because there is only one minimum. The experience with the minimization technique for two or more unknowns is that it always produced good results for synthetic data with or without random noise. In the case of field data, good results may only be obtained when using very good initial guesses on the model parameters. The optimization problem for the depth parameter is highly nonlinear, increasing the number of parameters to be solved simultaneously also increases the dimensionality of the energy surface, thereby greatly increasing the probability of the optimization stalling in a local minimum on that surface. Thus common sense dictates that the nonlinear optimization should be restricted to as few parameters as is consistent with obtaining useful results. This is why we propose a solution for only one unknown, z. Once z is known, the effective magnetization inclination, θ, can be determined from equation (4). Finally, knowing θ, the effective magnetization intensity, K, can be determined from equation (2). Then, we measure the goodness of fit between the observed and the computed magnetic data for each N value. The simplest way to compare two magnetic profiles is to compute the root-mean-square (RMS) of the differences between the observed and the fitted anomalies. The model parameters which give the least root-mean square error are the best. To this point we have assumed knowledge of the axes of the magnetic profile so that T(0) can be found. T(0) is determined using methods described by Stanley (1977). As illustrated in Figure 2, the line M-m intersects the anomaly profile at x_i = 0. The base line of the anomaly profile lies a distance M-T(0) above the minimum. A semi-automated interpretation scheme based on the above equations for analyzing field data is illustrated in Figure 3.

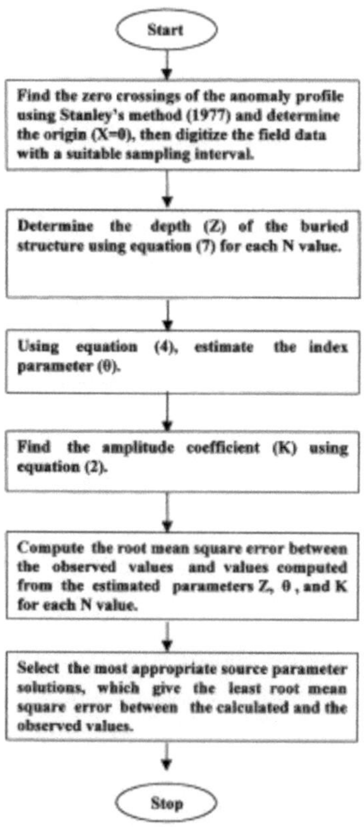

Fig. 3. Generalized scheme for semi-automatic depth, index parameter and amplitude coefficient estimations.

2.1.1 Synthetic example

Synthetic examples of spheres and horizontal cylinders buried at different depths (profile length = 40 units, K = 100 units, θ = 60°, sampling interval = 1unit) were interpreted using the present method [equations (7), (4), and (2)] to determine depth, index parameter, and amplitude coefficient, respectively. In each case, the starting depth was 5 units. In all cases examined, the exact values of z, θ, and K were obtained. However, in studying the error response of the least-squares method, synthetic examples contaminated with 5% random errors were considered. Following the interpretation scheme, values of the most appropriate model parameters (z, θ, K) were computed and percentage of error in model parameters was plotted against the model depth for comparison (Fig. 4). We verified numerically that the depth obtained is within 3% for horizontal cylinders and 4% for spheres. The index parameter obtained is within 5.5% whereas the amplitude coefficient is within 8% (Fig. 4). Good results are obtained by using the present algorithm, particularly for depth estimation, which is a primary concern in magnetic prospecting and other geophysical work.

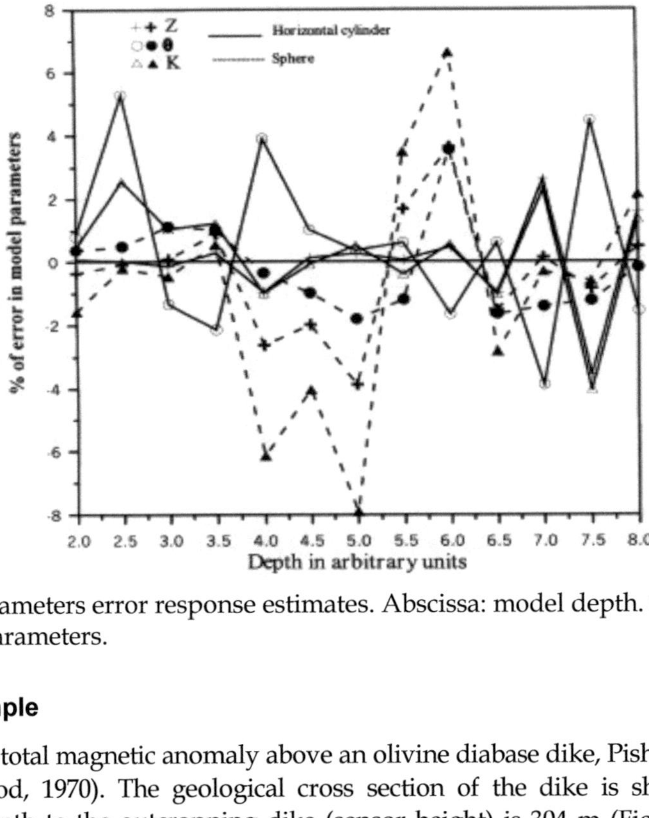

Fig. 4. Model parameters error response estimates. Abscissa: model depth. Ordinate: percent error in model parameters.

2.1.2 Field example

Figure 5 shows a total magnetic anomaly above an olivine diabase dike, Pishabo Lake, Ontario (McGrath & Hood, 1970). The geological cross section of the dike is shown beneath the anomaly. The depth to the outcropping dike (sensor height) is 304 m (Fig. 5). The anomaly profile was digitized at an interval of 100 m. Equations (7), (4), and (2) were used to determine depth, index parameter and amplitude coefficient, respectively, using all possible cases on N values. The starting depth used in this field example was 50 m. Then, we computed the root-mean-square of the differences between the observed and the fitted anomalies. The best fit model parameters are: z = 306 m, θ = 38° and K = 1422 nT*Z (Table 2). McGrath & Hood (1970) applied a computer curve-matching method to the same magnetic data employing a least-squares method and obtained a depth of 301 m. Moreover, Figure 5 shows that there is a lack of "thinness" because the thickness of the dike is only slightly less than the depth. However, because many of the characteristics of the thick dike anomalies are similar to those for thin dike (Gay, 1963), our method can be applied not only to magnetic anomalies due to thin dikes, but also to those due to thick dikes to obtain reliable depth estimates.

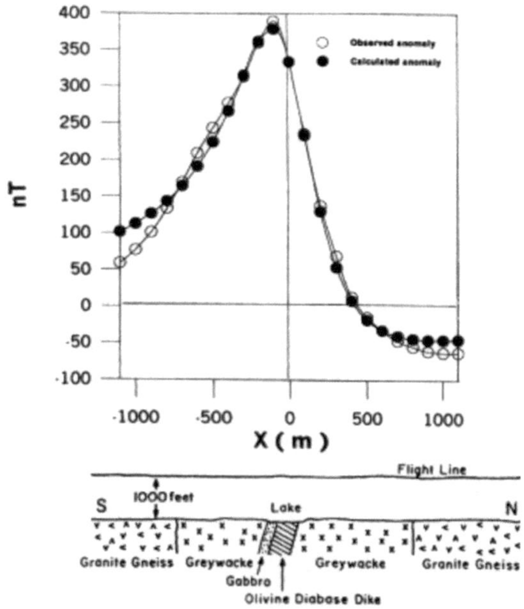

Fig. 5. Total magnetic anomaly (above) over an outcropping diabase dike (below), Pishabo Lake, Ontario, Canada (McGrath & Hood, 1970). The base line and zero crossing shown are determined using Stanley's method (1977).

N (m)	Depth, Z (m)	Index parameter, θ (deg.)	Amplitude coefficient, K (nT * Z)	Root-mean-square error (nT)
−200	305	−37.9	1411.7	16.4
−100	290	−42.1	1435.9	21.9
100	306	−37.9	1422.8	16.5
200	314	−37.5	1443.5	16.6
300	321	−36.7	1460.1	17.6
400	311	−37.6	1434.5	16.5
500	310	−37.7	1429.3	16.4
600	**306**	**−38.3**	**1422.7**	**16.3**
700	299	−39.8	1425.4	17.5
800	297	−41.3	1447.6	20.1
900	298	−43.0	1491.1	24.6
1000	302	−43.9	1533.8	27.9
1100	307	−44.5	1570.6	30.1

Table 2. Numerical results of the Pishabo field example.

2.2 Three least-squares minimization approach to model parameters estimation from magnetic anomaly due to thin dikes

From equation (1), the magnetic anomaly expression of a thin dike which extends to infinity in both strike direction and down dip (2-D) is given by:

$$T(x_i, z, \theta) = zK\frac{x_i \sin\theta + z\cos\theta}{x_i^2 + z^2}, \; i = 1,2,3,\ldots\ldots,N \qquad (9)$$

where z is the depth to the top of the body, x_i is a distance point along x-axis where the observed anomaly is located, K is the amplitude coefficient, and θ is the index parameter.

The values of K and θ for the anomalies in total, vertical, and horizontal fields are given in Table 3. The index parameter θ is related to the effective direction of the resultant magnetization, the dip and strike of the dike.

Anomaly (T)	Index parameter (θ)	Amplitude coefficient (K)
ΔV	$I'_0 - d$	$2kt\, T'_0 / z$
ΔH	$I'_0 - d - 90°$	$2kt\, T'_0 \sin\alpha / z$
ΔT	$2I'_0 - d - 90°$	$2kt\, T'_0 \sin I_0 / z \sin I'_0$

Table 3. Characteristic index parameter θ and amplitude coefficient K in vertical (ΔV), horizontal (ΔH), and total (ΔT) magnetic anomalies due to thin dikes where t is the thickness, k is the magnetic susceptibility contrast, d is the dip angle, T'o and Ío are the values of effective total intensity and effective inclination of magnetic polarization, and α is the strike (after Gay, 1963).

Following Griffin (1949), the moving average residual magnetic anomaly (R) is defined as

$$R = T(0) - \bar{T}(s), \tag{10}$$

where T(0) is the magnetic anomaly value at a given point on the magnetic anomaly map, and

$$\bar{T}(s) = \frac{1}{2\pi} \int_0^{2\pi} T(s,\beta)\, d\beta \tag{11}$$

is the average magnetic value at the radial distance s from the point where the magnetic value is T(0). The geometrical interpretation of equation (11) is that the average value, $\bar{T}(s)$, is obtained by going around a circle of radius s, having T(0) as its centre, and forming the sum of the products of T(s, β) and dβ for an infinite number of infinitesimally small dβ's. The result of the above is then divided by 2π. Since an integral form of $\bar{T}(s)$ is not know, Griffin (1949) has adopted a simple numerical method to obtain the average magnetic anomaly value. The method is to form the arithmetical average of a finite number of points about the circumference of a circle of radius s, or

$$\bar{T}(s) = \frac{[T_1(s) + T_2(s) + T_3(s) + \ldots\ldots + T_n(s)]}{n}. \tag{12}$$

for n points.

Along a profile, the moving average residual magnetic anomaly is then defined as

$$R(x_i, s) = \frac{2T(x_i) - T(x_i + s) - T(x_i - s)}{2}, \tag{13}$$

where s = 1, 2, 3,…., M sample spacing units and is defined here as the window length or graticule spacing.

Applying equation (13) to equation (9), the moving average residual magnetic anomaly due to a thin dike is obtained as

$$R(x_i, z, \theta, s) = \frac{zK}{2}\left\{ \frac{2x_i \sin\theta + 2z\cos\theta}{x_i^2 + z^2} - \frac{(x_i + s)\sin\theta + z\cos\theta}{(x_i + s)^2 + z^2} - \frac{(x_i - s)\sin\theta + z\cos\theta}{(x_i - s)^2 + z^2} \right\}. \tag{14}$$

The specific value $R(x_i, z, \theta, s)$ at $x_i = 0$ is given by

$$R(0) = \frac{Ks^2 \cos\theta}{s^2 + z^2}. \tag{15}$$

Using equation (15), equation (14) can now be written as

$$R(x_i, z, \theta, s) = \frac{z(s^2 + z^2)R(0)}{2s^2} \left\{ \frac{2x_i \tan\theta + 2z}{x_i^2 + z^2} - \frac{(x_i + s)\tan\theta + z}{(x_i + s)^2 + z^2} - \frac{(x_i - s)\tan\theta + z}{(x_i - s)^2 + z^2} \right\}. \tag{16}$$

Using equation (16), we can obtain the following equations at $x_i = \pm s$

$$\frac{R(s)}{R(0)} = \frac{z^2 - 2s^2 + 3zs\tan\theta}{4s^2 + z^2}, x_i = s \tag{17}$$

And

$$\frac{R(-s)}{R(0)} = \frac{z^2 - 2s^2 - 3zs\tan\theta}{4s^2 + z^2}, x_i = -s. \tag{18}$$

From equations (17) and (18), we obtain

$$\tan\theta = \frac{F(4s^2 + z^2)}{6zs}, \tag{19}$$

where

$$F = \frac{R(s) - R(-s)}{R(0)},$$

which is computed directly from observed anomaly. In all cases, the maximum value of s is less than or equal (N-1)/4, where N is the number of the data points along the anomaly profile whose centre is located at $x_i = 0$. Using equation (19), equation (16) can be written as

$$R(x_i, z, s) = R(0)W(x_i, z, s), \tag{20}$$

where

$$W(x_i, z, s) = \frac{(s^2 + z^2)}{12s^3} \left\{ \frac{2x_i F(4s^2 + z^2) + 12z^2 s}{x_i^2 + z^2} - \frac{(x_i + s)F(4s^2 + z^2) + 6z^2 s}{(x_i + s)^2 + z^2} - \frac{(x_i - s)F(4s^2 + z^2) + 6z^2 s}{(x_i - s)^2 + z^2} \right\}.$$

The unknown depth (z) in equation (20) can be obtained by minimizing

$$\varphi(z) = \sum_{i=1}^{N} \left[L(x_i) - R(0)W(x_i, z, s) \right]^2, \tag{21}$$

where L(xi) denotes the moving average residual magnetic anomalies at xi. Setting the derivative of φ(z) to zero with respect to z leads to

$$f(z)=\sum_{i=1}^{N}[L(x_i)-R(0)W(x_i,z,s)]W^*(x_i z,s)=0. \qquad (22)$$

where W*(xi, z, s) is evaluated by analytically differentiating W(xi, z, s) and is given by the following equation

$$W^*(x_i,z,s)=(s^2+z^2)\left[\begin{array}{l}\dfrac{(x_i^2+z^2)(4x_iFz+24zs)-2z(2x_iF(4s^2+z^2)+12z^2s)}{(x_i^2+z^2)^2}-\\ \dfrac{((x_i+s)^2+z^2)(2(x_i+s)Fz+12zs)-2z((x_i+s)F(4s^2+z^2)+6z^2s)}{((x_i+s)^2+z^2)^2}-\\ \dfrac{((x_i-s)^2+z^2)(2(x_i-s)Fz+12zs)-2z((x_i-s)F(4s^2+z^2)+6z^2s)}{((x_i-s)^2+z^2)^2}\end{array}\right]$$

$$+2z\left[\frac{2x_iF(4s^2+z^2)+12z^2s}{(x_i^2+z^2)}-\frac{(x_i+s)F(4s^2+z^2)+6z^2s}{(x_i+s)^2+z^2}-\frac{(x_i-s)F(4s^2+z^2)+6z^2s}{(x_i-s)^2+z^2}\right]$$

Equation (22) can be solved for z using standard methods for solving nonlinear equations. Here, it is solved by an iterative method (Demidovich & Maron, 1973).

Substituting the computed depth (z_c) as a fixed parameter in equation (16), we obtain

$$R(x_i,\theta,s)=\frac{z_c(s^2+z_c^2)R(0)}{2s^2}P(x_i,\theta,s), \qquad (23)$$

where

$$P(x_i,\theta,s)=\left\{\frac{2x_i\tan\theta+2z_c}{x_i^2+z_c^2}-\frac{(x_i+s)\tan\theta+z_c}{(x_i+s)^2+z_c^2}-\frac{(x_i-s)\tan\theta+z_c}{(x_i-s)^2+z_c^2}\right\}.$$

The unknown index parameter (θ) in equation (23) can be obtained by minimizing

$$\psi(\theta)=\sum_{i=1}^{N}\left[L(x_i)-\frac{z_c(s^2+z_c^2)R(0)}{2s^2}P(x_i,\theta,s)\right]^2. \qquad (24)$$

Setting the derivative of ψ(θ) to zero with respect to θ leads to

$$\lambda(\theta)=\sum_{i=1}^{N}\left[L(x_i)-\frac{z_c(s^2+z_c^2)R(0)}{2s^2}P(x_i,\theta,s)\right]P^*(x_i,\theta,s)=0, \qquad (25)$$

where P*(x_i, θ, s) is obtained by differentiating P(x_i, θ, s) with respect to θ and is given by the following equation

$$P^*(x_i,\theta,s)=Sec^2\theta\left[\frac{2x_i}{x_i^2+z_c^2}-\frac{(x_i+s)}{(x_i+s)^2+z_c^2}-\frac{(x_i-s)}{(x_i-s)^2+z_c^2}\right].$$

Equation (25) can be solved using a simple iterative method (Demodivch & Maron, 1973). Substituting the computed depth (z_c) and index parameter (θ_c) in equation (14) as fixed parameters, we obtain

$$R(x_i, s) = KD(x_i, s), \tag{26}$$

where

$$D(x_i, s) = \frac{z_c}{2}\left\{\frac{2x_i \sin\theta_c + 2z_c \cos\theta_c}{x_i^2 + z_c^2} - \frac{(x_i + s)\sin\theta_c + z_c \cos\theta_c}{(x_i + s)^2 + z_c^2} - \frac{(x_i - s)\sin\theta_c + z_c \cos\theta_c}{(x_i - s)^2 + z_c^2}\right\}.$$

Finally, applying the least-squares method to equation (26), the unknown amplitude coefficient (K_c) can be determined from

$$K_c = \frac{\sum_{i=1}^{N} L(x_i) D(x_i, s)}{\sum_{i=1}^{N} [D(x_i, s)]^2}. \tag{27}$$

Theoretically, one value of s is enough to determine the model parameters (z, θ, K) from equations (22), (25), and (27), respectively. In practice, more than one value of s is desirable because of the presence of noise, interference from neighbouring magnetic rocks, and complicated regional in data. Moreover, the advantage of this method over continuous modelling methods is that it requires neither magnetic susceptibility contrast nor depth information, and it can be applied if little or no factual information other than the magnetic data is available. The method may be automated to solve for a number of anomalies along a magnetic profile when the origin of each dike is known from other geophysical and/or geological data.

2.2.1 Synthetic example

The composite vertical magnetic anomaly in nanoteslas of Figure 6 consisting of the combined effect of an intermediate structure of interest (thin dike with K = 200 nT, z = 5 km, and θ = 40°), an interference neighbouring magnetic rocks (horizontal cylinder with K = 40 nT, z = 2 km, and θ = 30°), and a deep-seated structure (sphere with K = 80000 nT, z = 15 km, and θ = 10°) was computed by the following expression:

$$\begin{aligned}
\Delta T(x_i) = {} & 40\frac{(4-(x_i-10)^2)\cos 30^\circ + 4(x_i-10)\sin 30^\circ}{((x_i-10)^2+2^2)^2} \\
& \textit{Horizontal cylinder (vertical component)} \\
& + 200\frac{x_i \sin 40^\circ + 5\cos 40^\circ}{x_i^2 + 5^2} \\
& \textit{Thin dike (vertical component)} \\
& + 80000\frac{(450-(x_i+60)^2)\sin 10^\circ - 45(x_i+60)\cos 10^\circ}{((x_i+60)^2+15^2)^{5/2}}. \\
& \textit{Sphere (vertical component)}
\end{aligned} \tag{28}$$

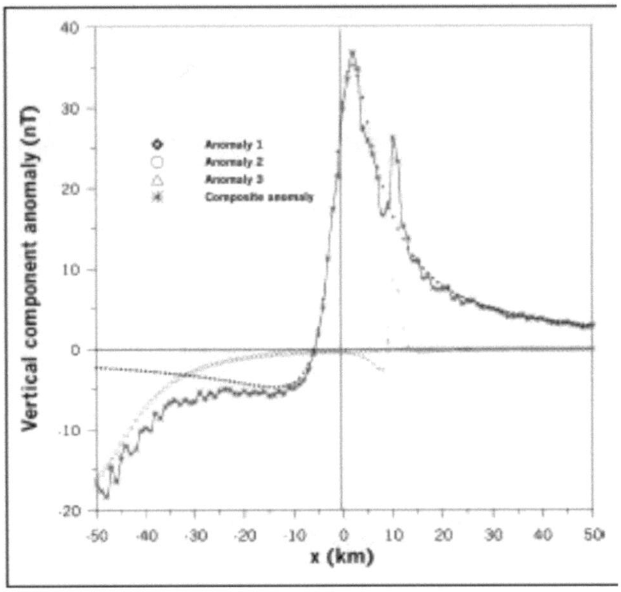

Fig. 6. Noisy composite vertical magnetic anomaly consisting of the combined effect of an intermediate structure (thin dike with K = 200 nT, z = 5 km, and θ = 40°) [anomaly 1], a deep-seated structure (sphere with K = 80000 nT, z = 15 km, and θ = 10°) [anomaly 2], and an interference from neighbouring magnetic rocks (horizontal cylinder with K = 40 nT, z = 2 km, and θ = 30°) [anomaly 3].

In Figure 6, anomaly 1 is the anomaly due to the intermediate structure of our interest, anomaly 2 is the anomaly due to the interference from neighbouring magnetic rocks, and anomaly 3 is the anomaly due to the deep-seated structure. The composite magnetic anomaly $\Delta T(x_i)$ is contaminated with 20% random error using the following equation:

$$\Delta T_{rnd}(x_i) = \Delta T(x_i)\left[1 + (RND(i) - 0.5) * 0.2\right], \tag{29}$$

where $\Delta T_{rnd}(x_i)$ is the contaminated anomaly value at x_i, and RND(i) is a pseudo random number whose range is [0, 1]. The interval of the pseudo random number is an open interval, i.e. it does not include the extremes 0 and 1.

The noisy composite magnetic anomaly (ΔT) is subjected to a separation technique using the moving-average method. Nine successive moving-average graticule spacings were applied to each set of input data. Equations (22), (25), and (27) were applied to each of the nine moving-average residual profiles, yielding depth, index parameter, and amplitude coefficient solutions. The results and their average values are given in Table 4.

We verified numerically that the depth obtained is within 0.38%. The index parameter obtained is within 0.4%, whereas the error in the amplitude coefficient is within 4.7%. Good results are obtained by using the present algorithm for depth, index parameter, and amplitude coefficient, which are a primary concern in magnetic prospecting and other geophysical work. On the other hand, the Euler deconvolution method (Thompson, 1982) has been applied to the same noisy composite magnetic anomaly to determine the depth to the buried dike. The result in this case is shown in Figure 7. The average depth obtained is 3.27 km compared with 4.99 km using the least-squares method. This result indicates that the present least-squares method is less sensitive to errors in the magnetic anomaly than Euler deconvolution method.

Graticule spacings (km)	Computed depth from equation (14)(km)	Computed index parameter from equation (17) (degree)	Computed amplitude coefficient from equation (19)(nT)
2	4.85	48.86	199.12
3	4.47	46.14	166.79
4	4.65	36.39	180.44
5	4.04	37.57	139.59
6	4.80	36.62	179.01
7	5.15	36.27	193.66
8	5.09	35.67	188.35
9	5.38	38.53	203.05
10	6.54	45.44	265.61
Average values	4.99 ± 0.7	40.16 ± 5.1	190.62 ± 34.1
% of error	0.38%	0.4%	4.7%

Table 4. Theoretical results obtained from the noisy composite magnetic anomaly. The depth, index parameter, and amplitude coefficient were computed from equations (22), (25), and (27), respectively. The actual dike parameters are: $z = 5$ km, $\theta = 40°$ and $K = 200$ nT.

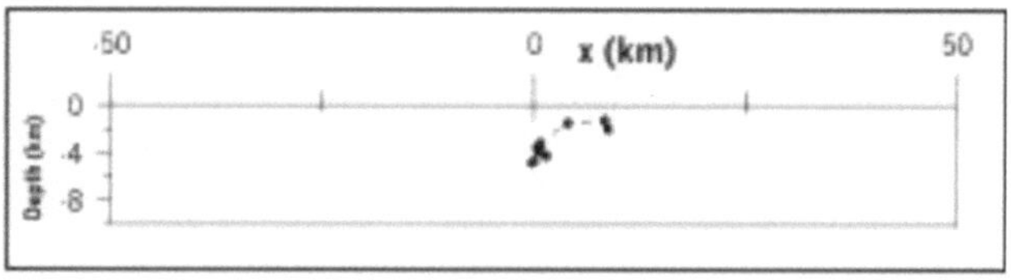

Fig. 7. Euler deconvolution depth estimates from the data of the noisy composite vertical magnetic anomaly of Figure 6 using a thin dike model. The average depth obtained is 3.27 km.

2.2.2 Field example

Figure 8 shows a total-field magnetic anomaly above a Mesozoic diabase dike (2 m thick) intruded into Palaeozoic sediments of the Parnaiba Basin, Brazil (Silva, 1989; Fig. 10). The total-field data were collected along profile perpendicular to the dike, with sampling interval of about 2 m. The depth to the outcropping dike (sensor height) is 1.9 m. This anomaly profile of 24.64 m was digitised at an interval of 0.385 m.

Graticule spacings (km)	Computed depth from equation (14)(km)	Computed index parameter from equation (17) (degree)	Computed amplitude coefficient from equation (19)(nT)
0.77	2.09	36.58	-205.11
1.155	2.24	43.16	-247.27
1.54	2.38	48.29	-282.20
1.925	2.23	53.69	-250.60
2.31	2.15	56.78	-236.57
2.695	2.18	58.19	-245.18
3.08	2.24	58.79	-258.43
3.465	2.25	58.96	-265.55
3.85	2.26	58.77	-273.72
Average values	2.23 ± 0.1	52.58 ± 8.2	-215.60 ± 22.7

Table 5. Computed values of depth, index parameter, and amplitude coefficient obtained from a total-field magnetic anomaly above a Mesozoic diabase dike intruded into Palaeozoic sediments from the Parnaiba Basin, Brazil.

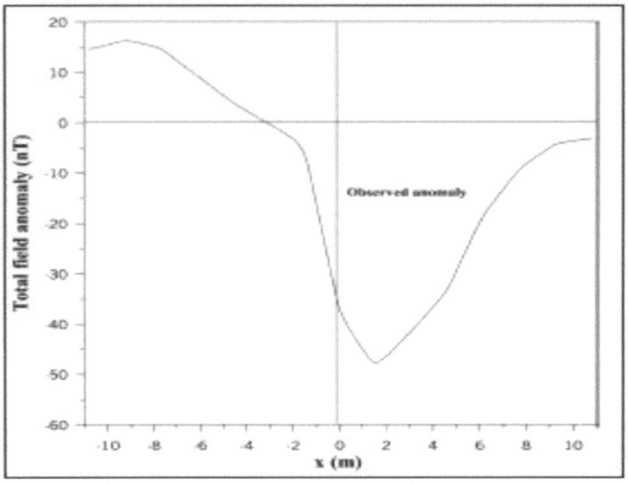

Fig. 8. A total field magnetic anomaly above a Mesozoic diabase dike (2 m thick) intruded into Palaeozoic sediments from the Parnaiba Basin, Brazil (after Silva, 1989).

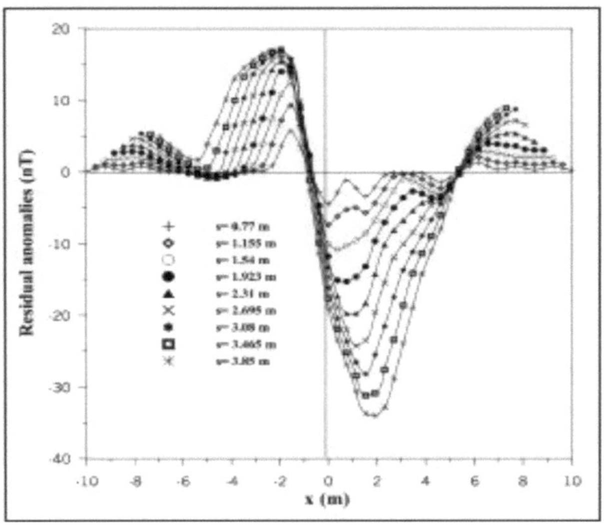

Fig. 9. Moving average residual magnetic anomalies above a Mesozoic diabase dike intruded into Palaeozoic sediments from the Parnaiba Basin, Brazil.

Adopting the same technique applied in the example above, we found the results given in Figure 9 and Table 5. The model parameters obtained are: $z = 2.22$ m, $\theta = 53^{o}$, and $K = -215$ nT. Silva (1989) applied a method to the same magnetic data employing the M-fitting technique and obtained a depth of about 3.5 m. The depth to the top is overestimated by both methods. This is reasonable because the upper part of the dike was weathered and the magnetite present was oxidized, losing most of its magnetic property (Silva, 1989). This field example shows also that there is a lack of "thinness". However, because many of the characteristics of wide dike anomalies are similar to those for thin dikes (Gay, 1963), our method can be applied not only to magnetic anomalies due to thin dikes, but also to those due to wide dikes to obtain reliable estimates of model parameters.

2.3 A window curves method to shape and depth solutions from third moving average residual gravity anomalies

Following Abdelrahman & El-Araby (1993), the general gravity anomaly expression produced by most geological structures is given by the following equation:

$$g(x_i,z,q)=\frac{A}{(x_i^2+z^2)^q}, \tag{30}$$

where q is the shape (shape factor), z is the depth, x is the position coordinate and A is amplitude coefficient related to the radius and density contrast of the buried structures. Examples of the shape factor for the semi-infinite vertical cylinder (3D), horizontal cylinder (2D), and sphere (3D) are 0.5, 1.0, and 1.5, respectively. Also, the shape factor for the finite vertical cylinder is approximately 1.0. The shape factor (q) approaches zero as the structure becomes a flat slab, and approaches 1.5 as the structure becomes a perfect sphere (point mass). The moving average (grid) method is an important yet simple technique for the separation of gravity anomalies into residual and regional components. The basic theory of the moving average methods is described by Griffin (1949), Agocs (1951), Abdelrahman & El-Araby (1996) and Abdelrahman et al. (2001).

Let us consider seven observation points (x_i-3s, x_i-2s, x_i - s, x_i, x_i + s, x_i +2s, and x_i+3s) along the anomaly profile where s=1, 2,..., M spacing units and is called the window length. The first moving average regional gravity field $Z_1(x_i, z, q, s)$ is defined as the average of $g(x_i-s, z, q)$ and $g(x_i+s, z, q)$, which for the simple shapes mentioned above can be written as

$$Z_1(x_i,z,s)=\frac{A}{2}\left\{((x_i-s)^2+z^2)^{-q}+((x_i+s)^2+z^2)^{-q}\right\}. \tag{31}$$

The first moving average residual gravity anomaly $R_1(x_i, z, q, s)$ at the point x_i is defined as (Abdelrahman & El-Araby, 1996).

$$R_1(x_i,z,q,s)=\frac{A}{2}\left[2(x_i^2+z^2)^{-q}-\left(((x_i-s)^2+z^2)^{-q}+((x_i+s)^2+z^2)^{-q}\right)\right]. \tag{32}$$

The second moving average residual gravity anomaly $R_2(x_i, z, q, s)$ at the point x_i is (Abdelrahman et al. 2001)

$$\begin{aligned}R_2(x_i,z,q,s)=\frac{A}{4}\Big[&6(x_i^2+z^2)^{-q}-4((x_i-s)^2+z^2)^{-q}-4((x_i+s)^2+z^2)^{-q}\\&+((x_i-2s)^2+z^2)^{-q}+((x_i+2s)^2+z^2)^{-q}\Big].\end{aligned} \tag{33}$$

The third moving average residual gravity anomaly $R_3(x_i, z, q, s)$ at any point x_i is

$$\begin{aligned}R_3(x_i,z,q,s)=\frac{A}{8}\Big[&20(x_i^2+z^2)^{-q}-15((x_i+s)^2+z^2)^{-q}-15((x_i-s)^2+z^2)^{-q}\\&+6((x_i+2s)^2+z^2)^{-q}+6((x_i-2s)^2+z^2)^{-q}\\&-((x_i+3s)^2+z^2)^{-q}-((x_i-3s)^2+z^2)^{-q}\Big].\end{aligned} \tag{34}$$

For all shape factors (q), equation (34) gives the following value at x_i=0

$$A = \frac{4R_3(0)}{\left[10z^{-2q} - 15(s^2+z^2)^{-q} + 6(4s^2+z^2)^{-q} - (9s^2+z^2)^{-q}\right]} \cdot \tag{35}$$

Using equation (34) and equation (35), we obtain the following normalized equation at x_i=s

$$\frac{R_3(s)}{R_3(0)} = \frac{\left[26(s^2+z^2)^{-q} - 16(4s^2+z^2)^{-q} + 6(9s^2+z^2)^{-q} - 15z^{-2q} - (16s^2+z^2)^{-q}\right]}{\left[20z^{-2q} - 30(s^2+z^2)^{-q} + 12(4s^2+z^2)^{-q} - 2(9s^2+z^2)^{-q}\right]}. \tag{36}$$

Equation (36) is independent on the amplitude coefficient (A). Let F= $R_3(s)/R_3(0)$ then from equation (36) we obtain

$$z = \left[\frac{-15}{F * P(s,z,q) + W(s,z,q)}\right]^{1/2q}, \tag{37}$$

Where

$$P(s,z,q) = \left[20z^{-2q} - 30(s^2+z^2)^{-q} + 12(4s^2+z^2)^{-q} - 2(9s^2+z^2)^{-q}\right],$$

and

$$W(s,z,q) = \left[16(4s^2+z^2)^{-q} - 6(9s^2+z^2)^{-q} + (16s^2+z^2)^{-q} - 26(s^2+z^2)^{-q}\right].$$

Equation (37) can be used not only to determine the depth, but also to simultaneously estimate the shape of the buried structure as will be illustrated by theoretical and field examples.

2.3.1 Synthetic example

The composite gravity anomaly in mGals of Figure 10 consisting of the combined effect of a shallow structure (horizontal cylinder with 3 km depth) and a deep-seated structure (dipping fault with depth to upper faulted slab = 15 km, depth to lower faulted slab = 20 km, and dip angle of fault plane = 50 degrees) was computed by the above expression:

$$\Delta g(x_i) = \frac{200}{(x_i^2 + 3^2)} + 100(\tan^{-1}(\frac{x_i - 5}{15} + \cot(50^o)) - \tan^{-1}(\frac{x_i - 5}{20} + \cot(50^o))), \tag{38}$$

(Horizontal cylinder + dipping fault)

The composite gravity field (Δg) was subjected to our third moving average technique. The third moving average residual value at point x_i is computed from the input gravity data $g(x_i)$ using the following equation:

$$R_3(x_i) = \frac{20g(x_i) - 15g(x_i+s) - 15g(x_i-s) + 6g(x_i-2s) + 6g(x_i+2s) - g(x_i+3s) - g(x_i-3s)}{8}. \tag{39}$$

Five successive third moving average windows (s= 2, 3, 4, 5, and 6 km) were applied to input data. Each of the resulting moving average profiles was analyzed by equation (37). For each window length, a depth value was determined iteratively for all shape values, and a window curve was plotted, illustrating the relation between the depth and shape.

We have also applied the first moving average method (Abdelrahman & El-Araby 1996) and the second moving average method (Abdelrahman et al. 2001) to the same input data generated from equation (38). The results in the case of using first, second, and third moving average techniques are shown in Figures 11, 12, and 13, respectively.

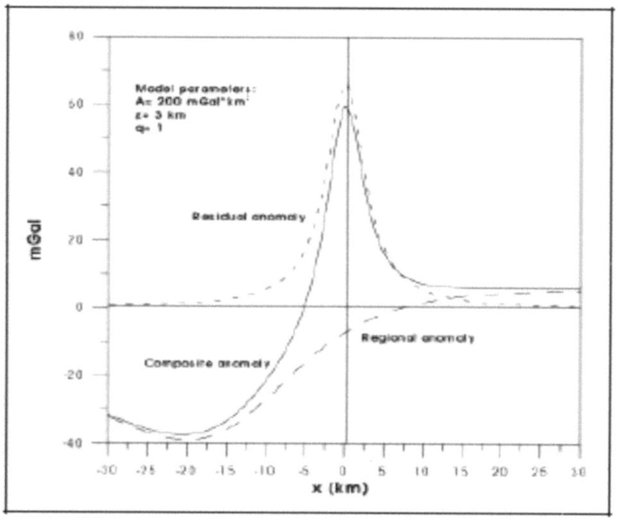

Fig. 10. Composite gravity anomaly (Δg) of a buried horizontal cylinder and a dipping fault as obtained from equation (38).

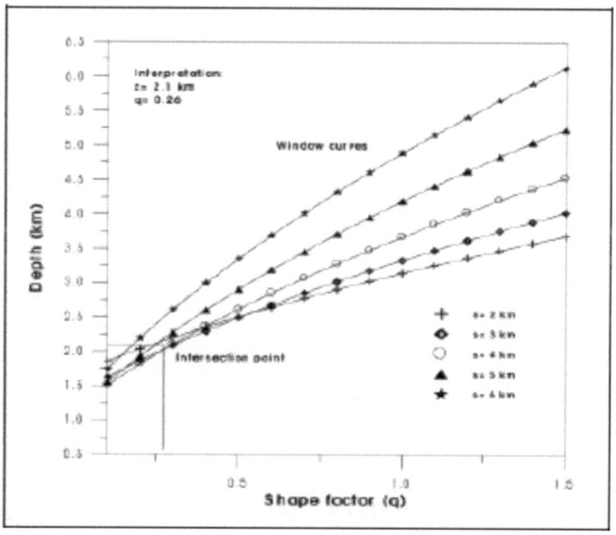

Fig. 11. Family of window curves of z as a function of q for s= 2, 3, 4, 5, and 6 km as obtained from gravity anomaly (Δg) using the first moving average method. Estimates of q and z are, respectively, 0.26, and 2.1 km.

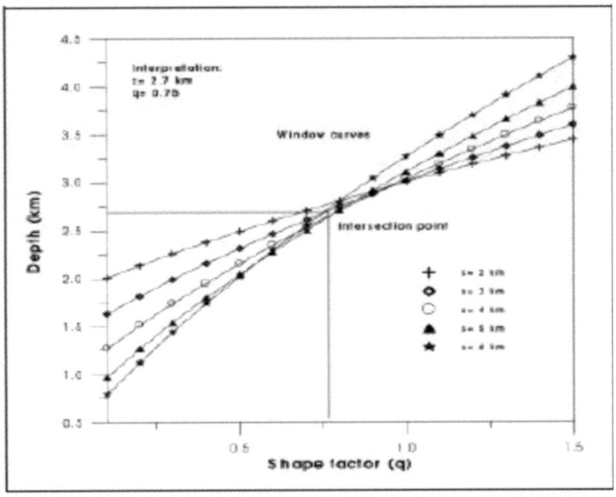

Fig. 12. Family of window curves of z as a function of q for s= 2, 3, 4, 5, and 6 km as obtained from gravity anomaly (Δg) using the second moving average method. Estimates of q and z are, respectively, 0.75, and 2.7 km.

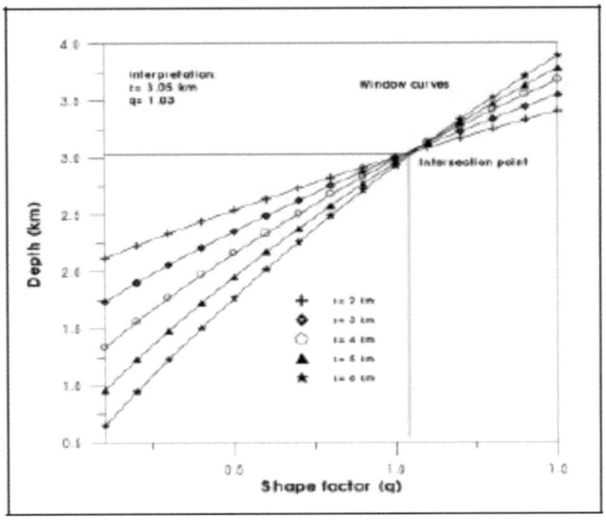

Fig. 13. Family of window curves of z as a function of q for s= 2, 3, 4, 5, and 6 km as obtained from gravity anomaly (Δg) using the present third moving average method. Estimates of q and z are, respectively, 1.03, and 3.05 km.

In the case of using first moving average technique, the window curves intersect at a narrow region where 0.55>q>0.05 and 2.5 km>z> 1.8 km (Fig. 11). The central point of this region occurs at the location q= 0.26 and z= 2.1 km. On the other hand, in the case of using second moving average technique, the curves intersect each other nearly at a point where q= 0.75 and z= 2.7 km (Fig. 12). Finally, in the case of using the third moving average method (present method), the curves intersect each other at the correct locations q= 1.03 and z= 3.05 km (Fig. 13). From these results (Figs. 11-13), it is evident that the first and the second moving average methods give erroneous results for the shape and depth solutions whereas the model parameters obtained from the third moving average method are in excellent agreement with the parameters of the shallow structure given in model [equation (38)]. The conclusion is inescapable that the third moving average technique gives the best result. Moreover, random errors were added to the composite gravity anomaly $\Delta g(x_i)$ to produce noisy anomaly. In this case, we choose a white random noise with amplitude being 5% of the maximum amplitude variation along the entire profile, so that the noise will be equally along the profile. The noisy anomaly is subjected to a separation technique using only the third moving average method. Five successive third moving average windows were applied to the noisy input data. Adapting the same interpretation procedure, the results are shown in Figure 14.

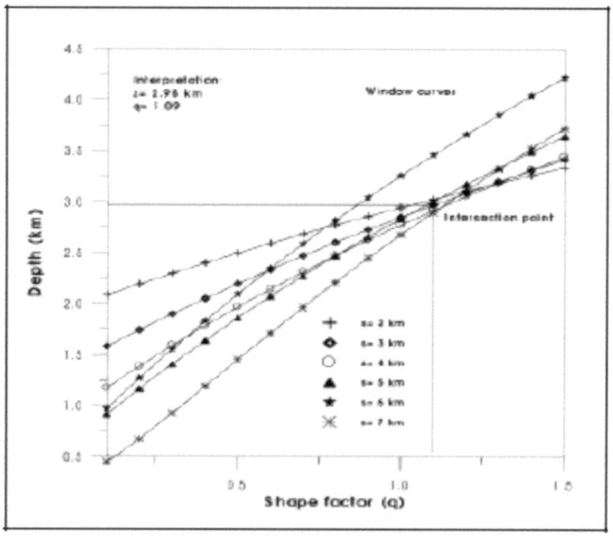

Fig. 14. Family of window curves of z as a function of q for s= 2, 3, 4, 5, 6, and 7 km as obtained from gravity anomaly (Δg) after adding 5% random errors to the data using the present third moving average method. Estimates of q and z are, respectively, 1.09, and 2.98 km.

Figure 14 shows the window curves intersect at approximately q= 1.09 and z= 2.98 km. The results (Fig. 14) are generally in very good agreement with the shape factor and depth parameters shown in [equation (38)]. This indicates that our method will produce reliable shape and depth estimates when applied to noisy gravity data.

2.3.2 Field example

A Bouguer gravity anomaly profile along AÁ of the gravity map of the Humble salt dome, near Houston (Nettleton, 1962, Fig. 22) is shown in Figure 15. The gravity profile has been digitized at an interval of 0.26 km. The Bouguer gravity anomalies thus obtained have been subjected to a separation technique using the third moving average method. Filters were applied in four successive windows (s= 2.60, 2.86, 3.12, and 3.38 km). In this way, four third moving average residual anomaly profiles were obtained (Fig. 16). The same procedure described for the synthetic examples was used to estimate the shape and the depth of the salt dome. The results are plotted in Figure 17. Figure 17 shows that the window curves intersect at a point where the depth is about 4.95 km and at q=1.41. This suggests that the shape of the salt dome resembles a sphere or, practically, a three-dimensional source with a hemi- spherical roof and root. This result is generally in good agreement with the dome form estimated from drilling top and contact (Nettleton, 1976; Fig. 8-16).

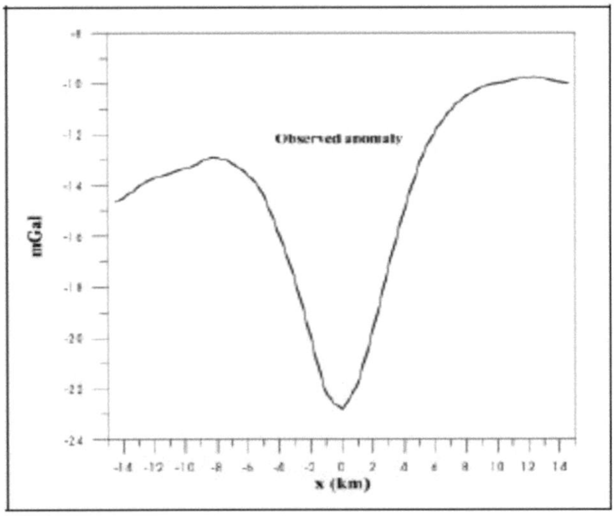

Fig. 15. Observed gravity profile of the Humble salt dome, near Houston, TX, USA (after Nettleton 1962)

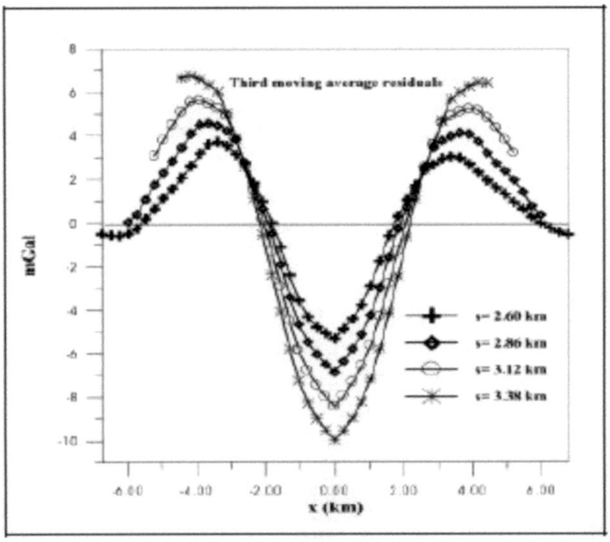

Fig. 16. Third moving average residual gravity anomalies on line AÁ of the Humble salt dome, for s= 2.60, 2. 86, 3.12 and 3.38 km.

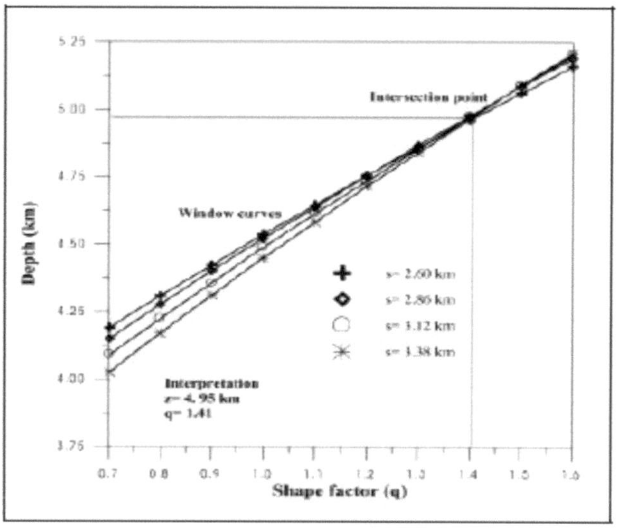

Fig. 17. Family of window curves of z as a function of q for s= 2.60, 2. 86, 3.12 and 3.38 km as obtained from the Humble gravity anomaly profile

3. Conclusion

Three different numerical algorithms to depth and other associated model parameters determination from magnetic and gravity data have been presented. In the first method, a least-squares minimization approach has been developed to determine the depth to a buried structure from magnetic data. The problem of determining the depth of a buried structure from the magnetic anomaly has been transformed into the problem of finding a solution of a nonlinear equation. The method presented is very simple to execute. The advantages of the present method over previous techniques, which use only a few points, distances, standardized curves, and nomograms are: (1) all observed values cal be used, (2) the method is automatic, and (3) the method is not sensitive to errors in the magnetic anomaly. Finally, the advantage of the proposed method over previous least-squares techniques is that any initial guess for the depth parameter works well. In the second method, the problem of determining the depth, index parameter, and amplitude coefficient of the buried thin dike from a magnetic anomaly profile has been transformed into the problem of solving two individual nonlinear equations and a linear equation, respectively. A scheme for interpreting the magnetic data to obtain the model parameters based on the least-squares method is developed. The method provides two advantages over previous least-squares techniques: (1) each model parameter is computed from all observed data, and (2) the method is less sensitive to errors in the magnetic anomaly. The method involves using a thin dike model convolved with the same moving average filter as applied to the observed data. As a result, the method can be applied not only to residuals, but also to measured magnetic data having noise, interference from neighboring magnetic rocks, and complicated regional. Synthetic and field studies demonstrate the efficiency of the present technique. Finally, the problem of determining the shape and depth from gravity data of long or short profile length can be solved using the window curves methods for simple anomalies. The window curves methods are very simple to execute and work well even when the data contains noise. The methods involve using simple models convolved with the same moving average filter as applied to the observed gravity data. As a result, the methods can be applied not only to residuals, but also to measured gravity data. Given their relative strength, the present techniques complement the existing methods of shape and depth determination from moving average residual gravity anomaly profiles and overcome some of their shortcomings. The present iterative methods converge to a depth solution for any shape factor value and when applying any window length. The methods are developed to derive shape–related parameter and depth and remove the regional trend from gravity data. These two parameters might be used to gain geological insight concerning the subsurface. However, the advantages of the third moving average method over the first and the second moving average techniques are: (1) the method removes adequately the regional field due to deep-seated structure from gravity data and (2) the method can be applied to large gridded data. Theoretical and field examples have illustrated the validity of the algorithms presented.

4. Acknowledgment

The author thanks Prof. Aleksander Lazinica, CEO, Prof. Hwee-San Lim, the Book editor and Mrs. Ana Skalamera, Publishing Process Manager for their excellent suggestions and collaborations. He would like to thank Prof. Sharf El Din Mahmoud, Chairman and Prof. El-Sayed Abdelrahman, Geophysics Department, Faulty of Science, Cairo University for their continuous support and encouragement.

5. References

Abdelrahman, E. M., & El-Araby, H. M. (1993). Shape and depth solutions from gravity data using correlation factors between successive least-squares residuals. *Geophysics,* Vol. 58, No. 12, pp. 1785-1791, ISSN 0016-8033.

Abdelrahman, E. M., & El-Araby, T. M. (1996). Shape and depth solutions from moving average residual gravity anomalies. *Journal of Applied Geophysics,* Vol. 36, No. 2-3, pp. 89-95, ISSN 0926-9851.

Abdelrahman, E. M., El-Araby, T. M., El-Araby, H. M., & Abo-Ezz, E. R. (2001). A new method for shape and depth determinations from gravity data. *Geophysics,* Vol. 66, No. 6, pp. 1774-1780, ISSN 0016-8033.

Abdelrahman, E. M., & Essa, K. S. (2005). Magnetic interpretation using a least-squares, depth-shape curves method. *Geophysics,* Vol. 70, No. 3, pp. L23–L40, ISSN 0016-8033.

Agocs, W. B. (1951). Least squares residual anomaly determination. *Geophysics,* Vol. 16, No. 4, pp. 686-696, ISSN 0016-8033.

Atchuta Rao, D. A., & Ram Babu, H. V. (1980). Properties of the Relation Figures between the Total, Vertical, and Horizontal Field Magnetic Anomalies over a Long Horizontal Cylinder Ore Body. *Current Science,* Vol. 49, No. 15, pp. 584–585, ISSN 001-3891.

Atchuta Rao, D., Ram Babu, H. V., & Sankar Narayan, P. V. (1980). Relationship of magnetic anomalies due to subsurface features and the interpretation of sloping contacts. *Geophysics,* Vol. 45, No. 1, pp. 32-36, ISSN 0016-8033.

Chai, Y., & Hinze, W. J. (1988). Gravity inversion of an interface above which the density contrast varies exponentially with depth. *Geophysics,* Vol. 53, No. 6, pp. 837–845, ISSN 0016-8033.

Demidovich, B., & Maron, I. A. (1973). *Computational mathematics,* MIR Publishers, ISBN B000R05KIU, Moscow.

Essa, K. S. (2007), A simple formula for shape and depth determination from residual gravity anomalies. *Acta Geophysica,* Vol. 55, No. 2, pp. 182-190, ISSN 1895-6572.

Essa, K. S. (2011), A new algorithm for gravity or self-potential data interpretation. *Journal of Geophysics and Engineering,* Vol. 8, No. 3, pp. 434-446, ISSN 1742-2132.

Gay, P. (1963). Standard curves for interpretation of magnetic anomalies over long tabular bodies. *Geophysics,* Vol. 28, No. 2, pp. 161-200, ISSN 0016-8033.

Griffin, W. R. (1949). Residual gravity in theory and practice. *Geophysics,* Vol. 14, No. 1, pp. 39-56, ISSN 0016-8033.

Keating, P., & Pilkington, M. (2004). Euler deconvolution of the analytic signal and its application to magnetic interpretation. Geophysical Prospecting, Vol. 52, No. 3, pp. 165 - 182, ISSN 1365-2478 .

Martin-Atienza, B., & Garcia-Abdeslem, J. (1999). 2-D gravity modeling with analytically defined geometry and quadratic polynomial density functions. *Geophysics,* Vol. 64, No. 6, pp. 1730–1734, ISSN 0016-8033.

Martins, C. M., Barbosa, V. C. F., & Silva, J. B. C. (2010). Simultaneous 3D depth-to-basement and density-contrast estimates using gravity data and depth control at few points. *Geophysics,* Vol. 75, No. 3, pp. I21–I28, ISSN 0016-8033.

McGrath, P. H., & Hood, P. J. (1973). An automatic least-squares multimodel method for magnetic interpretation. *Geophysics,* Vol. 38, No. 2, pp. 349-358, ISSN 0016-8033.

Mohan, N. L., Sundararajan, N. & Seshagiri Rao, S. V. (1982). Interpretation of some two-dimensional magnetic bodies using Hilbert transform. *Geophysics,* Vol. 47, No. 3, pp. 376-387, ISSN 0016-8033.

Nabighian, M. N. (1984). Toward a three-dimensional automatic interpretation of potential field data via generalized Hilbert transforms: Fundamental relations. *Geophysics,* Vol. 49, No. 6, pp. 780-786, ISSN 0016-8033.

Nettleton, L. L. (1962). Gravity and magnetic for geologists and seismologists. AAPG, Vol. 46, No. , pp. 1815-1838, ISSN 0149-1423.

Nettleton, L. L. (1976). *Gravity and magnetics in oil prospecting,* Mc-Graw Hill Book Co., ISBN 9780070463035, New York.

Nunes, T. M., Barbosa, V. C. F., & Silva, J. B. C. (2008). Magnetic basement depth inversion in the space domain. *Pure Applied Geophysics,* Vol. 165, No. 9-10, pp. 891–1911, ISSN 0033-4553.

Prakasa Rao, T. K. S. & Subrahmanyam, M. (1988). Characteristic curves for the inversion of magnetic anomalies of spherical ore bodies. *Pure and Applied Geophysics,* Vol. 126, No. 1, pp. 69-83, ISSN 0033-4553.

Prakasa Rao, T. K. S., Subrahmanyam, M., & Srikrishna Murthy, A. (1986). Nomogram for the direct interpretation of magnetic anomalies due to long horizontal cylinders. *Geophysics,* Vol. 51, No. 11, pp. 2156-2159, ISSN 0016-8033.

Press, W. H., Flannery, B. P., Teukolsky, S. A., & Vetterling, W. T. (1986). *Numerical recipes, the art of scientific computing,* Cambridge University Press, ISBN 0-521-43064-X, Cambridge.

Salem, A., Ravat, D., Mushayandebvu, M. F., & Ushijima, K. (2004). Linearized least-squares method for interpretation of potential field data from sources of simple geometry. *Geophysics,* Vol. 69, No. , pp. 783–788, ISSN 0016-8033.

Silva, J. B. C. (1989). Transformation of nonlinear problems into linear ones applied to the magnetic field of a two-dimensional prism. *Geophysics,* Vol. 54, No. 1, pp. 114-121, ISSN 0016-8033.

Silva, J. B. C., Oliveira, A. S., & Barbosa, V. C. F. (2010). Gravity inversion of 2D basement relief using entropic regularization. *Geophysics,* Vol. 75, No. 3, pp. I29–I35, ISSN 0016-8033.

Stanley, J. M. (1977) Simplified magnetic interpretation of the geologic contact and thin dike. *Geophysics,* Vol. 42, No. 6, pp. 1236-1240, ISSN 0016-8033.

Tarantola, A. (2005). *Inverse problem theory and methods for model parameter estimation,* SIAM, ISBN 978-0-898715-72-9, Philadelphia, USA.

Thompson, D. T. (1982). EULDPH: A new technique for making computer-assisted depth estimates from magnetic data. *Geophysics,* Vol. 47, No. 1, pp. 31-37, ISSN 0016-8033.

Tikhonov, A. N., & Arsenin, V. Y. (1977). *Solutions of ill-posed problems,* Winston & Sons, ISBN 0-521-43064-X, Washington, DC, USA.

Zhang, J., Zhong, B., Zhou, X., & Dai, Y. (2001). Gravity anomalies of 2-D bodies with variable density contrast. *Geophysics,* Vol. 66, No. 3, pp. 809–813, ISSN 0016-8033.

Zhang, C., Mushayandebvu, M. F., Reid, A. B., Fairhead, J. D., & Odegard, M. E. (2000). Euler deconvolution of gravity tensor gradient data. *Geophysics,* Vol. 65, No. 2, pp. 512-520, ISSN 0016-8033.

Zhdanov, M. S. (2002). *Geophysical inverse theory and regularization Problems,* Elsevier, ISBN 0-444-51089-3, Amsterdam.

3

Geophysics in Near-Surface Investigations

Jadwiga A. Jarzyna et al.*
AGH University of Science and Technology,
Faculty of Geology Geophysics and Environmental Protection, Krakow,
Poland

1. Introduction

Environmental and engineering geophysics are new branches of geophysics that focus on data collection and an analysis of the present day environment. Noninvasive, nondestructive, and inexpensive geophysical technologies are now capable of providing detailed 3D and 4D representations of the subsurface. These advances in geophysical survey techniques have not only enhanced traditional applications, such as prospecting geophysics, but also made way for development of new methodologies in near surface investigations. Environmental and engineering geophysics includes mapping of shallow geological formations, prospecting for resources located in the near surface, monitoring aquifers, mapping anthropogenic effects on the environment, and surveying for civil engineering and archaeological projects. Environmental and engineering geophysics measures parameters such as salinity of subsurface fluids, levels of radioactivity, and other soil properties affected by industrial activity.

The increasing demand for sustainable development as well as community efforts to preserve and conserve the environment have fostered the development of engineering technologies that are more environmentally friendly than traditional geophysical tools. Newer techniques are often used to monitor and protect the environment, as well as evaluate geotechnical risks of various natural and human - made structures. The increasing number of these modern geophysical applications, and of professionals specializing in them, bear witness to their emerging significance.

2. Petrophysical background for geophysical investigations (Jadwiga A. Jarzyna)

Petrophysics is a subdiscipline of geophysics that addresses the physical properties and other parameters of rock and other subsurface material. Applied geophysics combines the theoretical foundation of a given geophysical method with empirical understanding of the materials involved. In the case of petrophysics, theoretical parameters of rock and other subsurface materials (i.e. their physical properties) are used to interpret empirical

* Jerzy Dec, Jerzy Karczewski, Sławomir Porzucek, Sylwia Tomecka-Suchoń,
Anna Wojas and Jerzy Ziętek
AGH University of Science and Technology, Faculty of Geology Geophysics and Environmental Protection, Krakow, Poland

observations (i.e. quantities, subsurface extent etc.). For example, geoeletrical investigations use Archie's law (theoretical) to interpret voltage and current measurements (empirical) to gain insight into the effective porosity and resistivity of the materials under study.

Petrophysics provides empirical information on rocks from laboratory measurements and field investigations. Laboratory measurements offer a direct way to determine rock parameters, but the results are not always directly applicable to the real world. The relatively small size of samples studied in typical laboratory experiments, and the complexity of the natural environment bedevil integration of laboratory results not just in applied geophysics, but in nearly all Earth science disciplines. High costs of drilling and limited number of cores can also constrain the practicality of investigations that rely heavily on parameters gathered through laboratory experiments.

In spite of these constraints, petrophysics aims to consistently integrate information from laboratory and field measurements. In subsurface investigations for example, the accuracy and resolution of geological profiles depend heavily on parameters provided by field measurements. Field measurements such as resistivity, measured by surface geoelectric methods and resistivity well logs, in turn reflect both the heterogeneity of subsurface rock formations and operational characteristics of the device used to perform the measurement. Given the variability among these factors, statistical methods can be used to synthesize and interpret data obtained using different methodologies. Empirical formulas generated through petrophysical investigations can also be used to standardize, calibrate and scale geophysical devices.

In practice, petrophysics often provides geophysical models with estimates of subsurface parameters. Estimated parameters can include bulk density, resistivity, dielectric permittivity, velocity of elastic waves, and magnetic susceptibility. In reservoir characterization for example, petrophysics parameterizes porosity, permeability and filtration factors. Accurate and consistent estimation of these parameters requires understanding of mineral composition, porosity, and properties of the interstitial media, (i.e. salinity, temperature, etc.). In larger scale studies of anisotropy, petrophysics provides estimates of structure, texture and facies type for materials under investigation.

Petrophysical analysis begins with the assumptions that the subsurface is heterogeneous and may undergo change. Subsurface materials may vary in terms of their geometry, size, composition, structure and other properties. Petrophysical and petrochemical variables address heterogeneity in these properties by averaging them over unit volumes. Subsurface materials are also affected by temperature, pressure and the resonant frequencies of physical fields used to evaluate them. Near surface investigations are less dependent on subsurface pressure and temperature conditions than traditional (deeper) geophysical applications, but these factors still come into play. Lack of compaction or varying degrees of consolidation in near surface materials can result in differing resistivity and bulk density for rocks of identical lithology, but occurring at different depths. Physical fields are also affected by consolidation and compaction. Loose subsurface sands and gravels, as well as sandstones and mudstones in the aeration zone have anomalously high resistivity for example, due to the air content of their pore space.

Because most of the near surface is composed of sediments and sedimentary rock, this chapter focuses on these materials. The chapter first outlines the basics of petrophysical properties, and then describes their use in environmental and engineering geophysical applications. The most basic petrophysical property of rock is that of density. **Bulk density** is defined as the ratio of a rock's mass to its volume. It is lower or equal to matrix density (mineralogical density, specific density), which is defined as the ratio of the mass of a rock in powdered form, to its volume. Bulk density of sedimentary rocks, δ_b, depends on the density of minerals (matrix material), δ_m; porosity, Φ; and density of the pore space media, δ_f (2.1).

$$\delta_b = (1-\Phi)\delta_m + \Phi\delta_f = \delta_m - (\delta_m - \delta_f)\Phi \qquad (2.1)$$

Bulk density is intuitively inversely proportional to porosity, but it is specifically a function of matrix and pore media density. For example, gaseous hydrocarbons in the pore space of a reservoir rock will significantly decrease its bulk density. Most rock-forming minerals have bulk densities ranging from 2200 to 3500 kg/m^3. Ore bodies and associated minerals have average densities of 4000 to 8000 kg/m^3. Sedimentary rocks are composed of matrix materials that range in average density from 2500 to 2900 kg/m^3, of fluids ranging from 800 to 1240 kg/m^3, and of gases that have average densities of less than 1000 kg/m^3 (Kobranova, 1989). Given variation in the densities of sedimentary components, the range in bulk density can be quite broad from lignite (1000 - 1300 kg/m^3); clay (1300 - 2300 kg/m^3); sand and gravel (1400 - 2300 kg/m^3); loam (1500 - 2200 kg/m^3); sandstone (2000 - 2800 kg/m^3); shale (2300 - 2800 kg/m^3); limestone (2300 - 2900 kg/m^3); and dolomite (2400 - 2900 kg/m^3); (Schön, 2004). **Porosity** is an important property of sedimentary rocks that often enters into petrophysical considerations. Total porosity is defined as the volume of free space per given volume of rock. Effective porosity refers only to the volume of connected pores, fractures, fissures and vugs. Porosity can be modelled as spherical volumes of various radii (Kobranova, 1989) but is more reliably measured in the laboratory using rock samples or well log data. Porosity is related to texture, mineral composition, sedimentary fabric and

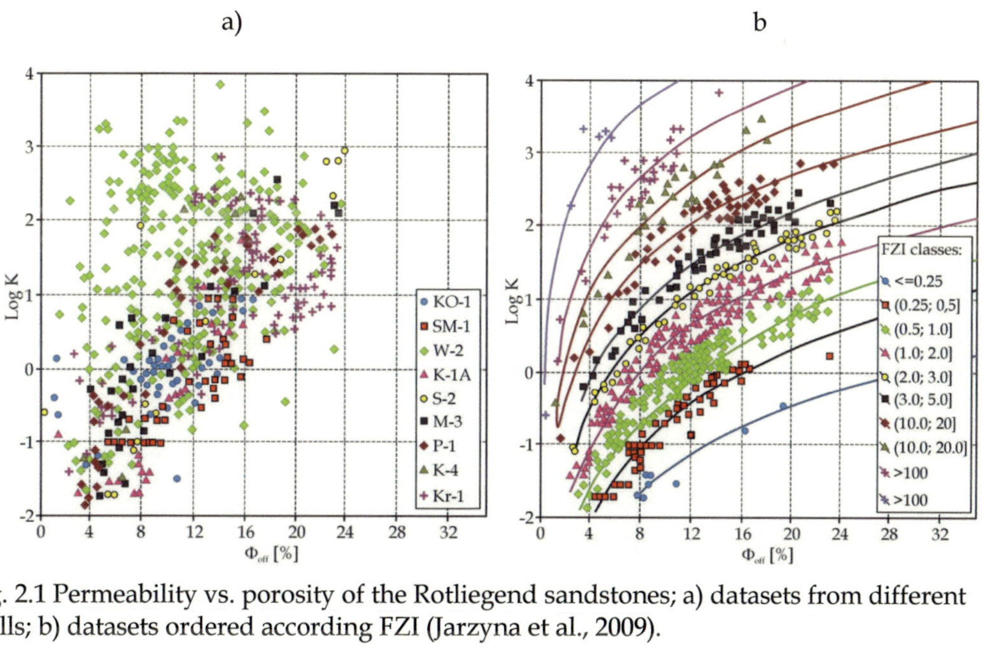

Fig. 2.1 Permeability vs. porosity of the Rotliegend sandstones; a) datasets from different wells; b) datasets ordered according FZI (Jarzyna et al., 2009).

degree of lithification. Clay content reduces porosity. Permeability is the ability of a material to transmit pore media, which depends heavily on the material's porosity. A functional alternative to calculating a rock's hydraulic properties as separate porosity and permeability variables is the Flow Zone Index (FZI). FZI depends on permeability and porosity, specifically considers tortuosity of pore space and specific surfaces without parameterizing either of these quantities. The advantages of using FZI over porosity and permeability are presented in figure 2.1 (Tiab & Donaldson, 2004; Jarzyna et al., 2009).

Resistivity and **dielectric permittivity** refer to electric properties of rocks. Both properties depend on mineral composition, porosity and properties of the pore space media. Dielectric permittivity is inversely proportional to the electromagnetic (EM) field frequencies often used in geophysical measurements. Factors such as water saturation may also influence the dielectric permittivity of subsurface materials. Frequency dependence makes it a dispersive parameter. Because the dielectric permittivity of water is relatively high compared to permittivity values for minerals and hydrocarbons, this parameter can be used to distinguish hydrocarbon plumes in water saturated materials. In terms of resistivity, silicates and carbonates have high specific resistivities (>10^9 Ohmm; Kobranova, 1989) and are thus classified as insulators. Sulfides and some oxides meanwhile are classified as mineral conductors (specific resistivities between 10^{-6} and 10^{-1} Ohmm), as are native metals (10^{-8} to 10^{-5} Ohmm). Variation in the specific resistivities of various minerals can be due to the presence of impurities, subsurface structural effects, and anisotropy. Relative dielectric permittivity, ε_r, is defined as the dielectric permittivity of rock, ε, divided by the dielectric permittivity of vacuum, ε_0. For most rock forming minerals, ε_r ranges between 4 and 10 (Table 2.1). Relative dielectric permittivity is correlated to bulk density (2.2). The generalized formula (2.2) applies to most minerals with the exception of hydrated forms of montmorillonite and members of the sulfide group (Schön, 2004).

$$\varepsilon_r = (1.93 \pm 0.17)^{\delta_b} \tag{2.2}$$

Mineral	*V_P [km/s]	*V_S [km/s]	*δ_b [kg/m^3]	**ε_r [MHz]	Rock	V_P [km/s]	E [GPa]	σ
Quartz	5.492	4.119	2650		Sandstone^	0.8-4.5	30-100	0.06-0.6
Calcite	6.403	3.436	2710	6.35; 7.5-8.7	Sandstone#	1.4-4.3	3-41	0.13-0.33
Orthoclase	5.690	3.260	2570	5.6; 4.5-6.2	Sandy shale^	1.45-5.18	5-69	0.12-0.21
Dolomite	7.007	4.293	2870	7.46; 6.3-8.2	Sand (dry)#	0.2-1.0	0.03-0.72	0.405
Anhydrite	6.096	3.126	2960	6.5; 5.7-6.7	Sand (wet)#	0.8-2.2	0.55-4.18	0.405
Siderite	6.959	3.590	3960	9.3; 5.2-7.4	Clay^	0.3-3	30	0.25-0.45
Pyrite	8.021	5.166	5020	33.7-81	Clay#	1.0-2.5	0.78-4.91	0.405
Hematite	6.626	4.233	5270	25	Shale (slate)^	2.3-6.65	24-72	0.17
Magnetite	4.175	1.966	5180	33.7-81	Limestone^	1.0-7.0	13-175	0.18-0.31
Kaolinite	1.438	0.929	2610	11.8; 9.1	Limestone#	5.9-6.1	55-63	0.34-0.354
Biotite	6.220	3.717	3010	6.3; 6.2-9.3	Marl^	1.3-4.5	10-135	0.11-0.23
Halite	4.549	2.628	2170	5.9; 5.7-6.2	Granite#	5.5-5.9	56-64	0.325

Table 2.1 Petrophysical properties of minerals and rocks (*after Halliburton, 1991; ** Schön, 2004; ^Kobranova, 1989; #calculated on the basis of averaged data).

The specific resistivity of air equals 10^{14} Ohmm, and is similar to that of hydrocarbons. Electrolytes are ionic conductors. Conductivity of porous, fractured rocks is caused by ionic movement in subsurface fluids. Under these conditions, resistivity depends primarily on mineral type, rock fabric, temperature of the subsurface fluid, and the volume of connected pores. The resistivity of pure water (containing only H^+ and OH^- ions) is very high (> 10^5 Ohms; Hearst et al., 2000). Conductivity, σ_w, and resistivity of subsurface water (ρ_w), depends on several factors (2.3; 2.4):

$$\sigma_w = \frac{1}{\rho_w} \cong \sum_{i=1}^{n} \alpha_i c_i z_i v_i \qquad (2.3)$$

$$\rho_w(t_2) = \rho_w(t_1)\frac{t_1 + 21.5}{t_2 + 21.5} \qquad (2.4)$$

In equations 2.3 and 2.4, n is the number of components in the subsurface water, α is the degree of dissociation, c is the concentration, z is the valence, and v is the mobility. Factors of α and v depend on temperature. Models often assume that pore fluid is a simple sodium-chloride solution. Formula 2.4 shows that pore fluid resistivity is temperature dependent. The variables t_1, t_2 are temperatures given in Celsius degrees. The relative dielectric permittivity of pore water, ε_r=81 (at t=21°C) differs from that of oil and gas (ε_r < 3). Salt concentration, c_{mol} (given in moles per liter), exerts a relatively small influence on the relative dielectric permittivity of pore fluids. The relative dialectric permittivity of pore fluids, ε_r, is related to c_{mol} as follows (Schön, 2004):

$$\varepsilon_r = \varepsilon_{r\_pure\_water} - 13 \cdot c_{mol} + 1.065 \cdot c^2_{mol} - 0.03006 \cdot c^3_{mol} \qquad (2.5)$$

Dielectric permittivity, ε_r, the velocity of electromagnetic waves, V_{EM}, and EM attenuation, A_{EM}, are used in georadar investigations. The dielectric permittivity of porous rocks strongly depends on the degree of water saturation. The salinity of pore water does not strongly affect ε_r (2.5), but does influence conductivity and attenuation of electromagnetic waves. Selected values for the electromagnetic properties of different pore media as well as surface and subsurface materials are presented below in Table 2.2.

Material	etha-nol	glyce-rine	oil	petro-leum	ice	ice *	ice **	snow ^	snow ^^
ε_r	25.8	56.2	2.0-2.7	2.0-2.2	3.1	4,15	3.2	1.2	1.55
Material	water (fresh)	water (saline)	ice #	silts #	sand (dry)	sand (wet)	air	clay	peat
ε_r	80	80	3-4	5-30	4	25	1	5-40	60-80
V_{EM} [m/ns]	0.03	0.03	0.16	0.07	0.15	0.06	0.30	0.05-0.13	0.03-0.04
A_{EM} [dB/m]	0.1	1000	0.01	1-100	0.01	0.03-0.3	0	1-300	0.3

Table 2.2 Relative dielectric permittivity of selected materials; *from pure distilled water (-12°C, 10^6 Hz), **from pure distilled water (-12° C, 10^9 Hz), ^freshly fallen, hard packed, followed by rain (-20° C, 10^6 - 10^9 Hz), ^^freshly fallen, hard packed, followed by rain (-6° C, 10^6 Hz) (after Schön, 2004; # from http://www.physics.utoronto.ca/~exploration/courses).

For rocks that have no clay minerals, Archie's laws (2.6 and 2.7) describe the relationships among the resistivity of fully saturated rock (R_0), the resistivity of rock partially saturated with water and hydrocarbons (R_t), effective porosity (Φ_{eff}), resistivity of formation water (R_w), and water saturation (S_w) or hydrocarbon saturation ($1-S_w$). Relationship I vs. S_w (2.7; Archie law) may be nonlinear on a bi-logarithmic scale, especially if there are several different brine or porosity systems within the subsurface formation, or if the formation has a significant fine-grained lithological component (≤ mud sized particles).

$$F = \frac{R_0}{R_w} = \frac{1}{\Phi_{ef}{}^{m}} \tag{2.6}$$

$$I = \frac{R_t}{R_0} = \frac{1}{S_w{}^{n}} \tag{2.7}$$

In equations 2.6 and 2.7, F is the formation factor, I is the resistivity index, m is a cementation factor ranging from 1.3 for unconsolidated sediments, to 2.2 for lithified, non-porous, non-permeable rock, and n is the saturation exponent, ranging from 1.12 to 2.55 for sandstone, and from 1.1 to 2.38 for limestone. The variables m and n are empirically determined (Tiab & Donaldson, 2004).

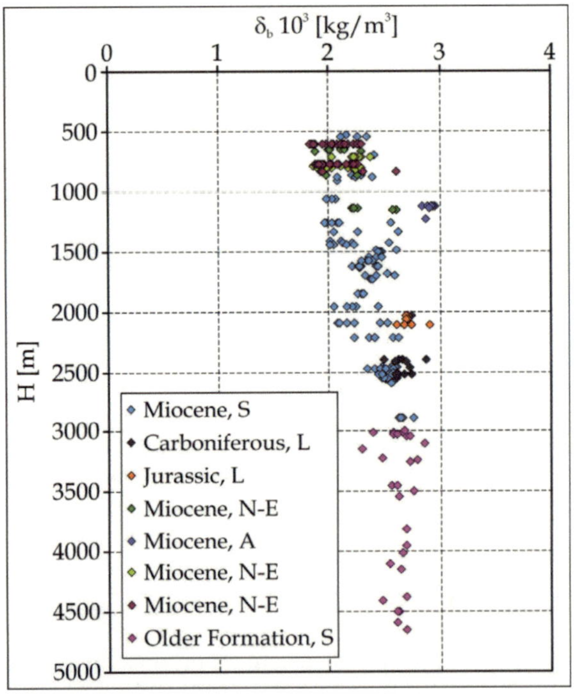

Fig. 2.2 Bulk density vs. depth; sandstone, S, limestone, L, anhydrite, A, Miocene N-E - north-eastern Carpathian Foredeep.

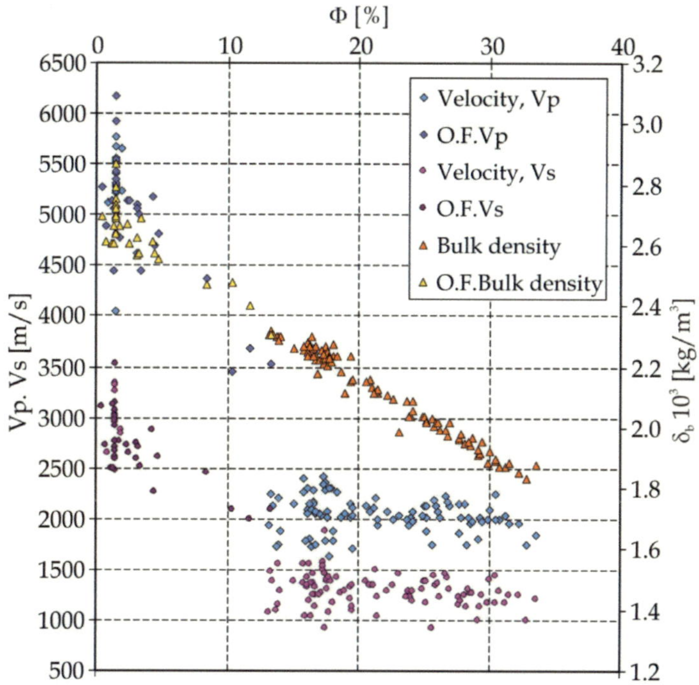

Fig. 2.3 Velocity and bulk density vs. porosity; Miocene sandstones and anhydrites; O.F.- units older than Permian.

Velocities of elastic waves used in seismic surveys (longitudinal P-waves, and shear S-waves) strongly depend on porosity and elastic moduli of the subsurface matrix, and on the type and properties of the pore space media. P- and S-wave velocities, along with bulk density and lithological information are used to calibrate seismic profiles. The velocity ratio, V_P/V_S, and acoustic impedance can indicate differences in lithology. Many petrophysical parameters are strongly depth dependent, and wave velocities in particular can be obscured by depth dependent changes in materials (Fig. 2.2). Porosity typically reduces velocity (Fig. 2.3.), as described by the Wyllie equation, the Raymer-Hunt-Gardner formula, and by others (Schön, 2004). A velocity correction for lack of compaction is necessary for empirical analysis of unconsolidated near-surface materials. Clays lower the elastic moduli of rocks and thus significantly reduce velocities of elastic waves. Velocities in clay-rich materials are lower than those in quartz- or calcite- rich materials (Table 2.1) while the compressibility of clay is higher than that of quartz and calcite. Clay properties are also strongly influenced by water volume. The above factors lead to complex relationships among the physical properties of subsurface materials that have high clay and water content.

Wave velocities in frozen unconsolidated sediments (permafrost) are similar to velocities in consolidated sediments, as ice can serve as a cement. Pronounced differences in the V_P/V_S ratio for gas- and water-saturated rocks are useful in energy exploration and cases of subsurface hydrocarbon contamination. Differences in the V_P/V_S ratio for gas and liquid saturated materials arise from the differential influence of pore space media on P- and S-wave velocities (differing effects on V_P and V_S, respectively). Velocities (for Poisson ratio, σ) and bulk density (for Young modulus, E, and others) can be used to calculate the dynamic elastic moduli of rocks. Sonic well logs record full, acoustic waveforms for *in situ* elastic wave velocity analysis. In unconsolidated, near-surface sediments, shear wave velocity can be lower than the velocity of borehole mud. Under these conditions, it is necessary to use a formula derived from the Stoneley wave equation in order to calculate S-wave velocity (Jarzyna et al., 2010). Sedimentary rocks are often anisotropic and well layered. When measured parallel to sedimentary layers, velocity, $V_{P//}$, and Young's modulus, $E_{//}$, are greater than $V_P{\perp}$ and $E{\perp}$, measured perpendicular to layers. The anisotropic coefficient of longitudinal wave velocity can reach values of 1.2 to 1.3. These values can exceed transverse wave coefficients and anisotropic coefficients pertaining to Young's modulus (~1.1 to 2.0; Kobranova, 1989). In finely laminated rocks, such as shale-rich sandstones, elastic waves having wavelengths longer than the thickness of the individual laminae must be normalized in order to calculate the effective properties of the material.

Magnetic susceptibility is an important parameter for soils and shallow subsurface rocks containing ferrimagnetic minerals. The magnetic susceptibility of soil and rock depends on shape, size and concentration of susceptible minerals. Most minerals are dia- and paramagnetic. Antiferrimagnetic minerals are relatively rare and ferromagnetic minerals are very rare (Kobranova, 1989). Specific volumetric susceptibility, κ, and mass susceptibility, χ, respectively, are proportionality coefficients in the presented below mutual relationships:

$$\vec{J}_v = \kappa \cdot \vec{H} \tag{2.8}$$

$$\vec{J}_m = \chi \cdot \vec{H} \tag{2.9}$$

$$\chi = \kappa / \delta_m \tag{2.10}$$

In equations 2.8 and 2.9, $\vec{H}$ is magnetic field, $\vec{J}_v$ is specific volume, and $\vec{J}_m$ is specific mass magnetization.

The magnetic susceptibility of dia- and paramagnetic native elements is governed by their chemical properties. For multi-element minerals, magnetic susceptibility is governed by stoichiometric composition and the arrangement of elements in the crystal lattice. The specific volumetric magnetic susceptibility of diamagnetic minerals is negligible and negative in value, ranging from -5.024 to -1.63 x 10^{-4} SI units (Kobranova, 1989). The most common minerals found in sedimentary rocks are diamagnetic (i.e. quartz, calcite, feldspar, dolomite, gypsum, halite, etc). Trace minerals such as biotite, pyrite, ilmenite, siderite, chlorite, and clays are paramagnetic or paraferrimagnetic. Accessory minerals such as magnetite, maghemite, hematite, and goethite are ferro- or ferrimagnetic. Magnetite is one of the most susceptible minerals, with 1.25 to 25 or more SI units. Water and oil are diamagnetic ($\kappa_w=\chi_w=$ -0.9 10^{-5} and $\kappa_o=\chi_o=$ -1.04 10^{-5} SI units, respectively). The influence of the mineral matrix on the susceptibility of the pore fluid is insignificant. Gases, including gaseous hydrocarbons have much lower magnetic susceptibility than liquids. Oxygen is paramagnetic with a relatively low, κ_{oxyg} = 0.17 x 10^{-5} SI units. The magnetic susceptibility of air, κ_{air}, equals 0.04 x 10^{-5} SI units. The ferromagnetic properties of magnetized rocks are the basis of magnetic surveys, paleogeographic reconstructions, and the paleomagnetic record.

Iron-oxide minerals are often used as paleoenvironmental proxies due to their high magnetic susceptibility, and because environmental conditions can be interpreted from iron oxide composition and grain size distribution.

3. Near surface georadar investigations (Jerzy Karczewski & Jerzy Ziętek)

In recent years, ground penetrating radar (GPR) has become the most popular geophysical method for near surface investigations. GPR belongs to a group of radio wave methods which evaluate electromagnetic wave propagation within a geological medium. The most popular GPR apparatus is the impulse type, which uses transmitting antennas to emit short electromagnetic pulses of 0.5 to 10 ns in the 10MHz to 6GHz frequency range (f). Under favorable conditions, low frequency antennas (f < 50 MHz) can be used to map the structure of the subsurface down to depths of a few dozen meters. Higher frequency pulses increase the resolution of shallow features, but decrease the depth range of the survey. EM waves from GPR devices can be attenuated, reflected or refracted. Reflection or refraction coefficients of EM waves for a given subsurface boundary layer will depend on different dielectric permittivities (ε) of the materials involved. Dielectric permittivity also influences velocities of the EM waves. Wave attenuation depends on wave frequency and the conductivity of the media, which in turn is strongly influenced by depth. While GPR devices are cheap and relatively easy to operate, GPR echograms are difficult to interpret and require knowledge of subsurface structure and petrophysical properties. Methods for processing GPR field data can enhance signal to noise ratios and the accuracy of reflector images. The most common GPR processing procedures include DC-Bias, time gain, filtration in the frequency domain, f/k filtration, deconvolution, migration and others (Annan, 2001; Daniels, 2004; Jol, 2009).

3.1 Landslide investigations

Landslides can damage infrastructure and pose a variety of other environmental risks. Mass wasting during a landslide event strongly depends on topography, hydrology, the structure of underlying bedrock, soil and bedrock type, and other factors. Strategies for landslide risk mitigation require a detailed understanding of the internal structure of the landslides, especially the slide surface. GPR surveys performed to image a landslide surface should profile the main landslide axis, as well as an axis perpendicular to the slide if possible.

For landslides that cut through inclined layers, a topographic correction is necessary to properly locate and map the slide surface. Stacking of multiple signals during data acquisition and maximizing the number stacked signals improves the significance and clarity of the echograms. If the terrain permits, several parallel profiles can be recorded to render a 3D map of the main slide surface. A small landside that destroyed a relatively new house in the south of Poland provides a case study of the methods described above. GPR was used to locate and map the landslide surface, which developed following a period of heavy rain in 1997 (Fig. 3.1). No landslide activity had been observed in the area prior to 1997, leading investigators to suspect that construction of the house and it adjacent swimming pool in 1995 may have disturbed the geotechnical balance of the property. Later drilling revealed the lithological composition of the subsurface rocks, and confirmed the presence of slide surface.

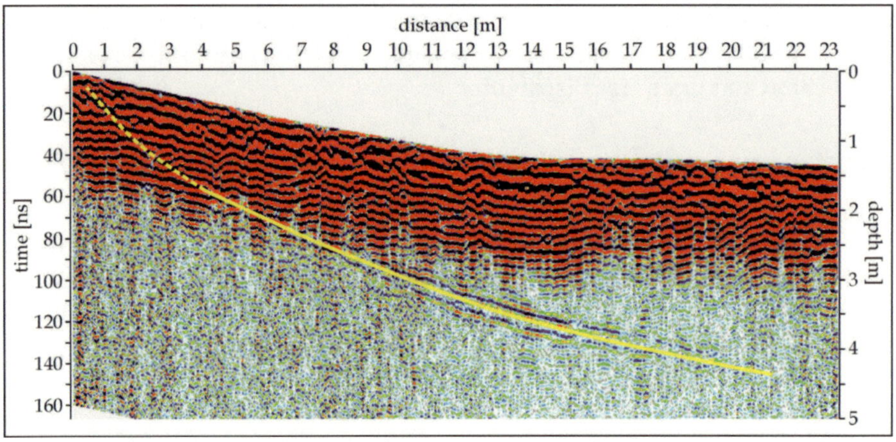

Fig. 3.1 GPR echogram of a landslide. RAMAC/GPR apparatus, 200 MHz antennas.

3.2 GPR in archeology

More and more archeological investigations utilize geophysical methods of magnetometry, GPR, and geoelectrics. Features such as old moats, ramparts, walls, basements, cemeteries, residual strongholds, foundations, and tumuli have all been located using GPR methods. Archeological objects can be successfully identified using medium frequency antennas, as they are buried at depths of less than 10 m. GPR is non-invasive and thus offers the advantages of speed and precision in archeology, as well as the possibility of object recognition without digging an exploratory trench. Archeological interpretations can be optimized by doing GPR profiles of an entire grid system if terrain conditions permit. Under these conditions a GPR survey can significantly reduce the time and effort necessary to locate subsurface objects, thus lowering the costs of excavation. GPR can also help optimize design of research trenches and excavation plans for specific objects. Higher frequency GPR records can help differentiate small artifacts from other, less significant fragments of rock in the subsurface.

GPR is an especially effective, low impact method for surveys carried out inside historic buildings. Void space such as crypts, basements and tunnels can be detected as a large increase in EM wave velocity. As a result, the boundaries of void spaces are visible in echograms at apparent depths that are below the actual boundaries or floors of these features. Voids appear in echograms as low frequency impulses zones. Shielded antennas are often used for GPR measurements inside of buildings to reduce feedback from walls, ceilings and other objects (i.e. reflex anomalies), but unshielded antennas can also provide adequate results. Echograms collected inside buildings can contain a significant amount of noise from reflex anomalies, but these datasets have nevertheless facilitated numerous archeological discoveries.

GPR surveys of Saint Margaret's collegiate in central Poland for example, were used to determine whether a Romanesque basement excavated outside of the church continued beneath the floor of the historic structure above. A GPR survey using shielded antennas (800 MHz) confirmed that the basement extended beneath the church (Fig. 3.2). Due to the architectural continuity of many historical structures, archeological GPR profiles can be planned according to known subsurface patterns, or the layout of the overlying structure. GPR surveys of the churches of Saint Peter and Saint Paul in Krakow were planned using the surface structure as a template (Fig. 3.3). In this example, the survey used shielded 250 MHz antennas and followed an aisle in the church that lay above known and accessible crypts. The longitudinal axes of the last two crypts (Fig. 3.3) are parallel to the profile. The echogram from the aisle was then used to interpret depth profiles of the nave of the church where unknown and inaccessible crypts were suspected to exist.

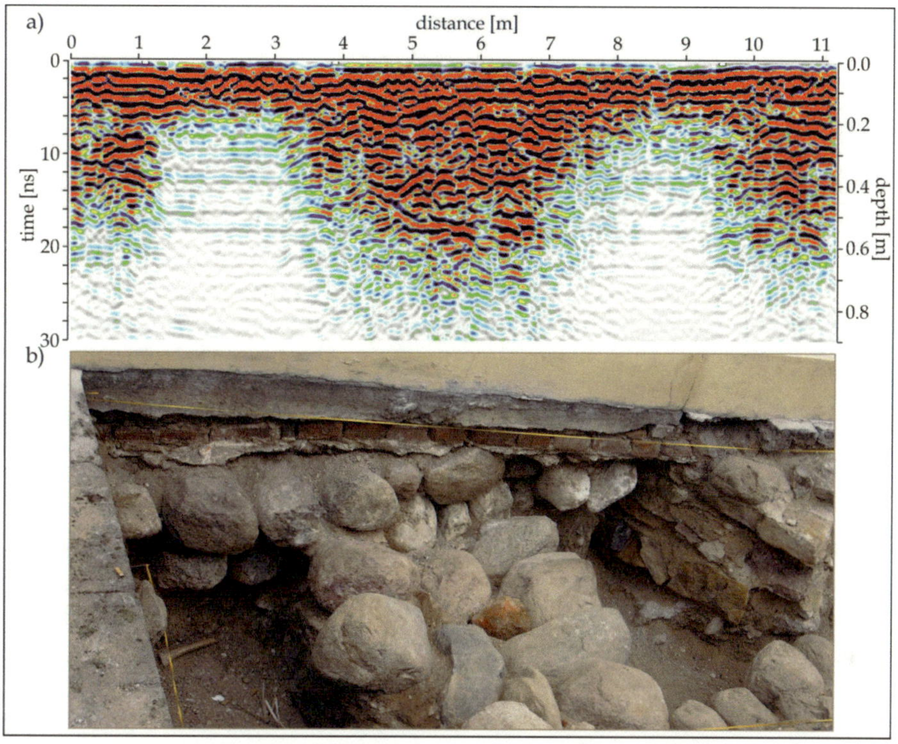

Fig. 3.2 GPR of Saint Margaret's collegiate in central Poland; a) echogram, b) ruins of the Romanesque basement. RAMAC/GPR apparatus, 800 MHz antennas.

3.3 GPR for controlling embankments

River embankments, levies and dams that secure reservoirs, holding ponds and other bodies of water, require periodic maintenance and upgrade. Even robustly constructed embankments degrade over time, usually from infiltration at the embankment's base. Water filtration across the body of the embankment results in soil suffusion and liquefaction.

Animal burrows, vegetation, and other biological modifications can also degrade embankments. Assessment of embankments is usually performed using geotechnical methods. These include drilling wells, conducting well tests, testing of well plugs in the laboratory, and *in situ* ground testing. Geomorphological factors affecting embankments can be ascertained via larger scale surveys of the area, using maps, photointerpretation, photogrammetry.

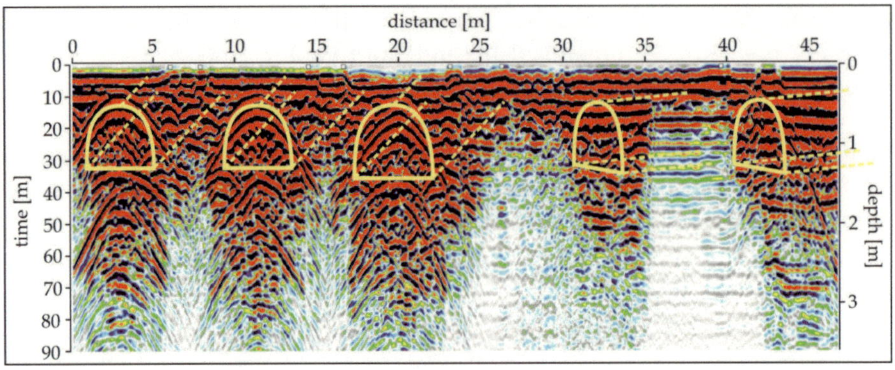

Fig. 3.3 Echogram in the Saint Peter and Saint Paul church in Krakow. RAMAC/CU II apparatus, shielded 250 MHz antennas.

GPR is highly effective in locating zones of weakness in river embankments. This method is useful for performing qualitative investigations of the embankment's inner structure and basal materials. GPR measurement profiles are best carried out along the rim of an embankment, along inner and outer shelves, and also within intervening areas between the river and the embankment. GPR signals can render the subsurface in sufficiently high resolution as to show relatively small anomalies in the embankment's subsurface structure. Varying degrees of cementation for example can be detected by GPR. Lower frequency antennas are generally used to query an embankment's basal structure.

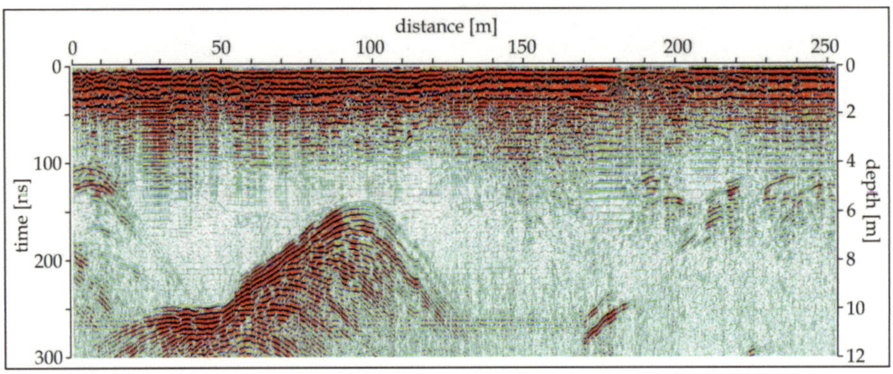

Fig. 3.4 Echogram of the Odra river embankment. RAMAC/GPR apparatus, unshielded 200 MHz antennas.

Frequently, river embankments are composed of impermeable clays and silts, materials that have relatively low resistivity. These materials attenuate EM signals, allowing GPR surveys to penetrate only the shallowest zones of the subsurface. GPR signals can also be obscured by high water, flooding, or heavy rains which saturate embankment material, and thus attenuate EM waves (Nguyen et al., 2005). Trees growing on or near the embankment can also interfere with signals, obscuring recognition of heterogeneities.

In spite of these limitations, the high measurement efficiency and low cost of GPR relative to other methods recommend it as a first order survey strategy for assessing embankments. GPR anomalies can then be confirmed with second order geotechnical methods. Used in this way, GPR can optimize the more expensive assessment techniques, limiting their use to zones where GPR anomalies are apparent. The echogram shown in Fig. 3.4 was recorded on the Odra river embankment using 200 MHz unshielded antennas. The height of the embankment along the section analyzed was about 4 m. The goal of the investigation was to identify heterogeneities within the embankment and its basal materials. No meaningful heterogeneities were observed within the embankment but a significant anomaly appeared at its base. Drilling revealed that the anomaly was a permeable layer of coarse-grained gravel. Further study of the anomaly identified this area of the embankment as being at risk of hydraulic puncture during high water events.

4. Georadar for monitoring soil contamination (Sylwia Tomecka-Suchoń)

Georadar methods (GPR) are a valuable tool in detection and remediation of soil pollution due to their sensitivity to the electrical properties of subsurface materials. Hydrocarbons and salts leaking from containment at industrial sites are the two most common classes of soil pollutants. Hydrocarbon contamination is usually a consequence of leakage from collecting tanks and pipelines used in various petroleum operations. Mining operations with their sizeable holding ponds, debris, and chemical dumps can also pollute soil and ground water. Identification and remediation of hazardous chemicals is complicated by the fact that waste material can be distributed at various depths, and may migrate in any direction, at any speed. The GPR method is a particularly sensitive method under conditions of pore water saturation, due to the high relative dielectric permittivity of water. In saturated soils for example, GPR can be used to perform real time monitoring of contaminant plume migration. As with applications described in the previous section, GPR is non-invasive and allows parties involved to forgo the more expensive activities of digging wells and trenches (Gołębiowski et al., 2010a). Two case studies of GPR use in soil contamination studies are described below. These include both types of liquid pollutants: low-conductivity hydrocarbons (a fuel station in Krakow) and high-conductivity chemical solutions (former military installation at B-S (Poland), and D waste dump).

4.1 Low-conductivity hydrocarbon contamination

The conductivity of soil media depends on 1) the relative volumes of the components (e.g., articles, water, air) and 2) their conductivity. The former volumetric factors will in turn affect the soil's overall dielectric permittivity. Owing to its very high dielectric permittivity, (ε_r =81) water content greatly facilitates imaging of Light Non-Aqueous Phase Liquid (LNAPL) contamination in sandy subsurface materials. Hydrocarbons have a low ε_r (Table 2.1). Detection of a LNAPL plume in dry, aerated soil horizons is difficult, but in water saturated soils, phase separation occurs between the LNAPL plume and uncontaminated surroundings. Differences in permittivity mean that higher concentrations of immiscible hydrocarbons (e.g. gasoline) can enhance detection and mapping of plumes. LNAPL plumes tend migrate in a downward direction due to gravity and capillary action. When a plume reaches the capillary fringe zone, contrasting electromagnetic properties between contaminants and the surrounding medium resolve the plume in GPR echograms.

A GPR survey was carried out at a fuel station in Krakow (Gołębiowski et al., 2010a). The study area was contaminated by repeated spills where hydrocarbon mixtures soaked and infiltrated the ground surface. GPR measurements were carried out in a constant-offset reflection mode along two lines that transected the site. A RAMAC/GPR device was used with 200 MHz antennas. Traces were collected every 0.05 m to generate a total of 64 stacked signals. All echograms were processed with the ReflexW program using phase correlation, time zero correction, amplitude declipping, dewowing, DC-shift, background removal, gain, the Butterworth filter, and smoothing procedures. Identification of contaminated zones was based on energy distribution analysis, counted from Hilbert transformation of raw signals. Areas where pockets of high energy (Fig. 4.1) overlapped anomalies in power spectra were interpreted as contaminated zones (Fig. 4.2). The energy envelope was normalized to the maximum value of the direct air wave.

Analysis of well cores (Fig. 4.1) found neither anisotropy nor other compositional anomalies in the subsurface of the study area. Parts of high energy in the echograms (violet area between 1 m and 2.5 m in Fig. 4.1) were therefore interpreted as contaminated zones. Analysis of GPR power spectra from the profile (Fig. 4.2) showed phase shifts and other irregularities that likewise indicated high concentrations of hydrocarbons. Power spectra irregularities correlated well with the zones of high energy in areas located at 15 and 32 m along the transect (Figs 4.1 and 4.2). The location of a high energy zone at ~2.5 m depth highlighted depression or downward bowing of the water table due to the gravitational pressure of the hydrocarbon plume.

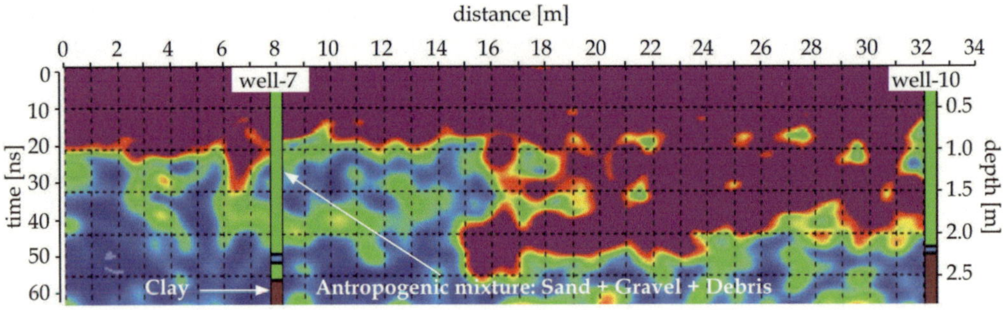

Fig. 4.1 GPR energy distribution (Gołębiowski et al., 2010a)

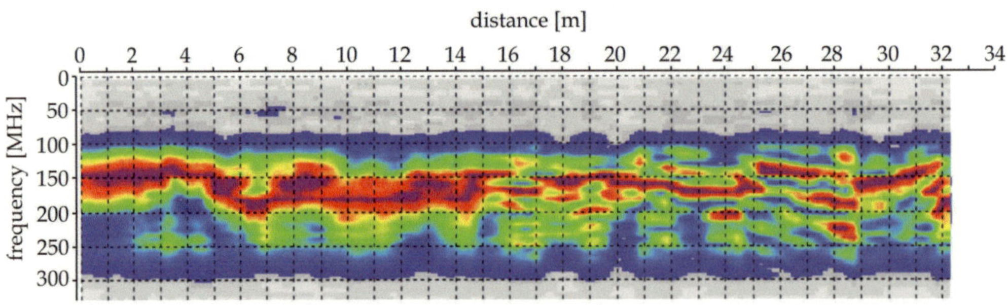

Fig. 4.2 GPR power spectra (Gołębiowski et al., 2010a)

4.2 High-conductivity chemical contamination

GPR can be used to detect subsurface chemical contamination through analysis of electrical conductivity profiles. Conductivity strongly affects attenuation of EM waves such that contaminated versus uncontaminated areas can be differentiated according to signal amplitudes in echograms (Marcak & Tomecka-Suchoń, 2010). The higher conductivity of some pollutants leads to greater attenuation and in echograms the absence of reflexes may be expected. Real time monitoring at the B-S military site revealed immediate changes in the GPR signals following injection of a high-conductivity NaCl solution into a zone of concentrated hydrocarbon contamination. The salt solution (5 kg of NaCl in 30 liters of water) was injected into a well located at point C along the profile (Fig. 4.3). Twenty liters of gasoline were simultaneously poured into the ground surface at point B. The injection well passes through sands of mixed grain size, overlying a clay layer. The clay layer is impermeable, separating the contamination experiment from the top of the water table at ~9 m depth.

GPR recordings were performed within the aeration zone of a lossless geological medium using high resolution antennas (500 and 800 MHz) at low depth, and operating in time-monitoring mode in order to track migration of contaminants (Fig. 4.3). A mean velocity of V_{mean}= 10 cm/ns, estimated from well core materials (i.e. dry sands) was used for time-depth conversions. GPR results show that attenuation increases dramatically around the borehole such that the lateral boundary of the contaminant plume was easily identified. The depth of contamination however was more difficult to locate due to the high conductivity and high attenuation of the NaCl solution. GPR time-monitoring of the gasoline spill showed that the horizontal length of the salt-contaminated area (~30 - 32.5 m; Fig. 4.3a-d) exceeded the original area of the spill, and did not change significantly. Twenty four hours later the contaminant plume had moved down on the profile and a hyperbola located at 30.5 m which probably originated from root reappeared on GPR echograms (Fig. 4.3d).

Different areas of the same water table can be contaminated from different local point sources, leading to varying levels of contamination along a given horizontal transect. In these situations, a single attenuation coefficient cannot be used to map the contamination. Changes detected in the coefficients themselves however may be used to locate boundaries of various contaminated areas. For porous media, reflection coefficients for EM waves depend not only on pore fluid type, but also on its degree of saturation (Schön, 1996). The effects of conductivity variation (ionic content of pore fluids) are evident in the GPR profile of an area of D waste dump contaminated by a high-conductivity solution, shown in Fig. 4.4. GPR data was winnowed according to an echogram time window of 20 and 35 ns to optimize recognition of the near surface aquifer and the horizon of salt contamination.

The energy component of the GPR signal reflected from the salt-contaminated horizon was subjected to a Hilbert transform, and then normalized along both horizontal and vertical axes to determine the dominant trend of energy changes (Fig. 4.4a, blue dots). Energy changes were then projected, using a cubic polynomial function, along the entire profile (Fig. 4.4a, red line). GPR results were merged with hydrological data on the basis of pore fluid conductivity (green line). Near the piezometer Pz-2, concentrated pore fluids corresponded to high energy GPR signals. To the right of Pz-2, GPR signal energy decreases rapidly until well site S-640A. The trend in energy change (Fig. 4.4a, red line) corresponds to trends in water conductivity change (Fig. 4.4a, green line).

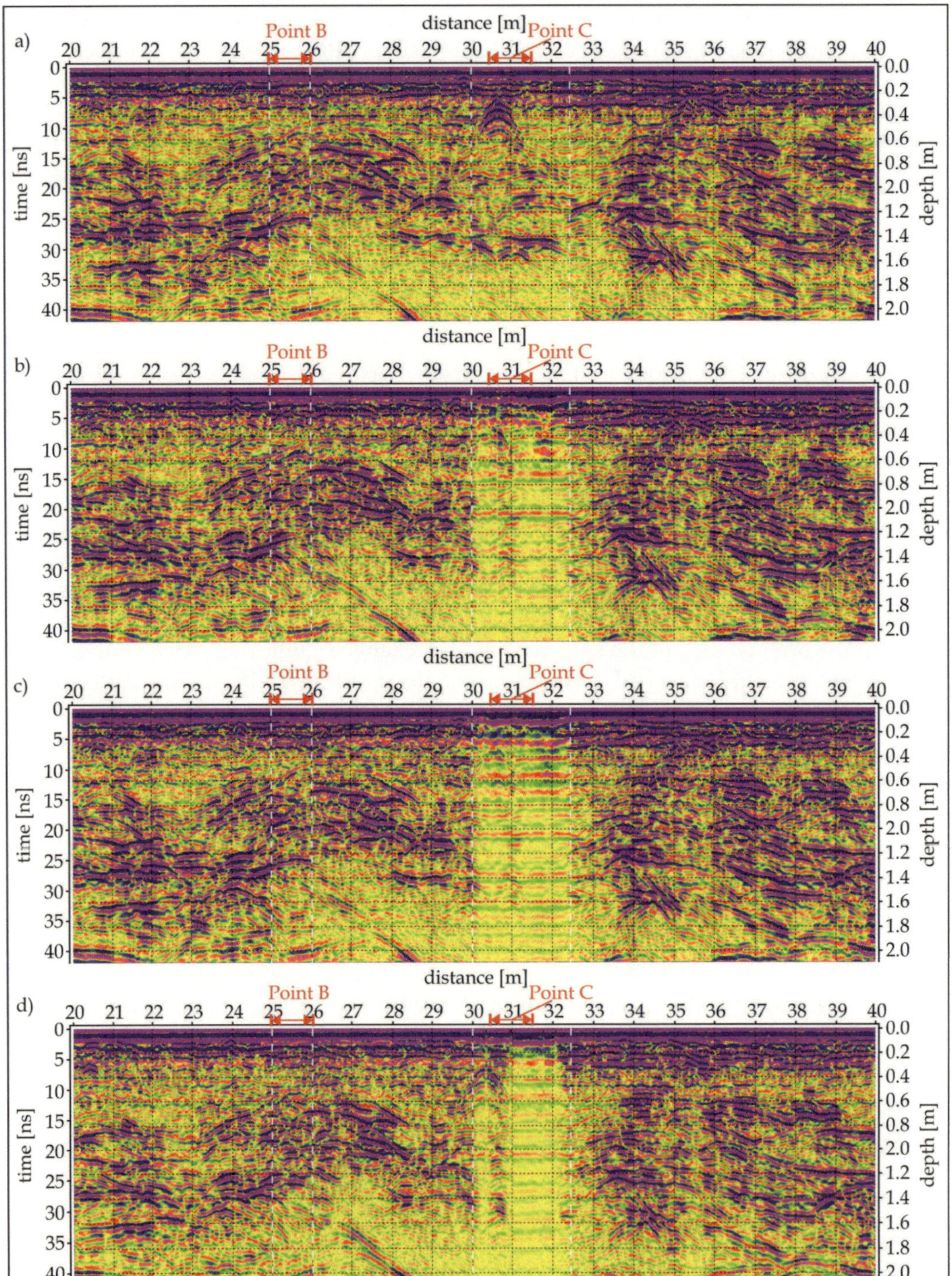

Fig. 4.3 Echograms (800 MHz antennas) from the B-S site a) before simulated gasoline spill, b) immediately after the spill, c) 3 hours after d), 24 hours after (Gołębiowski *et al.*, 2010a)

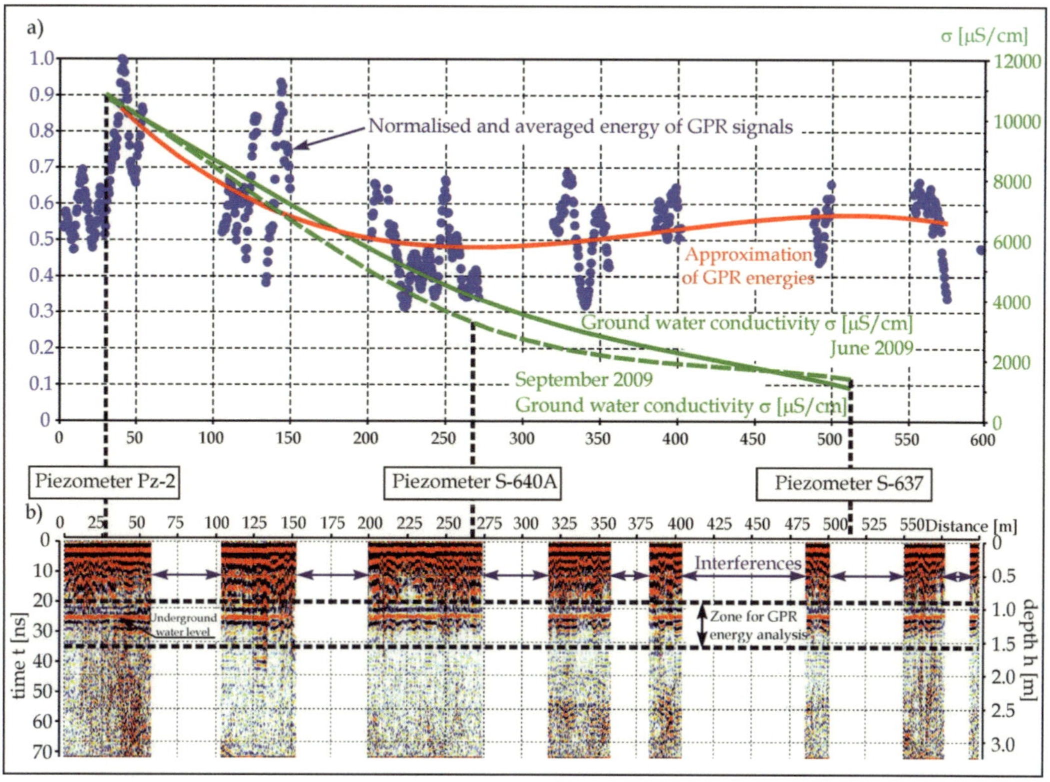

Fig. 4.4 Results of GPR investigations of an area surrounding a mining dump; a) energy distributions of GPR signals along the measurement profile and changes of ground water conductivity according to hydrogeological results; b) modified echogram - white perpendicular stripes are zones of high interference that were removed from GPR signals (Gołębiowski et al., 2010b)

5. Microgravity investigations (Sławomir Porzucek)

Micro-scale gravity measurements are the newest type of the gravimetric investigations in geophysics. Anomalies measured in the microgravity range are much smaller than those measured by traditional gravimetric methods. Microgravity amplitudes are only few times higher than average measurement errors for typical gravimetric devices. The newest generation of gravity meters have measurement accuracies of ~5 x 10^{-8} m/s^2 (0.005 mGal). Gravity field values for a given object depend primarily on how its bulk density is spatially distributed. For subsurface materials, shapes, sizes and the depth of objects may effect gravity measurements, but other petrophysical properties do not. Microgravity measurements can be influenced by strong winds or ground vibrations from traffic, both of which can lead to large errors. Special methods can be used to reduce the microgravity measurement error and enhance a signal/error ratio. These include shortening the time intervals between measurements to eliminate instrument drift (i.e. measurement intervals of no more than 1-3 hrs), performing replicate measurements to reduce random measurement error, and using relatively short distances between measurement stations in order to image anomalies as precisely as possible. Gravimetric modelling of the geological or human - made structures under investigation can assist in selecting the optimal distance interval between measurement stations. The height of the tripod on which the instrument is mounted should be measured as precisely as possible. An error of 0.03 m in tripod height estimation can change gravity measurements by as much as 0.01 mGal. Station location and terrain corrections are also extremely important. Improper processing of microgravity data (e.g. poorly chosen filters and so forth) may mask subsurface anomalies or create artefacts in the data. A poorly executed microgravity survey may reveal little more than the obvious Bouguer anomalies. For a carefully executed survey, microgravity anomalies with amplitudes of as little as 1.5 times the measurement error can be identified from as few as four stations, with a 99.9% confidence level (Liu, 2007). Microgravity methods are highly sensitive to the density of subsurface objects and thus are generally used for investigating small objects at relatively shallow depths. Microgravimetric information combined with borehole data can provide 2- and 3-D density models of the subsurface. Bulk density differences apparent in microgravity data can be to used locate and identify loosened debris, such as slag heaps, salts or other industrial waste materials. Microgravity surveys are also useful in searching for natural cavities caused by karst processes. In both active and abandoned subsurface mine sites, microgravity surveys can show underground shafts, chambers and other low density features. Microgravity surveys can thus identify risks to surface activities posed by old, closed underground workings and their potential deformation.

A microgravity survey of the Wieliczka salt mine provides an example of the application described above. A 600 x 350 m area was surveyed for microgravity anomalies to indentify zones in the rock mass that posed a structural risk to surface activities (Madej *et al.*, 2001). Residual anomalies (Fig. 5.1) reflected both the geological structure of the subsurface and structures related to former mining operations. Negative concentric anomalies revealed chambers excavated by historical mining operations. The negative, concentric anomaly denoted as '1' in Fig. 5.1, was the result of weakened materials around an abandoned 16th century structure, the Lois shaft. Negative anomalies 2, 3 and 4 (Fig. 5.1) were interpreted as the main excavation area of the 16th century operation. Collapsing mine chambers and corresponding surface subsidence were recorded as early as the 16th century, indicating an on-going risk to modern surface activities. Microgravimetric surveys and modelling based on historical information enabled location of the weakened zones between level I of the mine and the surface. The negative, concentric anomalies 5 and 6 (Fig. 5.1) corresponded to excavation chambers located on level II. The relatively low amplitudes of these anomalies indicated insignificant dewatering of the strata above the chambers, and thus a slower than expected advance of the weakened zones towards the surface. A WWN-EES trending negative anomaly 7; (Fig. 5.1) is situated near a local road. This anomaly corresponded to a system of chambers that was undergoing uplift and deformation. The collapse of roof material caused structurally weakened zones above the chambers and posed significant risk to the road.

Microgravity surveys have also been conducted near a calamine deposit that was exploited during the first half of the 19th century. These investigations sought to identify shafts that posed risk to planned construction of a nearby road. The calamine deposit occurred in a Lower Triassic dolomite unit directly overlain by relatively thin (< 1 m) Quaternary deposits. The exploited calamine horizon was about 2 m thick, and located at depths of ~20-30 m. The mining operation consisted of a system of narrow shafts having diameters of 1.5

m, and occurring at depths of several meters up to several tens of meters. Archival maps provided the approximate location of these shafts. The survey was conducted using a main grid of 50 x 50 m, with stations spaced every 5 m. Additional stations were implemented within the main grid, creating a secondary 17.5 x 17.5 m grid with stations every 2.5 m.

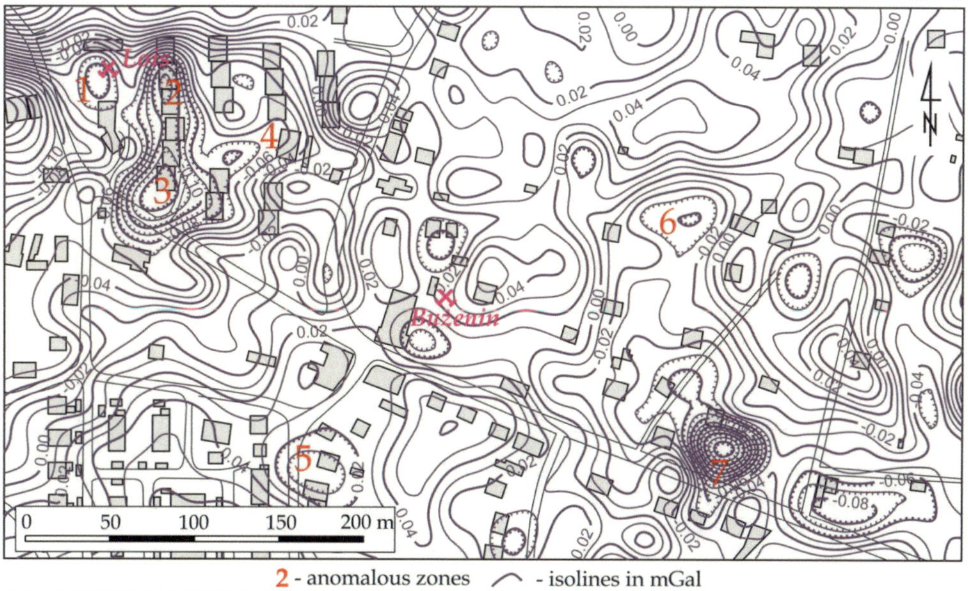

Fig. 5.1 Microgravity residual anomalies in the eastern part of the Wieliczka salt mine

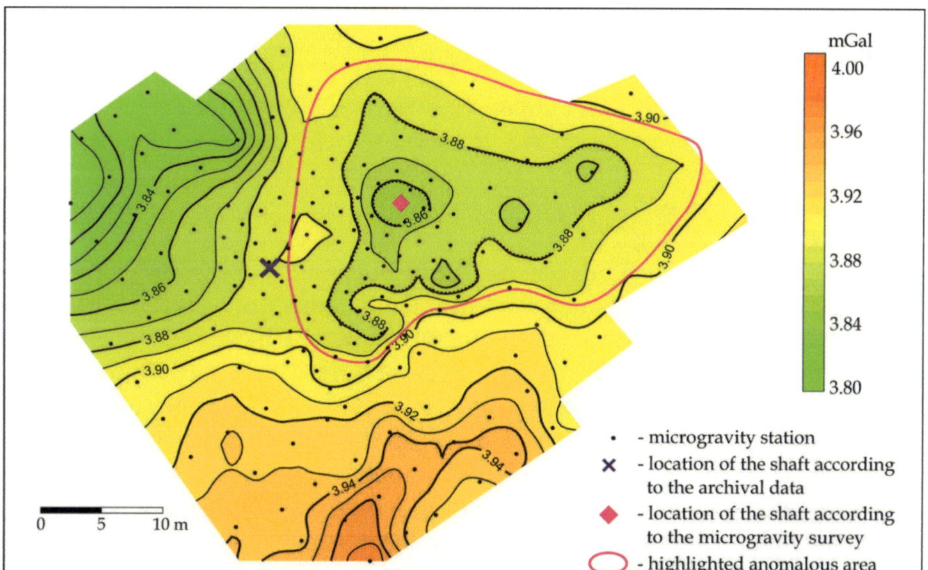

Fig. 5.2 Bouguer anomaly in the area where the shaft was suspected to be located; irregular positioning of stations is due to terrain, as well as surface and underground infrastructure

The number of stations used in this two grid survey strategy provided the subsurface resolution necessary to locate several shafts and surrounding zones of weakness that were at risk of collapse. The highlighted area shown along with the Bouguer anomalies in Fig. 5.2, is a 40 x 20 m zone that exhibited a distinct, negative gravity anomaly. This area encircled a smaller anomaly with a diameter of about 10 m. Old mining maps indicated uncertainty concerning the location of a shaft that might have been moved from its original position in the central part of the study area. The microgravimetric anomalies clearly identified the shaft as the smaller, encircled anomaly. The larger oval-shaped anomaly surrounding the shaft was interpreted as a weakened zone, and identified as a possible risk to road construction and other surface activities.

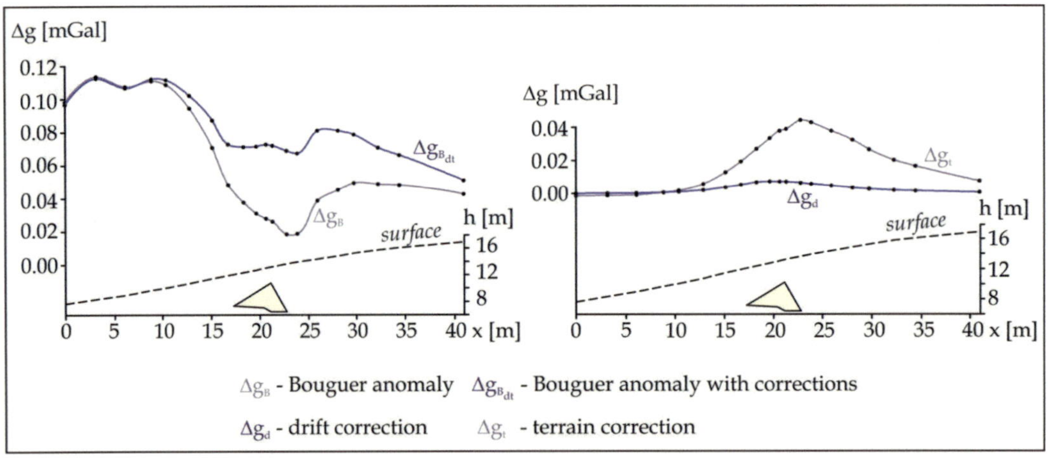

Fig. 5.3 Gravity anomalies and corrections over the subsurface drift

Microgravity surveys conducted prior to land development projects have also confirmed usefulness of the technique in risk mitigation. Results shown in figure 5.3 were collected from an area overlying a sealed mine drift, tunneled out of the local sandstone formation at the end of the 19th century. The tunnel had a total length of about 15.5 m, a triangular cross-section with a 3-5 m base, and a height of 3-4 m. The microgravity measurement stations were positioned across the drift at intervals of 1-4 m. The drift is visible as a relatively negative anomaly (Fig. 5.3, Δg_B) but the amplitude of the anomaly is difficult to determine due to the bulk density properties of the surrounding media. Excavation of the entrance to the drift increased the difficulty in detecting the structure by gravimetric methods. To minimize the impact of irregularities in the overlying terrain, the triangulation method was used to calculate a terrain correction, Δg_t (Wójcicki, 1993). Investigators also corrected for three dimensional gravitational effects of the drift, Δg_d, using geodesy measurements to subtract the drift from the local gravity anomaly. Together, these corrections yielded a final anomaly, Δg_{Bdt} (Fig. 5.3). The final anomaly was based on points 8–9, and bore some similarity to the initial distribution. The shape of the anomaly however did not correspond to the shape of the applied corrections. The final anomaly was thus interpreted as a fractured zone, weakened by earlier construction of the drift. These findings demonstrated that the subsurface was not entirely stable.

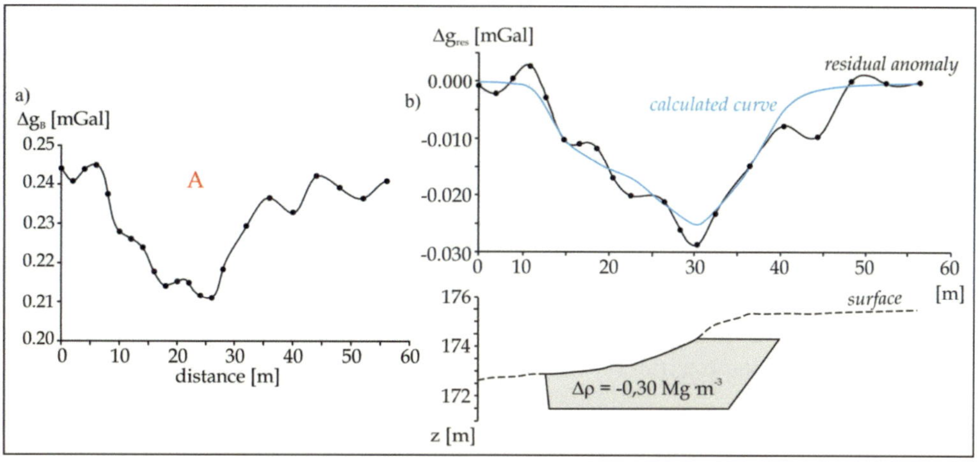

Fig. 5.4 a) Bouguer anomaly A, b) result of gravimetric modelling of the A anomaly

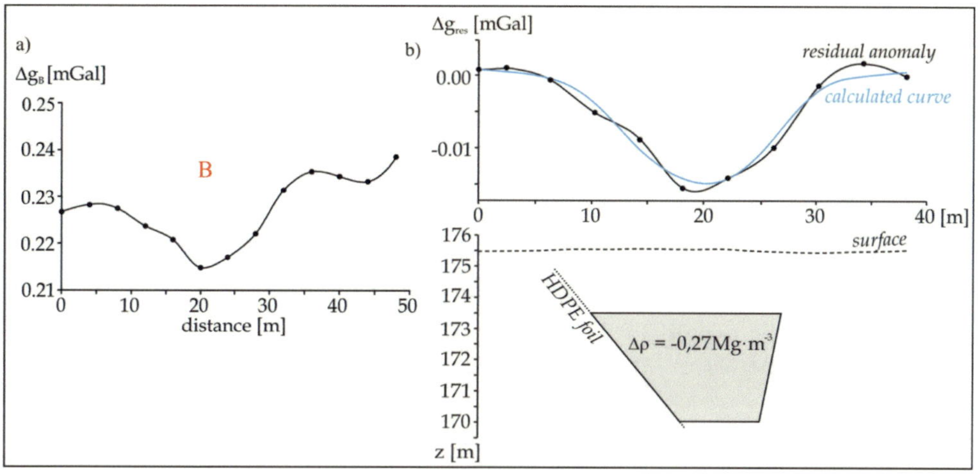

Fig. 5.5 a) Bouguer anomaly B, b) result of gravimetric modelling of the B anomaly

Microgravity can also be applied to investigate embankments and earthen dams that are at risk of water leakage and possible failure. Figures 5.4 and 5.5 show results from microgravity surveys conducted to detect zones of weakness along embankments that surround an underground water reservoir. Surveys were carried out along the embankment using a measurement interval of 2 m. A terrain correction was applied to the data to address irregularities in the topography of the study area. The survey revealed some relatively negative anomalies, and identified several zones of weakness within the embankment. The anomaly labeled A is based on 14 measurement stations and had an amplitude of about 0.03 mGal (Fig. 5.4a), a factor of 3 times higher than the overall measurement error of the study. The next anomaly was a separate residual anomaly, which was then modelled (results shown in Fig. 5.4b). One zone identified in the survey was found to have a density of 0.30 Mg/m^3 less than the average density of surrounding material in the embankment. This zone begins 2 m below the surface, has a thickness of about 3 m, and contains piping that carries water from the reservoir. Poor sealing of the pipes had likely caused leakage and soil suffusion, resulting in this lower density zone. The second anomaly marked B had a much smaller amplitude (Fig 5.5a) and was located beneath the causeway crossing the reservoir. The anomaly is based on 6 - 7 measurement stations, and was identified with a 99.99% confidence level (Liu, 2007).

The B anomaly was modelled (Fig. 5.5b) under the assumptions that it was a zone of poorly consolidated material that extended from the bottom of the reservoir to a depth of about 2 m below the surface. The top of the zone of weakness was also analyzed using GPR echograms, allowing for a consistent and unambiguous model solution. The left edge of the unconsolidated zone corresponded to the location of an HDPE reservoir liner, embedded within the slope of the main embankment. In this case, microgravimetric identification of the zone of weakness demonstrated poor construction of the embankment in the study area.

6. Magnetometry in the investigation of surface materials (Anna Wojas)

6.1 Origin of magnetic particles

In near surface investigations of the Earth's crust (shallow rock formations and soils), magnetometry can be used to identify magnetic minerals, determine their origin and reveal certain physicochemical processes that may be occurring in the soil. Soil magnetometry measures a given soil's magnetic susceptibility. The technique is commonly used to assess topsoil contamination by heavy metals and to track sources of magnetic particles in urban and industrial areas. Fly ash from incinerators is among others the source of magnetic particles in urban and industrial soils. The magnetic particles derive mostly from combustion of coals containing iron sulfides. High temperature oxidation transforms iron sulfides into magnetite, maghemite, and other less common types of ferrites. Both human activities and natural processes can influence the magnetic susceptibility of soils. Magnetic anomalies can thus be caused by industrial operations, weathering of ferrimagnetic and antiferromagnetic minerals, or other pedogenic processes.

6.2 The role of magnetic susceptibility in assessing topsoil contamination

Surface surveys of the magnetic susceptibility of topsoil were carried out in the vicinity of a major steel operation in Krakow. The soil surrounding the steel plant is contaminated by high levels of Zn, Pb, Cu, Cd and Hg.

Magnetic susceptibility surveys were performed only in the northeastern corner of the steel plant due to limitations in site accessibility. Prevailing winds blow emissions from the plant in a southeasterly direction. A nearby railroad track was also considered as a possible source of magnetic particles in study area soils. *In situ* measurements of magnetic susceptibility were performed using a MS2 meter and a MS2D sensor (Bartington Co.), working from principles of magnetic induction. Measured magnetic susceptibilities of soil were in the range of 20 to 435 x 10^{-5} SI units. The highest values were found closest to the steel plant (average values of 215 x 10^{-5} SI units). These results confirmed fall and nearby deposition of the heavy fraction of emission particles in the soil. The outlying regions of the study area had soil magnetic susceptibility ranging from 20 to 122 x 10^{-5} SI units, with an average value of 64 x 10^{-5} SI units (Fig. 6.1). An oval shaped magnetic susceptibility anomaly (isoline 150 x 10^{-5} SI units) having an easterly directed axis, highlights the direction of prevailing winds and magnetic particle transport. Magnetic and mineralogical analyses of the magnetic concentrate from soil samples revealed high concentrations of magnetite. Maghemite and iron sulfides were also identified. Subsequent studies have revealed a significant increase in the magnetic susceptibility of soil closest to the steel plant in the last four years (Rosowiecka & Nawrocki, 2010). Land near the steel plant is composed of fertile soil (chernozem, brown soils, loess) and is under cultivation. Magnetometry is especially suitable for further effective monitoring of the surrounding area.

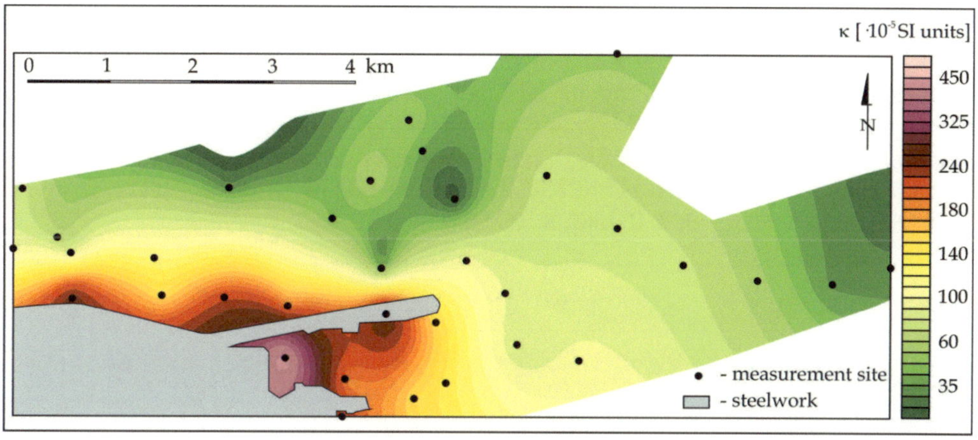

Fig. 6.1 Map of the magnetic susceptibility of topsoil in the vicinity of a Krakow steel plant

6.3 Magnetic susceptibility of historical sequence layers

A magnetic susceptibility study of historical sequence layers underlying the Main Market Square in Krakow revealed information about past human activity in the area. Subsurface samples collected during an archaeological excavation were analyzed in the laboratory using a MS2 meter and MS2B sensor. Analysis revealed contrasts in the magnetic susceptibility of underlying sands and overlying layers affected by anthropogenic activities (Fig. 6.2). Anthropogenic layers, which contained dark particles, fragments of bricks and sediment filled void space, often showed greater magnetic susceptibility. Overall, the historical layers had relatively weak magnetic properties.

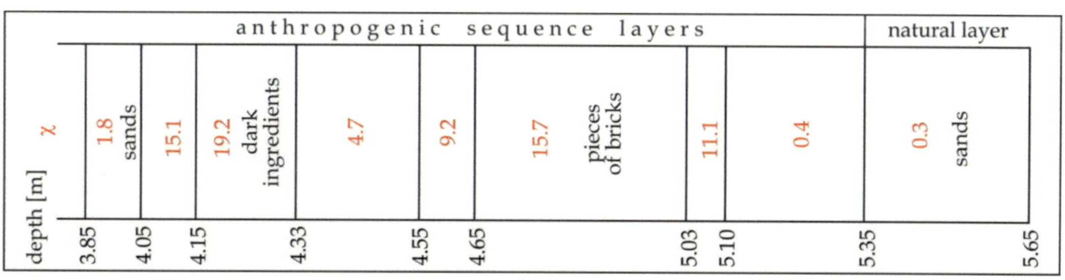

Fig. 6.2 Magnetic susceptibility of samples ($\chi \cdot 10^{-8}\ m^3kg^{-1}$) of historical sequence layers from the Main Market Square in Krakow.

6.4 Magnetic susceptibility studies of ochra deposit

Magnetometry studies of an ochre deposit in the Carpathians demonstrate the use of magnetometry in reconnaissance of potential economic deposits (Wojas, 2009). Certain iron oxy- hydroxide minerals weather to an ochre colour, which lends its name to ochre type deposits. These deposits are generally composed of goethite (yellow), hematite (red), and manganese oxides (dark hues). The magnetic susceptibility of the ochre deposit under investigation was attributable to the accumulation of secondary iron sulfides weathered from nearby rocks, and was also influenced by underlying bedrock. Magnetic susceptibility of the area was measured *in situ* using a MS2 meter and MS2F sensor. Low values of magnetic susceptibility (40 to 60 x 10^{-5} SI units) indicated soil devoid of ochre. The ochre deposit itself was recognizable by its stronger magnetic properties, and wide ranging magnetic susceptibility values. Magnetic susceptibility of yellow and rubiginous ochre ranged from 60 to 200 x 10^{-5} SI units. Crystalline rocks containing high concentrations of Fe_2O_3, and appearing as brown ochre had the highest magnetic susceptibility, from 200 to 900 x 10^{-5} SI units (Kotlarczyk & Ratajczak, 2002) (Fig. 6.3).

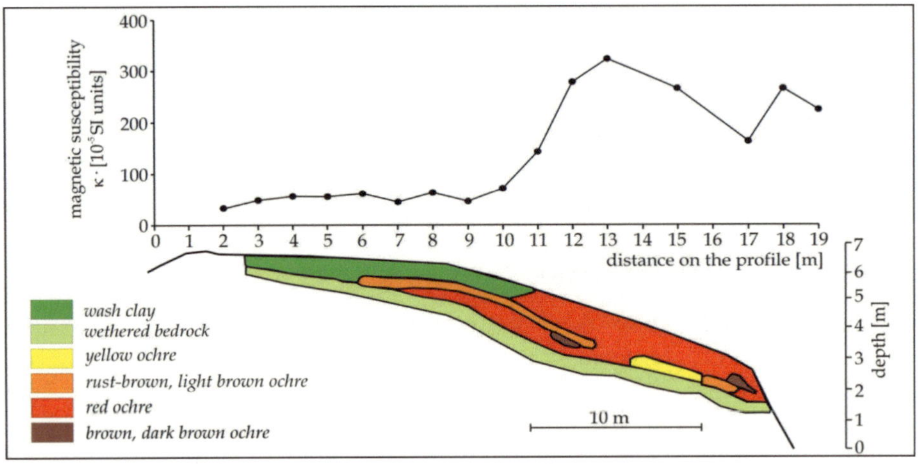

Fig. 6.3 Magnetic susceptibility values for an ochre deposit shown with a geological cross section of the deposit (Kotlarczyk & Ratajczak, 2002).

7. Environmental seismic investigations (Jerzy Dec)

7.1 Seismic evaluation of surface risks in developed areas overlying former mine sites

Closed or abandoned coal mines are often monitored using seismic surveys, which can locate and characterize workings a few meters below the surface, to depths of up to 60 m. Such workings may collapse or destabilize their surroundings, and thus pose a significant risk to surface activities. The stress field surrounding an abandoned working depends on the initial stress affecting the subsurface structure, which is in turn related to the depth and geometry of the working. The stress field can be evident as both continuous and discontinuous deformation (Mange & Kochonov, 1994). Continuous deformation occurs as elastic and plastic strain in layers, which deform but maintain their continuity. Discontinuous deformation manifests as fractures and displacement of bedrock, evident in small scale faulting and other discontinuities (Goodman & Shi, 1985). Discontinuous deformational features pose the greatest risk to the overlying surface. In solid rocks, seismic surveys can identify fractured zones and potential fault planes as velocity changes in seismic data. A seismic refraction survey of shallow workings at a former mine site within Carboniferous rocks, revealed differing behaviour and degrees of deformation (Table 7.1).

Terrain Category	Velocity, V_P, [m/s]	Structure of Carboniferous rocks	Fractures	Hazards to surface
A	>1900	contiguous rock	None	none
B	>1900	separate blocks of rocks, contiguous rock within a block	None	infiltration hazard at block contacts
C	1500-1900	continuous surface of uppermost Carboniferous	fractured rocks	infiltration hazard
C1	1500-1900	separate blocks of rocks, displaced	fractured blocks of rocks	discontinuous deformation hazard
C2	1100-1500	fractured rock, possible blocks of rocks	highly fractured rocks	infiltration and discontinuous deformation hazards
D	<1100	destruction of rock mass	highly fractured rocks, fissures	high risk of discontinuous deformation

Table 7.1 Seismic categories of Carboniferous rocks

7.1.1 Seismic surveys in developed areas

Standard seismic surveys often cannot be conducted at construction sites in developed areas due to their disruptiveness, risks to nearby inhabitants, and accessibility of the area surrounding the site.

In these cases, the uppermost body of bedrock beneath the surface can be imaged by seismic refraction tomography (Fig. 7.1). This technique interprets changes in wave velocity during P-wave propagation, in this example, through a near-surface Quaternary deposit (AB and CD in Fig. 7.1; velocity = V_0) and underlying bedrock (BC in Fig. 7.1; velocity = V_1). P-wave velocity for the Quaternary deposit was much smaller than that of the Carboniferous rocks ($V_0 < V_1$), which resulted in a relatively small critical angle of refraction ($i < 10^o$). The near surface Quaternary deposit was only a few meters thick. AB was thus insignificant, and the distance BC was assumed to equal that of AD. The errors for estimating unit thickness and their associated velocity changes were based on critical angle, i, and refractor depth, h.

Given an AD thickness of 100 m, refractor depth h of 2 - 8m, and i equal to 5 - 15°, errors were relatively small, ranging from 0.3 – 4.3 %. In the model described above, the initial arrival time for the refracted signal T_{ABCD} (traversing ABCD), could be approximated as the time necessary to traverse BC (Dec, 2004). The study used a seismic source located at point B, and receiver at point C, which reduced the initial arrival time to T_{AB}, or time necessary to traverse distance AB. Arrival times were used to construct a tomographic image of the P-wave velocity distribution for rocks beneath the Carboniferous surface. Velocity values were directly correlated to the composition and physical state of the subsurface material. Fractures and void space for example were evident as areas of reduced velocity.

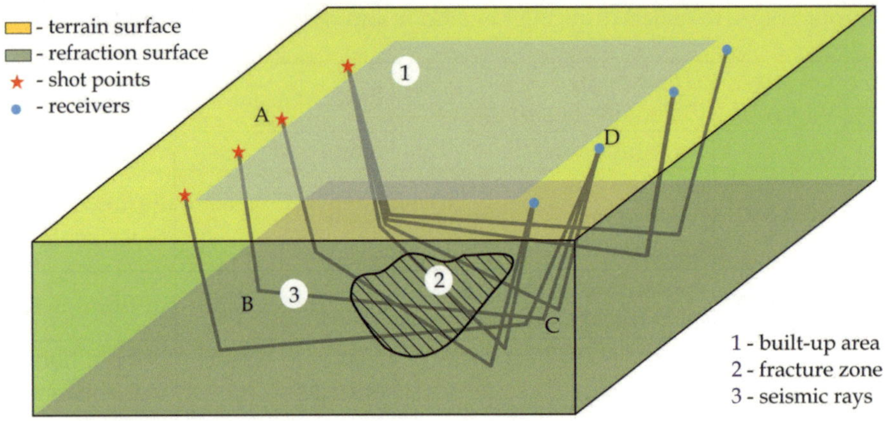

Fig. 7.1 Seismic refraction tomography: the study model

7.1.2 Evaluating ground surface stability risks

A survey carried out at the K-K coal mine in Katowice, Poland was used to image the continuity and stability of a subsurface Carboniferous unit, and the risk it posed to surface activities. The survey imaged old shallow (20-50 m) workings dating from the late 19th to the 20th century, which were located in upper parts of the Carboniferous unit. The workings have caused extensive fracturing of the unit (V_P <1200 m/s) and cavity migration towards the surface. Fractures from the working significantly increase the risk of ground surface deformation and failure. Due to urban development in the area, a standard seismic profile was not possible; refraction seismic tomography was used instead.

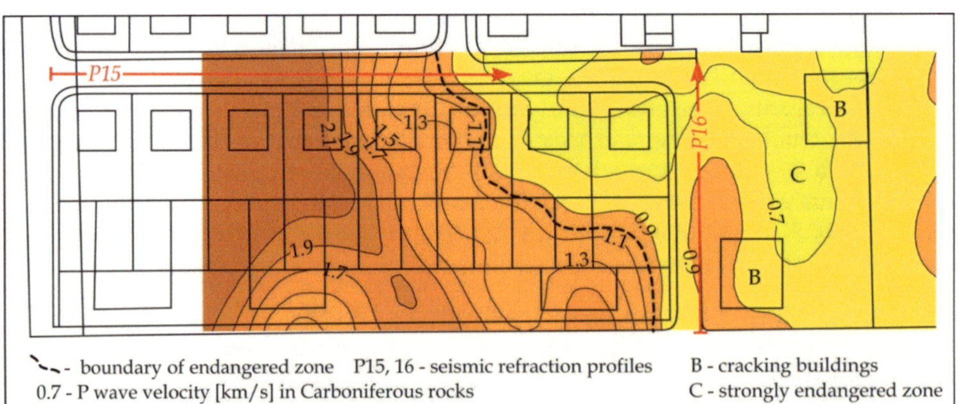

Fig. 7.2 Location and results (velocity distribution) of a tomographic study conducted in a developed area, Katowice-D, Poland; colours indicate velocity values [km/s] from the Carboniferous surface

Quaternary deposits that overly the Carboniferous unit include several meters of loamy sands and clays. The Carboniferous unit is comprised of sandstone, mudstone, shale and coal seams. The contact between the two units is an angular unconformity (lower units dip ~ 6 - 8°) with a pronounced erosional surface. The survey showed numerous locations where the rock mass was continuous and not fractured, as indicated by velocities of $V_P > 1900$ m/s (Table 7.1). Most of the subsurface in the study area however was strongly fractured and appeared as a discontinuous refraction boundary ($700 > V_P > 1800$ m/s). A representative location of the study area was the Katowice-D district (Fig. 7.2), where the Carboniferous unit exists in each of the different physical states or categories as described in Table 7.1.

Velocity, V_P, from profile 15 was equal to around ~2200 to 1100 m/s (Fig. 7.3), and the state of the Carboniferous surface (B-C2) was categorized as significantly fractured near the edges of the profile. The adjacent profile 16 had V_P=900 m/s, and rock structure category D, having no contiguous structure. Data from boreholes drilled in the survey area verified results of the seismic survey. Concrete was then injected into the top parts of the Carboniferous unit through the boreholes. The concrete was readily absorbed by the fractures and cavities.

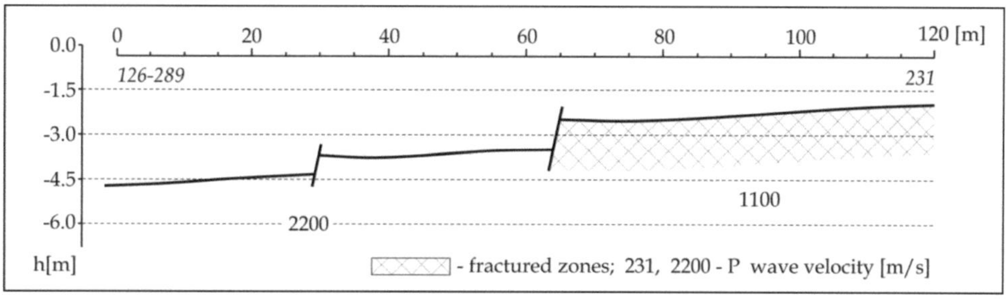

Fig. 7.3 Cross-sections along refraction profile 15, Katowice-D, Poland

The 160 x 70 m investigation area was located between two streets, and was thus divided into two sections (Fig. 7.2). Measurements were performed using a TERRALOC MK6 seismic system. Accelerated weight-drop source EWG-III and L-40B geophones (100 Hz) were used to record seismic sources. Descriptive results for the structure and physical state of the Carboniferous rocks are presented as categories in (Table 7.1). Figure 7.2 illustrates that the subsurface structure was unstable throughout most of study area (category B-D). Two different zones of the study area were distinguished based on the physical state of the subsurface (Fig. 7.2). The left-hand zone was characterized by fractured rock (category B-C2) and the right-hand zone was characterized by rocks that had been pulverized (category D). A contour outlining material with velocity V_P = 900 m/s, separates the zone with a fractured foundation. A low velocity anomaly (V_P < 700 m/s) is probably related to old workings (cavities) which connect to the ground surface through a system of fractures. Such zones are at risk of infiltration, suffusion and ground collapse. In conclusion, refraction tomography was effective in determining rock properties beneath a developed area, where traditional seismic methods could not be used. Fracture zones overlying old mine workings were evident as strong velocity anomalies, whose magnitude could be correlated to the degree of rock fracturing. Seismic refraction methods are thus suitable for evaluating surface risks of abandoned mines in developed areas.

7.2 High resolution seismic techniques for investigating the economic sulphur deposits

Exploitation of sulphur deposits using the underground 'melting' method (also referred to as well mining, or hydrodynamic process mining, or Frash method) poses significant environmental risks. A seismic investigation of the Osiek sulphur mine (Staszow district, Poland) illustrates the usefulness of seismic surveys in managing these risks. Sulphur is extracted at Osiek by pumping hot extraction fluids into the subsurface sulphur deposit. Melted, liquid sulphur is then brought to the surface through a series of production wells. Although the Osiek mine is the only operation in the world to use this technique, extraction at Osiek as well as similar oil and gas operations, put soils, ground water, and air quality at risk. Mine operators can manage risks by closely monitoring the melting process, and incorporating real time geophysical data into risk management decisions. Seismic surveys can specifically be used to monitor the spatial range of the melting zone (Dec, 2010) (i.e. its rate of expansion), and potential subsidence of the overburden (materials overlying the deposit). A seismic investigation of the Osiek mine site illustrates the application described above. Fluid removal of sulphur at the Osiek mine decreases the mechanical strength of porous limestone host rock, which can fail under the pressure of the overburden. Melting zones are not only at risk of failure, but melting may also cause changes in elastic properties of the areas surrounding the extraction zone. Such changes can induce subsidence and other disruptions to the subsurface stress field. High-resolution reflected seismic profiles provide accurate images of the structure of the deposit, its overburden, and signs of deformation in the surroundings. Seismic methods were also used to monitor the melting front and to watch for potential initiation of subsidence troughs, whose risks can be better managed by early detection (Al-Rawahy & Goulty, 1995). Changes in subsurface structure imparted by the Osiek melting method were detected by seismic surveys as early as a month after melting operations began at a particular well site. The extraction fluids used to control the size and expansion rate of the melt zone can also be used to manage subsidence troughs. Under these circumstances, pressure and flow direction of the extraction fluids were used to shape the subsidence trough in a way that prevented the destruction of adjacent wells located along the trough margins.

7.2.1 Field methods for the Osiek mine seismic survey

Seismic measurements imaged the Miocene unit being exploited by the mining operation and underlying sub-Tertiary strata, at depths of 20 - 200 m. The survey was conducted along lines defined by the positions of overlying production wells (Fig. 7.5). A short spread, high frequency seismic source (Elastic Wave Generator EWG-III, 250 kg accelerated weight-drop), and high frequency geophones (100 Hz) were used to obtain high vertical and horizontal resolution (Brouwer & Helbig, 1998). In order to avoid errors caused by heterogeneities in the subsurface, a split spread with a 100 m interval was applied. Geophones were positioned at 5 m intervals (trace spacing). Near offset was equal to 50 m and far offset was equal to 165 m. The 5 m trace spacing and 1 m shot interval gave CMP resolution equal to 2.5 m and a 24 fold CMP gather.

The majority of the profile was collected using the full 24 fold CMP gather. For parts of the study area that were difficult to access, a 24 channel end-off spread with a 12 fold CMP gather was used. Vertical stacking of signals helped reduce ambient noise caused by mine equipment. This procedure combines signals recorded from the same seismic source and at the same geophone location (Upadhyay, 2004). As many as 20 stacked signals were required to generate high resolution records for some areas. A model of the deposit and the seismic profile obtained from the 24 channel end-off spread are shown in figure 7.4. The following data processing procedures were applied to enhance profile resolution: field static correction, spherical divergence (spreading) correction, refraction static based on the first break picking, surface consistent scaling, surface consistent deconvolution, shaping filter based on the refraction wavelet, Ormsby filter 40/60-150/200 Hz, iterative velocity analysis, NMO, residual static correction, and CMP stacking. Several profiles, referred to as time sections, were recorded throughout the extraction process to monitor temporal changes associated with various phases of extraction.

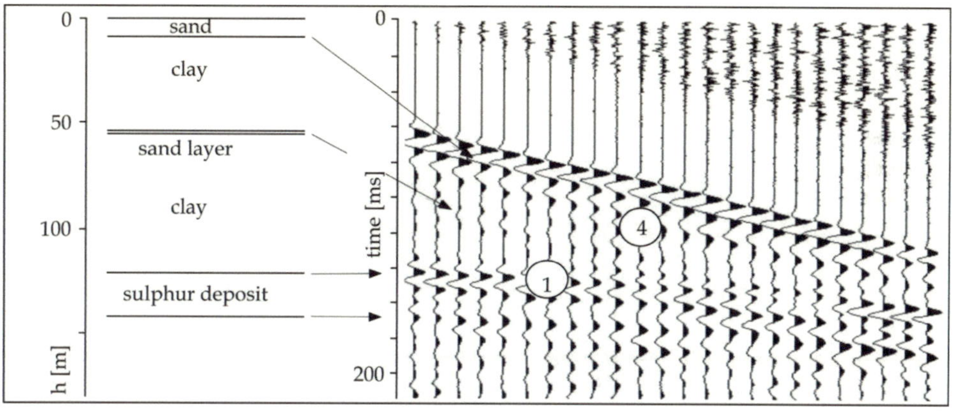

Fig. 7.4 Deposit model and seismic raw record; 1-top of the deposit, 4 sand horizon in the overburden

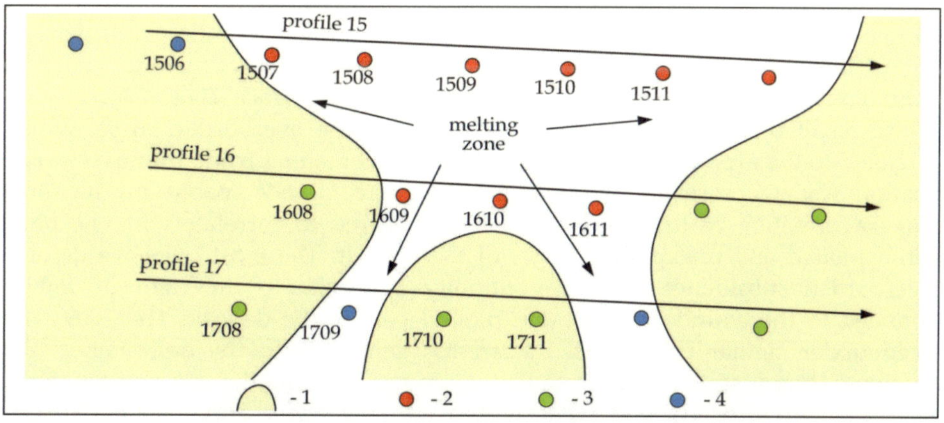

Fig. 7.5 Osiek mine, location of seismic profiles; 1-unexploited zone, 2-producing wells, 3-wells prepared for exploitation, 4-wells with increasing temperature

7.2.2 Influence of melting on the overburden

Time sections obtained prior to the initiation of melting revealed a number of continuous reflectors in the overburden. Profiles collected during extraction show distinct changes in the top surface of the deposit (above the producing formation), which usually appeared as discontinuities in previously continuous reflectors. These changes were also evident as changes in amplitudes and even disappearance of certain reflectors.

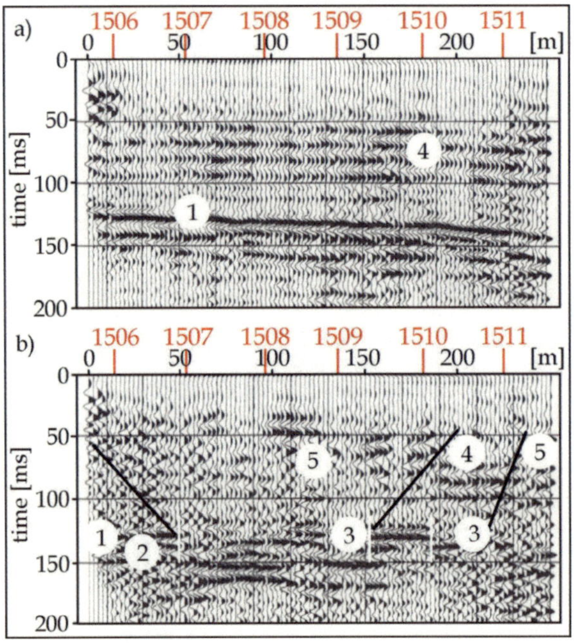

Fig. 7.6 Seismic sections of the sulphur –bearing horizon obtained before (a) and during exploitation (b); 1-top of the sulphur horizon, 2-unexploited zone, 3-exploitation zone, 4-horizons in overburden, 5-deformation zone

Anomalies were also evident in seismic images of the overburden material. These anomalies (50-100 ms) appeared as discontinuities in specific reflectors, indicating subsidence of the formation. Time sections of profile 15 collected prior to initiation of melting, and 3 months after initiation of melting, showed changes in the subsurface (Fig. 7.6). Asymmetric subsidence resulted from destruction and collapse of the overburden strata, which was inturn related to the expansion rate and direction of the melting front. Intuitively, observed deformation was more severe in areas having more closely spaced production wells (Fig. .7a). Deformation of the overburden also appeared to correlate with the boundary between exploited and unexploited zones of the deposit. The most intensive deformation and overburden subsidence (broken continuity of strata; to the right of 1609 well), corresponded to the boundary of an unexploited part of the deposit. The 1609 well was destroyed under similar conditions, by stresses generated in the deformation zone. If deformation-induced stresses generated along planes within the overburden formation attain critical values, production wells within the subsidence zone can be subjected to shearing and ultimate destruction. This scenario (Fig. 7.7b) is evident in the time section profile presented in figure 7.7a.

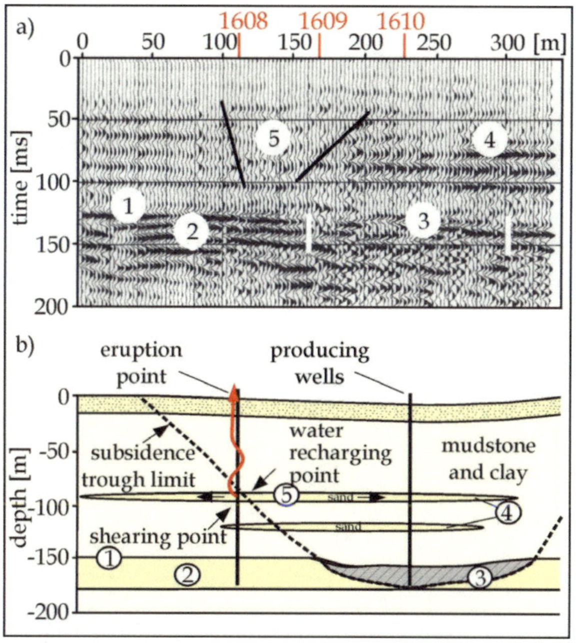

Fig. 7.7 a) Seismic image of overburden layers and associated deformation zone;
b) Subsidence effects around exploitation zone and the area of subsidence trough; numerical labels the same as those in Fig. 7.6

7.2.3 Studies of water flow in the Osiek subsurface

Infiltration of fluids into the overburden can change the elastic properties of the materials therein, and thus cause the disappearance of natural reflectors observed in previous surveys. For example, fluid migration into arenaceous horizons of the overburden caused the disappearance of a reflector, demonstrating the physical response of subsurface materials to infiltration. The water infiltration occurred within a zone of the overburden that allowed investigators to identify the source of the infiltration. Successful location of the leak allowed for repair of the failed component. Subsurface monitoring of infiltration is recommended if a given recharge point (e.g. the C-45 recharge well) is located a significant distance from its discharge (eruption) zone (Fig. 7.8). The unexploited part of the deposit is located to the left and the melting-induced destruction zone is located on the right side of figure 7.8. Strong subsidence above the exploited area caused failure of the C-45 well and fluid infiltration into the overburden strata.

The disappearance of seismic reflectors in profiles affected by the infiltration event demonstrates that the zone of infiltration can be precisely identified (Fig. 7.9). Near the F-71 well for example, all seismic reflectors found within the overburden strata disappeared. Fluid infiltration of the arenaceous horizon at 95 m depth was also visible. Post infiltration seismic profiles also indicated that subsidence associated with exploitation of this particular area may facilitate water migration not only within the arenaceous horizon, but also through horizons of the overburden. The broad infiltrated zone around the C-45 well narrowed into an elongated flow path directed to the F-71 well, a site of intensive surface discharge. The system of parallel seismic sections allowed precise imaging of the direction and width of the flow path.

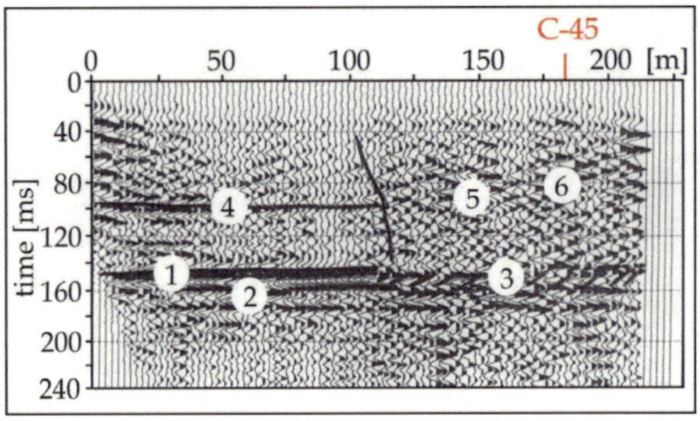

Fig. 7.8 Seismic section in the zone around the failed well; 1-top of the bed, 2-unexploited zone, 3-exploitation zone, 4-horizons in the overburden, 5-deformation zone, 6-recharge zone.

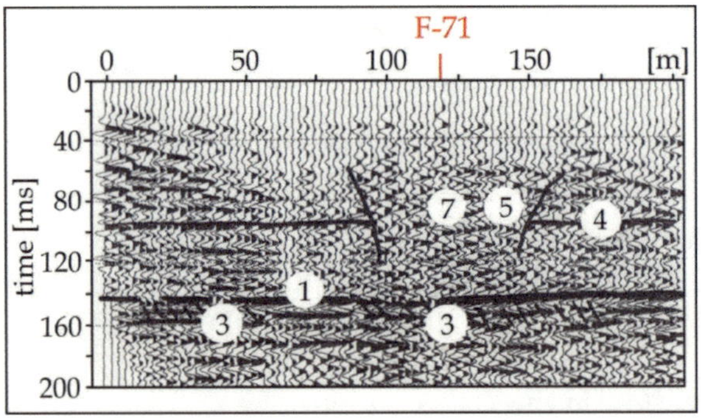

Fig. 7.9 Seismic profile of a zone affected by extraction fluid migration from the overburden to the surface; labels 1 through 6 are the same as in Fig. 7.8; 7-flow and eruption zone.

7.3 Evaluating landslide stability from seismic surveys

The current geophysical methods for investigating landslides assume that each landslide must be analyzed separately. Various geophysical methods, including seismic refraction, geoelectrical profiles, and GPR can be integrated in geophysical analysis of landslides. Integrated methods were used to study a landslide that occurred in the Krynica subunit of the Magura unit, Outer Carpathians. The landslide covered approximately 3 hectares, and was located on a slope that descends to a river. The slide surface was initiated in thin, weathered subsurface rocks, and in outcrops of the Eocene Piwniczna sandstones (the sandstone member of the Magura unit). The Piwniczna sandstone is a massive, thickly bedded (1.5 - 6 m) unit with conglomeratic horizons. Fractures are rare in the unit. Within the Piwniczna, packets of medium to fine grained, rhythmically bedded flysch occur. These beds are generally several meters thick. Thin-bedded sandstones and marly shales were also observed in the eastern part of landslide. Field measurements of the landslide were difficult due to the steep gradient of the slope, dense vegetation, and damage to slide surface caused by recent minor reactivation. Nevertheless, eight seismic refraction profiles, three geoelectrical profiles, and six GPR profiles were collected from accessible areas. The seismic boundary separating the overburden ($V_P \sim 1000 - 1400$ m/s) and basement ($V_P \sim 2200 - 2500$ m/s) were distinctly visible on seismic records (Fig. 7.10).

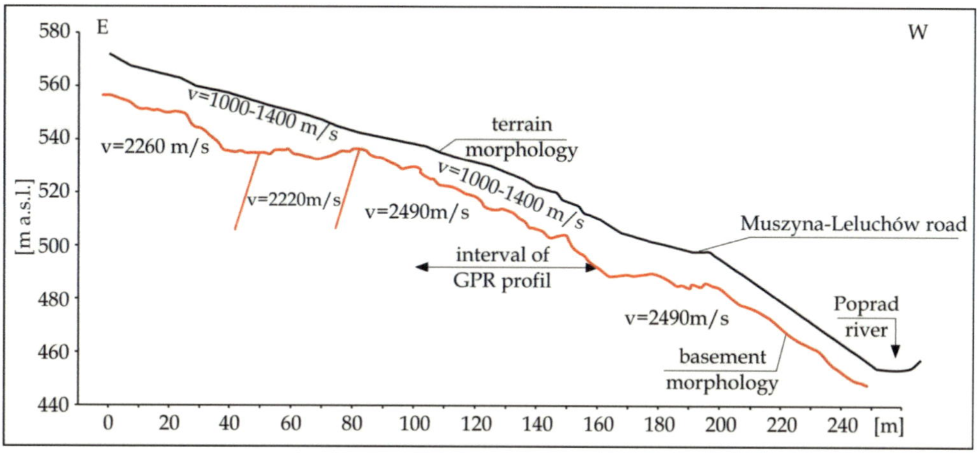

Fig. 7.10 Landslide - seismic refraction depth section.

Geoelectric profiles were strongly influenced by the water saturated overburden, and were consistent with seismic profiles. The GPR survey showed several areas of the overburden where shallow layers of differing inclination highlighted the landslide's progression. Large blocks of rocks within the slide probably caused the irregular movement observed in the overall landslide. Visible variations in bedrock morphology (Fig. 7.10) provided evidence that the landslide plane formed within flysch strata. Geophysical investigations enabled determination of the depth and horizontal range of the landslide. The integrated interpretation was used to select sites for geotechnical boreholes, and forecast the landslide's future stability. These interpretations suggested that even slight shifts in geotechnical features could reactivate the landslide and jeopardize a nearby road. The risk is a natural consequence of the geomorphology of the study area, where fine-grained units within the subsurface can easily become natural slide surfaces.

8. Conclusions

The examples presented above demonstrate how geophysical methods can be used in near surface applications such as resource development, engineering, archaeology, mitigation, remediation and environmental protection. In all cases, the available geophysical techniques were applied using modified methodologies, field measurements and processing strategies.

Contemporary catastrophic events and Earth processes in macro scale force application of current geophysical methods in recognition of their results influencing the environment. Heavy rains and floods in Poland in the last ten years gave rise to intensive works in identification and recognizing and mitigation of landslides. Results of geophysical surveys including GPR and shallow seismic refraction were integrated for investigation of landslides. The presented investigations enabled determination of the depth and horizontal range of the landslide. The integrated interpretation showed sites for geotechnical boreholes, and forecasted the landslide's future stability. Geophysical surveys were also useful in preparing the strategy for landslide risk mitigation and helped in the detailed understanding of the internal structure of the landslides, especially the slide surface.

Results of GPR and microgravity surveys turned out also to be highly effective in locating zones of weakness in river embankments. The results of field measurements enabled qualitative determination of the embankment's inner structure and basal materials. It was shown that GPR and microgravity methods were useful in the investigation of embankments and earthen dams that are at risk of water leakage and possible failure.

Well known applications of GPR in archaeology were also presented together with new area oriented to identification of low conductivity and high conductivity contaminations of subsurface formations.

Seismics in tomography mode and refraction mode and microgravity surveys were shown as methods useful in identification and recognition of weak zones in the industrial areas of former and contemporary mining activity. Fracture zones overlying old mine workings were evident as strong velocity anomalies and low density anomalies, whose magnitude could be correlated to the degree of rock fracturing suitable for evaluating surface risks of abandoned mines in developed areas.

Magnetic susceptibility measurements *in situ* and in laboratory were presented as an useful tool in effective monitoring of anthropogenic influences on the environment in the past (historical layers) and in tracking the contemporary pollution by iron compounds. Magnetic anomalies are also the evidence of weathering of ferrimagnetic and antiferromagnetic minerals, or other pedogenic processes.

Universal petrophysical formulas were demonstrated to combine empirical laboratory data with geophysical and well logging measurements and theoretical expectations and parameters forming relationships useful in a consistent, quantitative geophysical interpretation. Exemplary values of parameters for the lithology types were presented in tables and figures to illustrate the level of parameters values frequently found in the near surface formations.

9. Acknowledgments

Authors would like to thank Ms Teresa Staszowska for preparing the figures and text edition.

10. References

Al-Rawahy, S., Y., S. & Goulty, N., R. (1995). Effect of Mining Subsidence on Seismic Velocity Monitored by a Repeated Reflection Profile, *Geophysical Prospecting,* Vol. 43, pp. 191-201.

Annan, A., P. (2001). Ground Penetrating Radar. Workshop Notes, Sensors & Software, Canada.

Brouwer, J. & Helbig, K. (1998). *Shallow High-Resolution Reflection Seismics.* Elsevier Science Ltd., Amsterdam.

Daniels, D., J. (2004). *Ground Penetrating Radar* - 2nd Ed., The Institution of Electrical Engineers, ISBN 978-0-86341-360-5, GB.

Dec, J. (2004). Seismic Survey to Evaluate the Danger of Ground Surface Damage in Built-Up Terrain in Mining Areas. *Polish Journal of Environmental Studies,* Vol. 13, pp. 70-73.

Dec, J. (2010). High Resolution Seismic Investigations for the Determination of Water Flow Directions During Sulphur Deposits Exploitation. *Acta Geophysica*, Vol. 58, pp. 5-14.

Gołębiowski, T., Tomecka-Suchoń, S., Marcak, H. & Żogała, B. (2010a). Aiding of the GPR Method by the Other Measurement Techniques for the Liquid Contamination Detection, XIII International Conference on Ground Penetrating Radar, Lecce/ 21-25 June 2010.

Gołębiowski, T., Marcak, H., Tomecka-Suchoń, S., Zdechlik, R., Zuberek, W. (2010b). Use of Geophysical Methods for the Assessment of Migration of Contaminants from the Coal-Mining Waste Dumps, Extended Abstracts, abstract id: 317,,XXXVIII IAH Congress, Groundwater Quality Sustainability, Krakow, 12-17 September 2010, Eds: Kania J., Kmiecik, E., Zuber A.

Goodman, E., E. & SHI G., H. (1985). *Block Theory and its Application to Rock Engineering*. Englewood Clifs, N.J., Prentice-Hall, Inc.

Halliburton (1991). *Log Interpretation Charts*, Halliburton Logging Services, Inc., Houston, Texas, USA.

Hearst, J., R., Nelson, P., H. & Paillet, F., L. (2000). *Well Logging for Physical Properties. A Handbook for Geophysicists, Geologists and Engineers*. Wiley & Sons, ISBN 0-471-96305-4, Chichester, England.

Jarzyna, J., Bała, M. & Cichy, A. (2010). Elastic parameters of rocks from well logging in Near Surface Sediments. *Acta Geophysica*, Vol. 58, No 1, pp. 34-48.

Jarzyna, J., Puskarczyk, E., Bała, M. & Papiernik, B. (2009). Variability of the Rotliegend Sandstones in the Polish Part of the Southern Permian Basin - Permeability and Porosity Relationships. *Annales Societatis Geologorum Poloniae*, Vol. 79, pp. 13-26.

Jol , H., M. (2009). *Ground Penetrating Radar Theory and Applications*, Elsevier, ISBN 978-0-444-53348-7, Amsterdam.

Kobranova, V., N. (1986). (Eng. Translation 1989). *Petrophysics*. MIR Publish. Moscow, Springer-Verlag, ISBN 3-540-51524-0, Berlin, Heidelberg, New York.

Kotlarczyk, J. & Ratajczak, T. (2002). *Carpathian ochre from Czerwonki Hermanowskie near Tyczyn*, Mineral and Energy Economy Research Institute, ISBN 83-89174-60-X, Krakow (in Polish).

Liu Y.X., (2007). Evaluation and extraction of weak gravity and magnetic anomalies. *Applied Geophysics*, Vol. 4, No.4, pp. 288-293.

Madej J., Jakiel K. & Porzucek S. (2001). Microgravimetric assesment of possible surface deformations in some post-mining areas. Mineral Deposits at the Beginning of the 21st Century, Proc. of the Joint Sixth Biennial SGA-SEG Meeting, Krakow, 26-29 August 2001, A.A. Balkema Publishers, pp.1035-1038.

Mange, C. & Kochonov M. (1997). Anisotropic Material with Interacting Arbitrarily Oriented Cracks. Stress Intensity Factor and Crack-Microcrack Interactions. *International Journal of Fracture*, Vol. 65, pp. 115-141.

Marcak, H. & Tomecka-Suchoń S. (2010). Properties of Georadar Signals Used for an Estimation of the Mineralization of the Soil Waters, *Archives of Mining Sciences*, Vol. 55, No. 3, pp. 469-487.

Nguyen, van, G., Ziętek, J., Nguyen, Ba, D., Karczewski, J. & Gołębiowski, T. (2005). Study of Geological Sedimentary Structures of the Mekong River Banks by Ground Penetrating Radar: Forecasting Avulsion-Prone Zones, Acta Geophysica Polonica, Vol. 53, No. 2, pp. 167-181, ISSN 1895-6572.

Rosowiecka, O. & Nawrocki, J. (2010). Assesment of Soils Pollution Extent in Surroundings of Ironworks Based on Magnetic Analysis. *Studia Geophysica et Geodaetica*, Vol. 54, No.1, (December 2009), pp.185-194, ISSN 0039-3169.

Schön, J., K. (2004). *Physical Properties of Rocks. Fundamentals and Principles of Petrophysics.* Handbook of Geophysical Exploration, Seismic Exploration, Helbig K. and Treitel S., (Eds), v. 18, Elsevier, ISBN 0-08-044346-X, Oxford, UK.

Tiab, D., & Donaldson, E.,C. (2004). *Petrophysics. Theory and Practice of Measuring Reservoir Rock and Fluid Transport Properties.* (sec. ed.), Elsevier, ISBN 0-7506-7711-2, Amsterdam, Boston, Heidelberg, London, New York, Oxford, Paris, San Diego, San Francisco, Singapore, Sidney, Tokyo, Gulf Professional Publishing.

Upadhyay, S., K. (2004). *Seismic Reflection Processing*, Springer, Berlin.

Wojas, A. (2009). Magnetic Susceptibility of Soils, Including Iron Oxides of Anthropogenic and Natural Origin - Measurements Using the Bartington Instrument, Proceedings of InterTech - II International Interdisciplinary Technical Conference of Young Scientists, ISBN 978-83-926896-1-4, Poznań, Poland, May 2009.

Wójcicki, A. (1993). Approximation of the Gravity Attraction Caused by the Terrain Relief Forms Using a Polyhedron Method. *Acta Geophysica Polonica*, Vol. 41, No. 3, pp.1-24.

Magnetotelluric Tensor Decomposition: Insights from Linear Algebra and Mohr Diagrams

F.E.M.(Ted) Lilley
Research School of Earth Sciences, Australian National University, Canberra Australia

1. Introduction

The magnetotelluric (MT) method of geophysics exploits the phenomenon of natural electromagnetic induction which takes place at and near the surface of Earth. The purpose is to determine information about the electrical conductivity structure of Earth, upon which the process of electromagnetic induction depends. The MT method has been well described recently in books such as Simpson & Bahr (2005), Gubbins & Herrero-Bervera (2007), and Berdichevsky & Dmitriev (2008). The reader is referred to these for general information about the method and its results. Notable modern extensions of the method are observation at an array of sites simultaneously, and observation on the seafloor (both shallow and deep oceans).

In the most simple form of the method, data are observed at a single field site. Typically, three components (north, east and vertically downwards) of the fluctuating magnetic field are observed, and two components (north and east) of the fluctuating electric field. The magnetic field is measured using a variety of instruments such as fluxgates and induction coils. The electric field is measured more simply, between grounded electrodes typically several hundred metres apart.

The natural signals observed cover a frequency band from 0.001 to 1000 Hz. They have a variety of causes, the relative importance of which varies with position on the Earth, especially latitude. Recorded data are transformed to the frequency domain, and interpretation proceeds based on frequency-dependence.

The reduction of observed time-series to the frequency domain is thus fundamental to the MT method. In the frequency domain various transfer functions are determined, encapsulating the response of the observing site to the source fields causing the induction.

Another fundamental part of the data-reduction process, the focus of the present chapter, is "rotation" of observed data. By rotation is meant an examination of the observed transfer functions to see how they would vary were the observing axes rotated at the observing site. Rotation may reveal geologic dimensionality, and strike direction.

The MT tensor has a 2 x 2 form and is well-suited to analysis by the methods of linear algebra. This suitability was evident early in the development of MT, and it is common for a paper on MT to start with such an analysis. The papers of Eggers (1982), LaTorraca et al. (1986), and Yee & Paulson (1987), for example, specifically proceed with eigenvalue analysis and singular value decomposition (SVD), and see also Weaver (1994).

There appear however to be a number of aspects of this approach as yet unexplored, especially when the in-phase and quadrature parts of the tensor are analysed separately. This chapter now follows such a line of enquiry, investigating the significance of symmetric, antisymmetric and non-symmetric parts, eigenvalue analysis and SVD.

The Mohr diagram representation is found to be a useful way to display some of the results. Cases of 1D, 2D and 3D electrical conductivity structure are examined, including the 3D case where the MT phase is greater than 90° or "out of quadrant".

Taking the Smith (1995) treatment of telluric distortion, a particular case examined is that where SVD gives an angle which is frequency independent, implying that it contains geological information. Such angles arise when MT tensor data are nearly singular.

The analysis of in-phase and quadrature parts separately is seen to give further insight into the question of whether, as suspected by Lilley (1998), the determinants of the parts taken separately should always be positive.

2. Notation

The common representation of a magnetotelluric tensor **Z** is taken,

$$\mathbf{E} = \mathbf{Z}\mathbf{H} \tag{1}$$

of components

$$\begin{bmatrix} E_x \\ E_y \end{bmatrix} = \begin{bmatrix} Z_{xx} & Z_{xy} \\ Z_{yx} & Z_{yy} \end{bmatrix} \begin{bmatrix} H_x \\ H_y \end{bmatrix} \tag{2}$$

linking observed electric **E** and magnetic **H** fluctuations at an observing site on the surface of Earth.

All quantities are complex functions of frequency ω, and in Equations 1 and 2 a time dependence of $\exp(i\omega t)$ is understood.

In this chapter the subscripts p, q will be used to denote in-phase and quadrature parts. For example the complex quantity Z^σ is expressed

$$Z^\sigma = Z^\sigma_p + iZ^\sigma_q \tag{3}$$

Note that adopting a time-dependence of $\exp(-i\omega t)$ as recommended by authors such as Stratton (1941) and Hobbs (1992) would change the sign of Z^σ_q. Such a change may be misinterpreted, especially when distortion has caused the phase to be out of its expected first quadrant.

The subscripts p, q will also be used to denote quantities which are derived from the in-phase and quadrature parts, respectively, of a complex quantity, but which are themselves not recombined to give a further complex quantity. In Sections 2.1, 3, 5 and 11 below, where derivations apply equally to in-phase and quadrature cases, subscripts p, q are omitted for simplicity. Such derivations can be taken as applying to in-phase components, with similar derivations possible for quadrature components.

Also in this chapter, for compactness of text, a 2 x 2 matrix such as that for **Z** in Equation 2 will in places be written $[Z_{xx}, Z_{xy}; Z_{yx}, Z_{yy}]$. A rotation matrix $\mathbf{R}(\theta)$ will be introduced

$$\mathbf{R}(\theta) = [\cos\theta, \sin\theta; -\sin\theta, \cos\theta] \tag{4}$$

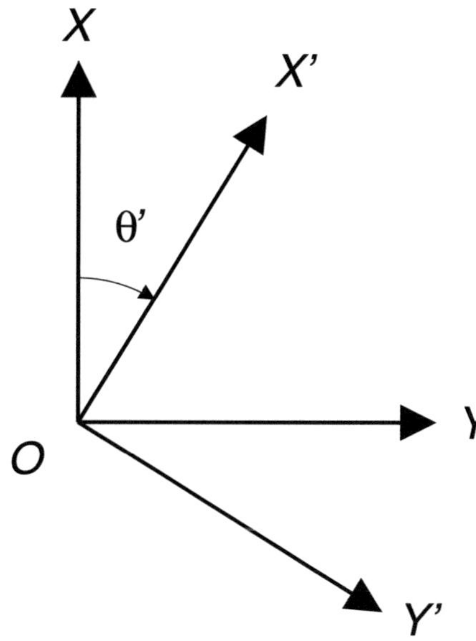

Fig. 1. The rotation of MT observing axes clockwise by angle θ', from OX and OY (north and east) to OX′ and OY′.

with $\mathbf{R^T}(\theta)$ the transpose of $\mathbf{R}(\theta)$. $\mathbf{R^T}(\theta)$ will sometimes be written $\mathbf{R}(-\theta)$.

2.1 Rotation of the horizontal axes

Upon rotation of the horizontal measuring axes clockwise by angle θ' as shown in Fig. 1, matrix $[Z_{xx}, Z_{xy}; Z_{yx}, Z_{yy}]$ changes to $[Z'_{xx}, Z'_{xy}; Z'_{yx}, Z'_{yy}]$ according to

$$\begin{bmatrix} Z'_{xx} & Z'_{xy} \\ Z'_{yx} & Z'_{yy} \end{bmatrix} = \mathbf{R}(\theta') \begin{bmatrix} Z_{xx} & Z_{xy} \\ Z_{yx} & Z_{yy} \end{bmatrix} \mathbf{R}(-\theta') \quad (5)$$

Thus the elements of the second matrix are related to the first by the following equations:

$$Z'_{xx} = (Z_{xx} + Z_{yy})/2 + C\sin(2\theta' + \beta) \quad (6)$$

$$Z'_{xy} = (Z_{xy} - Z_{yx})/2 + C\cos(2\theta' + \beta) \quad (7)$$

$$Z'_{yx} = -(Z_{xy} - Z_{yx})/2 + C\cos(2\theta' + \beta) \quad (8)$$

$$Z'_{yy} = (Z_{xx} + Z_{yy})/2 - C\sin(2\theta' + \beta) \quad (9)$$

where

$$C = [(Z_{xx} - Z_{yy})^2 + (Z_{xy} + Z_{yx})^2]^{\frac{1}{2}}/2 \quad (10)$$

and β is defined by

$$\tan\beta = (Z_{xx} - Z_{yy})/(Z_{xy} + Z_{yx}) \quad (11)$$

It is also useful to define an auxiliary angle β' by

$$\tan\beta' = (Z'_{xx} - Z'_{yy})/(Z'_{xy} + Z'_{yx}) \tag{12}$$

with

$$\theta' = (\beta' - \beta)/2 \tag{13}$$

and angle μ as

$$\tan\mu = (Z_{yy} + Z_{xx})/(Z_{xy} - Z_{yx}) \tag{14}$$

also Z^L as

$$Z^L = [(Z_{xx} + Z_{yy})^2 + (Z_{xy} - Z_{yx})^2]^{\frac{1}{2}}/2 \tag{15}$$

Then plotting Z'_{xx} against Z'_{xy} as the axes are rotated (i.e. θ' varies) defines a circle, known (with its axes) as a Mohr diagram. On such a figure axes for Z'_{yx} and Z'_{yy} may be included, to display the variation of these components also.

3. The depiction of MT tensors using Mohr diagrams

The Mohr diagram representation, straightforward for a 2 x 2 matrix, is an informative figure for the student of linear algebra. For the MT case it may be a useful way to display results, as will now be shown for the general case of 3D conductivity structure, and the particular cases of 2D and 1D conductivity structure to which the 3D case simplifies.

In this chapter in-phase and quadrature data will be presented in adjacent figures. There is an appeal in putting both in-phase and quadrature data on the same axes, as demonstrated by Szarka & Menvielle (1997) and Weaver et al. (2000), but when showing a full frequency range separate sets of axes are practical. Also at times it is helpful to add axes for Z'_{yx} and Z'_{yy} as on Fig. 2a, and such an addition is not possible for Z'_{xx} and Z'_{xy} axes common to both in-phase and quadrature parts.

3.1 The general case; 3D structure

Mohr diagrams for the general case of the MT tensor are shown in Fig. 2a. Different points on the diagram can be checked to confirm that Equations 6 to 15 for the rotation of axes are obeyed. The diagram is "Type 1" of Lilley (1998, p.1889), and shows how axes for Z'_{yx} and Z'_{yy} can be included. The diagram in Fig. 2a is drawn for relatively mild 3D characteristics. Diagrams for strong 3D characteristics are discussed below and shown in an example; however, what limits there may be to extreme 3D behaviour have not yet been fully explored. Some examples of 3D behaviour which appear to be prohibited will be discussed in this chapter.

3.2 2D structure

When the geologic structure is 2D and the axes are rotated to be along and across geologic strike, the MT tensor can be rotated to have the form $[0, Z^\sigma; -Z^\chi, 0]$, where it is expected both Z^σ and Z^χ are positive (a point discussed in Section 13.1 below). As shown in Fig. 2b, the diagram becomes a pair of circles with origins on the Z'_{xy} axes, and angles μ_p and μ_q are zero. The in-phase and quadrature radial arms are parallel. Z^σ and Z^χ are known as the *TE* and *TM* modes (or vice-versa).

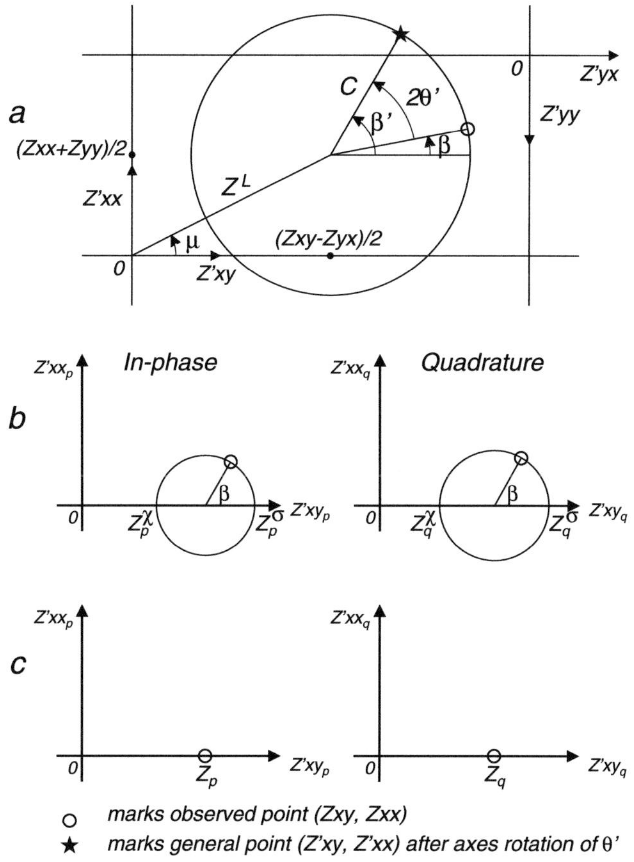

Fig. 2. Mohr diagrams for a: The general MT tensor; both in-phase and quadrature parts take this general form. b: The 2D case; radial arms are now parallel, with circle centres on the horizontal axes. The 2D values Z^{σ} and Z^{χ} could be interchanged. c: The 1D case, for **Z** observed as $[0, Z; -Z, 0]$.

3.3 1D structure

If a 2D case simplifies further to become a 1D case, $Z^{\sigma} = Z^{\chi} = Z$ say, and the tensor for all rotations has the form $[0, Z; -Z, 0]$. The length C of the radial arm vanishes, and the diagram reduces to a pair of points on the horizontal axes, as shown in Fig. 2c.

4. Invariants of rotation of the measuring axes

It can be seen from Fig. 2a that the MT observations can be expressed in terms of seven invariants of rotation. From different sets which are possible, and evident from formal analysis (Szarka & Menvielle, 1997; Weaver et al., 2000), this chapter adopts the invariants shown in Fig. 3. Thus the eight values of the MT tensor, all of which generally change upon axes rotation, become seven invariants plus one angle. That angle, θ' in Fig. 2a, is the angle which defines the direction of the measuring axes (for example, with respect to north).

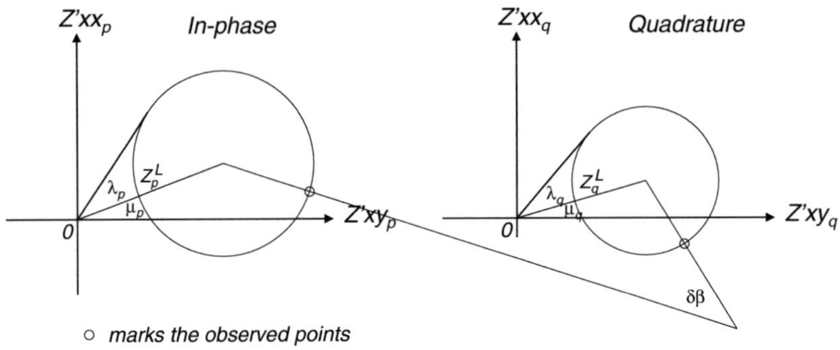

Fig. 3. Seven invariants of rotation for a general 3D MT tensor.

4.1 Two invariants summarising the 1D character of the tensor

The two invariants $Z^L_{p,q}$ summarising the 1D character are straightforward and were termed "central impedances" by Lilley (1993). Note that the second could be expressed normalised by the first, and so as a ratio (and then as an angle, taking the arctangent of that ratio).

4.2 Two invariants summarising the 2D character of the tensor

Two invariants $\lambda_{p,q}$ measure the 2D character, and are also straightforward. They are naturally angles, and were termed anisotropy angles by Lilley (1993). They may be expressed

$$\lambda_{p,q} = \arcsin(C_{p,q}/Z^L_{p,q}) \tag{16}$$

where $\lambda_{p,q}$ is in the range 0 to 90°. (Note this definition fails if $C_{p,q} > Z^L_{p,q}$ and the relevant circle encloses the origin.)

4.3 Two (of three) invariants summarising the 3D character of the tensor

Two angles $\mu_{p,q}$ characterising the 3D nature of the impedance tensor are also straightforward, and are shown in Fig. 3. It may be effective to express them as their mean and difference values, because certain mechanisms for causing 3D effects, especially static distortion, give the same μ contribution to both the in-phase and quadrature parts of a tensor (Lilley, 1993). In such cases, static distortion of a regional 2D structure is then measured by $(\mu_q + \mu_p)/2$, and the difference $(\mu_q - \mu_p)$ would be zero. Thus $(\mu_q - \mu_p)$, when non-zero, may be a measure of any 3D effects present, beyond static distortion.

4.4 The third 3D invariant

A third 3D invariant $(\delta\beta = \beta_q - \beta_p)$ can be seen from Fig. 3 to be the angle by which the two radial arms of the in-phase and quadrature circles are not parallel. It is significant as it alone, of the invariants chosen, links the in-phase and quadrature parts of an observed tensor. Another possibility for this seventh invariant is $(\delta\beta - \mu_q + \mu_p)$, where the difference $(\mu_q - \mu_p)$ is removed first from $\delta\beta$, and the departure of the radial arms from being parallel is then judged afresh.

However a related angle for the seventh invariant, derived by Weaver et al. (2000), has the great utility that it appears again in the Mohr diagram for the phase tensor, to be introduced in Section 8 below. There, as angle γ, it has a simple significance concerning geologic strike.

4.5 Angles or their sines?

Weaver et al. (2000) take sines of the angles to give measures in the range 0 to 1, and this technique is adopted by Marti et al. (2005). However, if ambiguity is to be avoided, such a procedure restricts the angles to being not greater than 90°. Because examples occur for which the values of $\mu_{p,q}$ are greater than 90°, it may be preferable to quote the invariants as angles, allowing them a 360° range (perhaps most usefully expressed in the range ±180°).

5. Some basic techniques of linear algebra applied to the analysis of the MT tensor

To gain familiarity with the MT tensor, it is instructive to explore some common steps taken in matrix analysis. The steps described form part of the history of MT.

5.1 Separation into symmetric and antisymmetric components

The MT tensor $\mathbf{Z}$, if split into symmetric $\mathbf{Z^s}$ and antisymmetric $\mathbf{Z^a}$ parts

$$\mathbf{Z} = \mathbf{Z^s} + \mathbf{Z^a} \tag{17}$$

may be written

$$\begin{bmatrix} Z_{xx} & Z_{xy} \\ Z_{yx} & Z_{yy} \end{bmatrix} = \begin{bmatrix} Z_{xx} & (Z_{xy}+Z_{yx})/2 \\ (Z_{xy}+Z_{yx})/2 & Z_{yy} \end{bmatrix} + \begin{bmatrix} 0 & (Z_{xy}-Z_{yx})/2 \\ (Z_{yx}-Z_{xy})/2 & 0 \end{bmatrix} \tag{18}$$

The second and antisymmetric part, $\mathbf{Z^a}$, is immediately recognised as being of the ideal 1D MT form described in Section 3.3.

In much MT interpretation, it has been common practice to obtain "1D estimates" from 2D or 3D data by taking the average of observed Z_{xy} and $(-)Z_{yx}$ as $(Z_{xy} - Z_{yx})/2$. Equation 18 demonstrates the approximations made in such a procedure. For in taking such an average, it can be seen that the information in the first matrix term $\mathbf{Z^s}$ is ignored, perhaps without justification. Diagrammatically, the procedure is equivalent, in Fig. 4a, to representing the circle on the left-hand side by the sum of the two circles on the right-hand side. The first circle on the right-hand side is then ignored, leaving the second circle which, reduced to its central point, is a 1D case.

5.2 Separation into symmetric and 2D components

The exercise in Section 5.1 may be regarded as a separation into symmetric and 1D components, and suggests a similar separation of the observed tensor into two parts of which the second, $\mathbf{Z^{2D}}$, is chosen to be of ideal 2D form. The first part, $\mathbf{Z^{s2}}$, is found to again be symmetric.

Thus the MT tensor is expressed

$$\mathbf{Z} = \mathbf{Z^{s2}} + \mathbf{Z^{2D}} \tag{19}$$

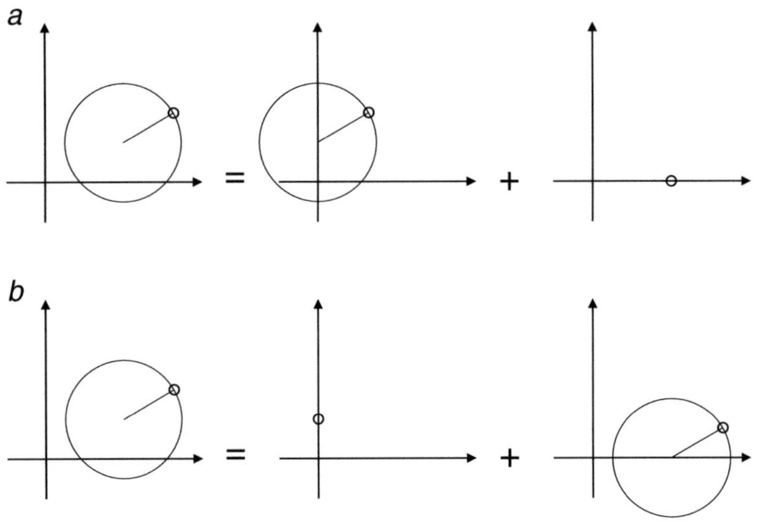

Fig. 4. Separation of a 2 x 2 matrix, a: into symmetric and 1D parts as in Equation 18; b: into symmetric and 2D parts as in Equation 20. The diagrams are drawn for $\mathbf{Z} = [1.75, 3.30; -0.70, 0.25]$. Thus case a is: $\mathbf{Z} = [1.75, 1.3; 1.3, 0.25] + [0, 2.0; -2.0, 0]$, and case b is: $\mathbf{Z} = [1.0, 0; 0, 1.0] + [0.75, 3.3; -0.7, -0.75]$.

and by partition as

$$\begin{bmatrix} Z_{xx} & Z_{xy} \\ Z_{yx} & Z_{yy} \end{bmatrix} = \begin{bmatrix} (Z_{xx} + Z_{yy})/2 & 0 \\ 0 & (Z_{xx} + Z_{yy})/2 \end{bmatrix} + \begin{bmatrix} (Z_{xx} - Z_{yy})/2 & Z_{xy} \\ Z_{yx} & (-Z_{xx} + Z_{yy})/2 \end{bmatrix} \quad (20)$$

where the second part is now of ideal 2D form.

Equation 20 is expressed in Mohr diagrams in Fig. 4b, where it can be seen that the first part is a point on the vertical axis, and the second part, $\mathbf{Z^{2D}}$, taken by itself plots as an ideal 2D circle with centre on the horizontal Z'_{xy} axis. The intersections of the circle with the axis give the values of the TE and TM impedances for this (now artificially) ideal tensor.

Thus the common practice with MT data, when seeking TE and TM values, of finding the maximum and minimum values of Z'_{xy} as the observing axes are rotated, can be seen to be tantamount to an assumption that the $\mathbf{Z^{s2}}$ part of the matrix be ignored, so that the circle for the $\mathbf{Z^{2D}}$ part indeed plots in ideal 2D form.

5.3 Eigenvalue analysis

Eigenvalues ζ_1 and ζ_2 of a matrix $\mathbf{Z}$ are found by solving the characteristic equation

$$\zeta^2 - (Z_{xx} + Z_{yy})\zeta + Z_{xx}Z_{yy} - Z_{xy}Z_{yx} = 0 \quad (21)$$

to obtain

$$\zeta_1, \zeta_2 = (Z_{xx} + Z_{yy})/2 \pm [(Z_{xx} + Z_{yy})^2 + 4(Z_{xy}Z_{yx} - Z_{xx}Z_{yy})]^{\frac{1}{2}}/2 \quad (22)$$

Three cases are possible and of interest. Each case will be discussed separately.

5.3.1 Eigenvalues are conjugate pairs

The first case occurs when

$$(Z_{xx} + Z_{yy})^2 + 4(Z_{xy}Z_{yx} - Z_{xx}Z_{yy}) < 0 \tag{23}$$

and the two roots of the conjugate equation form a conjugate pair. The product of the two roots, $\zeta_1\zeta_2$, will always be positive. An equivalent way of expressing Inequality 23 is as

$$(Z_{xx} - Z_{yy})^2 + 4Z_{xy}Z_{yx} < 0 \tag{24}$$

5.3.2 Eigenvalues are real and equal

The second case occurs when

$$(Z_{xx} + Z_{yy})^2 + 4(Z_{xy}Z_{yx} - Z_{xx}Z_{yy}) = 0 \tag{25}$$

The two roots of the conjugate equation are now both real (positive or negative), and equal. In fact

$$\zeta_1 = \zeta_2 = (Z_{xx} + Z_{yy})/2 \tag{26}$$

Again, the product of the two roots, $\zeta_1\zeta_2$, will be positive.

5.3.3 Eigenvalues are real and different

The third case occurs when

$$(Z_{xx} + Z_{yy})^2 + 4(Z_{xy}Z_{yx} - Z_{xx}Z_{yy}) > 0 \tag{27}$$

The two roots of the conjugate equation are now both real, different, and positive or negative depending on the signs of $(Z_{xx} + Z_{yy})$ and $(Z_{xx}Z_{yy} - Z_{xy}Z_{yx})$, the trace and determinant respectively of **Z**.

The product of the two eigenvalues is given by

$$\zeta_1\zeta_2 = \det \mathbf{Z} \tag{28}$$

and is positive if det**Z** is positive, and negative if det**Z** is negative.

5.4 Eigenvalues on Mohr diagrams

The three eigenvalue cases discussed in the preceding three subsections are clearly defined when the data are plotted on Mohr diagrams. Eigenvalues which are real are shown graphically. The directions of their eigenvectors may be read from the diagrams remembering, in Fig. 2a, the $2\theta'$ anticlockwise rotation of the radial arm for, in Fig. 1, the θ' clockwise rotation of the axes.

Thus for the case of Section 5.3.1, circles which (as in Fig. 5a) do not touch the vertical axes obey Inequality 23. Their eigenvalues are complex conjugate pairs, and real eigenvectors for them do not exist.

Secondly, for the case discussed in Section 5.3.2, Fig. 5b shows a circle which is just touching both vertical axes, $Z'_{xy} = 0$ and $Z'_{yx} = 0$. The eigenvalues may be read off the Z'_{xx} axis, as

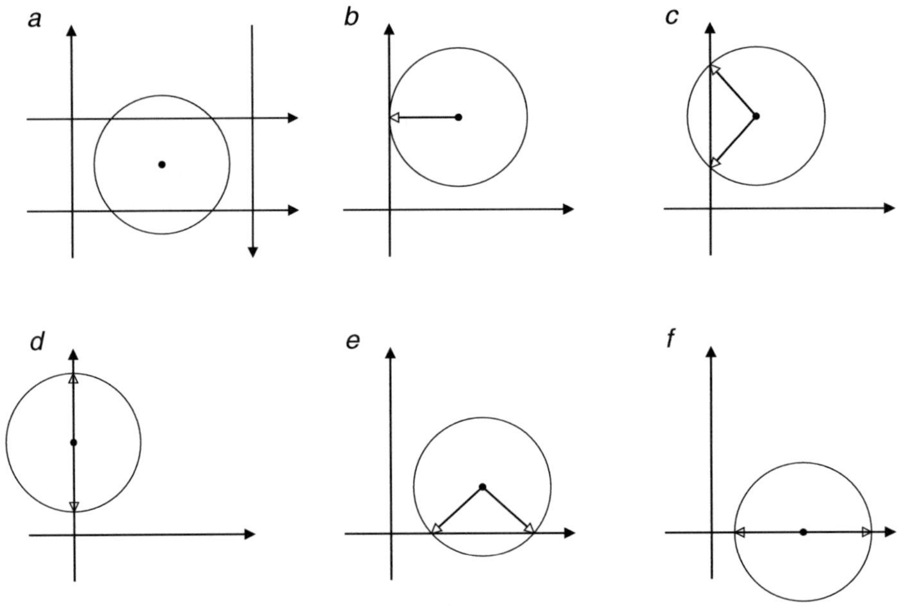

Fig. 5. Eigenvector directions (according to the radial arrows) and eigenvalues (where the arrowheads touch an axis) displayed on Mohr diagrams for a 2 x 2 matrix. a: Eigenvalues are complex conjugates, and are not evident on the diagram. b: Eigenvalues are real and equal; both eigenvectors are the same. c: Eigenvalues are real and different; eigenvectors are not orthogonal. d: Eigenvalues are real and different; eigenvectors are orthogonal. e: Associated MT eigenvalues are real and different; eigenvectors are not orthogonal. f: Associated MT eigenvalues are real and different; eigenvectors are orthogonal (the 2D case). In Fig. 5a axes are also drawn for Z'_{yx} and Z'_{yy}. Between the Z'_{xx} and Z'_{yy} axes, the product $Z'_{xy}.Z'_{yx} < 0$. On the Z'_{xx} and Z'_{yy} axes, $Z'_{xy}.Z'_{yx} = 0$. Outside the Z'_{xx} and Z'_{yy} axes, $Z'_{xy}.Z'_{yx} > 0$.

$(Z_{xx} + Z_{yy})/2$. The direction of the repeated eigenvector corresponds to the direction of a radial arm which is horizontal in the diagram, as shown.

Thirdly, for the case discussed in Section 5.3.3, in Fig. 5c a circle is shown which, like examples to be discussed below, now crosses the vertical axes. The two eigenvalues may again be read off the Z'_{xx} axis where the circle cuts this axis, and the two eigenvector directions are given by the θ' values for the radial arms to these points.

The example in Fig. 5c demonstrates that real eigenvalues correspond to the MT case of "phase going out of quadrant". Also it can be seen, by an extension of the discussion in Section 5.3.3 above, that for a Mohr circle to not capture the origin the product of the two eigenvalues must be positive; i.e. det**Z** must be positive.

Of equal interest in MT are the eigenvectors of the associated problem, which correspond to intersections of a circle with the horizontal Z'_{xy} axis. As shown in Fig. 5e, these might be regarded as an approximation to 2D TE and TM values, and for the true 2D case (Fig. 5f) they will indeed be so. These eigenvalues may be found by similar formalism, or by elementary trigonometry based on Fig. 2a. They may be evident from inspection, as in Lilley (1993).

6. Singular value decomposition

It is instructive also to apply SVD to an MT tensor, taking the in-phase and quadrature parts of the tensor separately.

As described for example by Strang (2005), decomposition by SVD factors a matrix $\mathbf{A}$ into

$$\mathbf{A} = \mathbf{U}\boldsymbol{\Sigma}\mathbf{V}^{\mathbf{T}} \tag{29}$$

where the columns of $\mathbf{V}$ are eigenvectors of $\mathbf{A}^{\mathbf{T}}\mathbf{A}$, and $\mathbf{A}^{\mathbf{T}}$ denotes the transpose of $\mathbf{A}$. The columns of $\mathbf{U}$ (which are eigenvectors of $\mathbf{A}\mathbf{A}^{\mathbf{T}}$) may be found by multiplying $\mathbf{A}$ by the columns of $\mathbf{V}$. The singular values on the diagonal of $\boldsymbol{\Sigma}$ are the square roots (taken positive by convention, a most important point in the present context) of the non-zero eigenvalues of $\mathbf{A}\mathbf{A}^{\mathbf{T}}$. As $\boldsymbol{\Sigma}$ is diagonal, it is straightforward to convert it to the antidiagonal form of an ideal 2D tensor, for example by expressing Equation 29 as

$$\mathbf{A} = \mathbf{U}\boldsymbol{\Sigma}\mathbf{W}\mathbf{W}^{-1}\mathbf{V}^{\mathbf{T}} \tag{30}$$

where $\mathbf{W} = [0, 1; -1, 0]$ and $\mathbf{V}^{\mathbf{T}}$ is pre-multiplied by $\mathbf{W}^{-1}$, with $\mathbf{W}^{-1}$ denoting the inverse of $\mathbf{W}$.

Equations for this form of the SVD of an MT tensor were derived in an earlier paper (Lilley, 1998). Taking phase relative to the $\mathbf{H}$ signal, so that $\mathbf{H}_q$ is zero, Equation 1 is written in its in-phase and quadrature parts as

$$\mathbf{E}_{p,q} = \mathbf{R}(-\theta e_{p,q}) \begin{bmatrix} 0 & \Upsilon_{p,q} \\ -\Psi_{p,q} & 0 \end{bmatrix} \mathbf{R}(\theta h_{p,q})\mathbf{H} \tag{31}$$

where the electric and magnetic observation axes are rotated clockwise independently, the electric axes by $\theta e_{p,q}$ and the magnetic axes by $\theta h_{p,q}$. Thus the in-phase part of the tensor, $\mathbf{Z}_p$, is factored into

$$\begin{bmatrix} Z_{xx_p} & Z_{xy_p} \\ Z_{yx_p} & Z_{yy_p} \end{bmatrix} = \mathbf{R}(-\theta e_p) \begin{bmatrix} 0 & \Upsilon_p \\ -\Psi_p & 0 \end{bmatrix} \mathbf{R}(\theta h_p) \tag{32}$$

where

$$\theta e_p = \frac{1}{2}\left[\arctan\frac{Z_{yy_p} - Z_{xx_p}}{Z_{xy_p} + Z_{yx_p}} + \arctan\frac{Z_{yy_p} + Z_{xx_p}}{Z_{xy_p} - Z_{yx_p}}\right] \tag{33}$$

$$\theta h_p = \frac{1}{2}\left[\arctan\frac{Z_{yy_p} - Z_{xx_p}}{Z_{xy_p} + Z_{yx_p}} - \arctan\frac{Z_{yy_p} + Z_{xx_p}}{Z_{xy_p} - Z_{yx_p}}\right] \tag{34}$$

$$\Upsilon_p - \Psi_p = \cos(\theta e_p + \theta h_p)[(Z_{xy_p} + Z_{yx_p}) - \tan(\theta e_p + \theta h_p)(Z_{xx_p} - Z_{yy_p})] \tag{35}$$

and

$$\Upsilon_p + \Psi_p = \cos(\theta e_p - \theta h_p)[(Z_{xy_p} - Z_{yx_p}) + \tan(\theta e_p - \theta h_p)(Z_{xx_p} + Z_{yy_p})] \tag{36}$$

(Equation 36 corrects equation 22 of Lilley (1998), where a negative sign is missing.) A similar set to Equations 32, 33, 34, 35 and 36 applies for the quadrature part of a tensor, with subscript q replacing p.

The quantities $\Upsilon_{p,q}$ and $\Psi_{p,q}$ are termed principal values. An examination of the possibility that one or both of $\Psi_{p,q}$ may be zero or negative is addressed in Section 13.3.

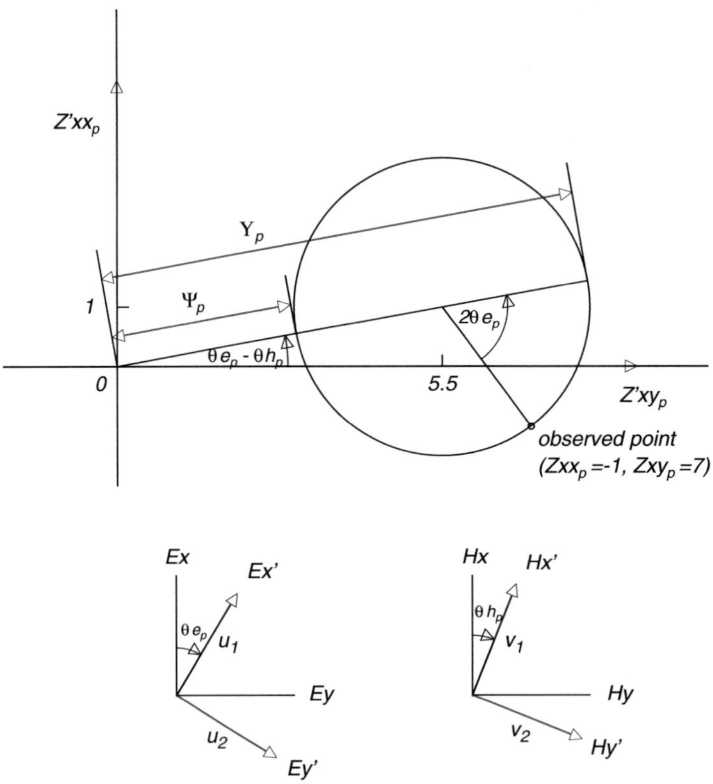

Fig. 6. Diagram showing the Mohr representation and the SVD of the matrix in Equation 37. The values of θe_p, θh_p, Υ_p and Ψ_p are evident as $32°, 21°, 8.1$ and 3.1. The axes for E'_x, E'_y and H'_x, H'_y show the rotations, from E_x, E_y and H_x, H_y by θe_p and θh_p respectively, to give an ideal 2D antidiagonal response.

The quantities arising from the SVD may be displayed on a Mohr diagram as in Fig. 6, which has been drawn for a tensor

$$\mathbf{Z}_p = [-1,7;-4,3] \tag{37}$$

The values of θe_p, θh_p, Υ_p and Ψ_p may be evaluated by the equations above as $31.7°, 21.4°, 8.09$ and 3.09 respectively. These values may also be read off the figure.

When the matrix of Equation 37 is put into a standard computing routine, the SVD returned is commonly in the form of Equation 29:

$$\mathbf{Z}_p = \begin{bmatrix} -.8507 & -.5257 \\ -.5257 & .8507 \end{bmatrix} \begin{bmatrix} 8.0902 & 0 \\ 0 & 3.0902 \end{bmatrix} \begin{bmatrix} .3651 & -.9310 \\ -.9310 & -.3651 \end{bmatrix} \tag{38}$$

which may be given the form of Equation 32 by expressing it first as

$$\mathbf{Z}_p = \begin{bmatrix} .8507 & -.5257 \\ .5257 & .8507 \end{bmatrix} \begin{bmatrix} 0 & 8.0902 \\ -3.0902 & 0 \end{bmatrix} \begin{bmatrix} .9310 & .3651 \\ -.3651 & .9310 \end{bmatrix} \tag{39}$$

and then as

$$\mathbf{Z}_p = \begin{bmatrix} \cos 31.71^\circ & -\sin 31.71^\circ \\ \sin 31.71^\circ & \cos 31.71^\circ \end{bmatrix} \begin{bmatrix} 0 & 8.0902 \\ -3.0902 & 0 \end{bmatrix} \begin{bmatrix} \cos 21.41^\circ & \sin 21.41^\circ \\ -\sin 21.41^\circ & \cos 21.41^\circ \end{bmatrix} \quad (40)$$

The columns of the first matrix $[\cos 31.71^\circ, -\sin 31.71^\circ; \sin 31.71^\circ, \cos 31.71^\circ]$, which have been derived from $\mathbf{U}$ in Equation 29, can be seen to define unit vectors u_1 and u_2 which are as given for E'_x and E'_y in Fig. 6. The rows of the third matrix $[\cos 21.41^\circ, \sin 21.41^\circ; -\sin 21.41^\circ, \cos 21.41^\circ]$, which have been derived from $\mathbf{V}^\mathbf{T}$ in Equation 29, can be seen to define unit vectors v_1 and v_2 as given for H'_x and H'_y in Fig. 6. Also, as is evident, the singular values in the diagonal of matrix $\mathbf{\Sigma}$ (the second matrix) give the values of Υ_p and Ψ_p.

Note that by rotation of the electric and magnetic axes separately, the MT tensor has been reduced to an ideal 2D form. In a search for the nearest 2D model in a 3D situation the results of SVD may be valuable to bear in mind; the magnetic axes may be an indication of regional strike, with the electric axes showing the brunt of the distortion. A disadvantage of such an analysis however is that a rather artificial conductivity structure is required to simply twist the electric field at an observing site to explain the different rotations required of the E and H axes. A more specific (if simple) model for local distortion is widely accepted, and will be discussed in Section 11. With this model, in the case of near-singular MT data, it will be shown that a surficial strike direction is determined. However regional strike determination remains possible if regional anisotropy is high.

7. A condition number to measure singularity in an MT tensor

It is common experience to find very strong anisotropy in an observed tensor, both for distorted 2D cases, and indeed generally. As a consequence, the tensor approaches a condition of singularity, in both its in-phase and quadrature parts. In a Mohr diagram the condition of singularity is shown by a circle touching the origin. If for example $\mathbf{Z}_p$ is a singular tensor, then there is some rotation of axes for which both Z'_{xx_p} and Z'_{xy_p} are zero (or indistinguishable from zero, when error is taken into account). For the student of linear algebra, an example of a null space occurs: it is the line of the direction of nil electric field change, holding for all magnetic field changes.

A condition number may be used to warn that singularity is being approached (Strang, 2005). When the condition number becomes high in some sense, the matrix is said to be ill-conditioned (Press et al., 1989). The condition number suggested by Strang (2005) is the norm of the matrix (sometimes called the spectral norm) multiplied by the norm of the inverse of the matrix; or equivalently, the greater principal value of the matrix divided by the lesser principal value. For the 2 x 2 matrix $\mathbf{Z}$ in Equation 1 the condition numbers $\kappa_{p,q}$ are

$$\kappa_{p,q} = \Upsilon_{p,q} / |\Psi_{p,q}| \quad (41)$$

The greater and lesser principal values $\Upsilon_{p,q}$ and $\Psi_{p,q}$ are given by Equations 35 and 36 and, as discussed above, are the singular values of the matrix. Following the convention that singular values are never negative, a modulus sign is put into the denominator of Equation 41 to cover cases where computation of the lesser principal value produces a negative number. In terms

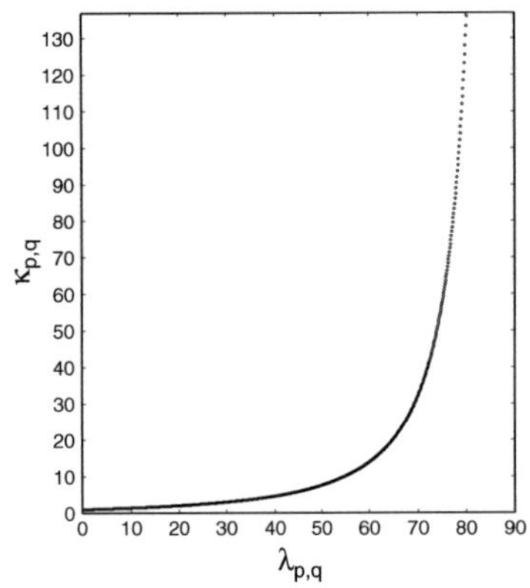

Fig. 7. The condition number $\kappa_{p,q}$ displayed as a function of the anisotropy angle $\lambda_{p,q}$ in degrees, using linear (left) and logarithmic (right) scales for $\kappa_{p,q}$.

of the Mohr representations in Figs 2, 3 and 6, from Equation 41 the condition numbers are given by

$$\kappa_{p,q} = (Z^L_{p,q} + C_{p,q})/|(Z^L_{p,q} - C_{p,q})| \tag{42}$$

that is

$$\kappa_{p,q} = 2/(\csc\lambda_{p,q} - 1) + 1 \tag{43}$$

remembering Equation 16. Fig. 7 shows $\kappa_{p,q}$ as a function of $\lambda_{p,q}$. While $\lambda_{p,q}$ itself is a measure of condition, Fig. 7 demonstrates that $\kappa_{p,q}$ is more sensitive than $\lambda_{p,q}$ as ill-condition is approached. Note that $\kappa_{p,q} \geq 1$.

An example is given in Fig. 8 of condition numbers determined for a set of data from a recent MT site, NQ142, in north Queensland, Australia. An increase in condition number above 10 monitors the decrease of the lesser principle value $\Psi_{p,q}$, which becomes negligible. The variation of angle $\theta e_{p,q}$ with period (T) stabilizes as condition numbers rise, an effect discussed below in Section 11.

8. The phase tensor

Caldwell et al. (2004) introduced a "phase tensor" based on the magnetotelluric tensor, and the concept was further developed by Bibby et al. (2005). The phase tensor is a real matrix $\mathbf{\Phi}$, defined by

$$\mathbf{\Phi} = \mathbf{Z_p}^{-1}\mathbf{Z_q} \tag{44}$$

and it has the property that it is unaffected by the in-phase distortion (described in Section 11.1 below) which is recognised as common in MT data. Note, however, that the computation of phase tensor values may encounter difficulties for high condition numbers and singularities in $\mathbf{Z_p}$ and $\mathbf{Z_q}$, which can be caused by strong distortion.

8.1 Mohr diagram for the phase tensor

The phase tensor, a 2 x 2 matrix, can also be represented by a Mohr diagram, as described by Weaver et al. (2003; 2006). The general form is shown in Fig. 9, following the convention for axes of Fig. 2a. Equations 6 to 15 can be adapted to apply to the quantities shown in Fig. 9.

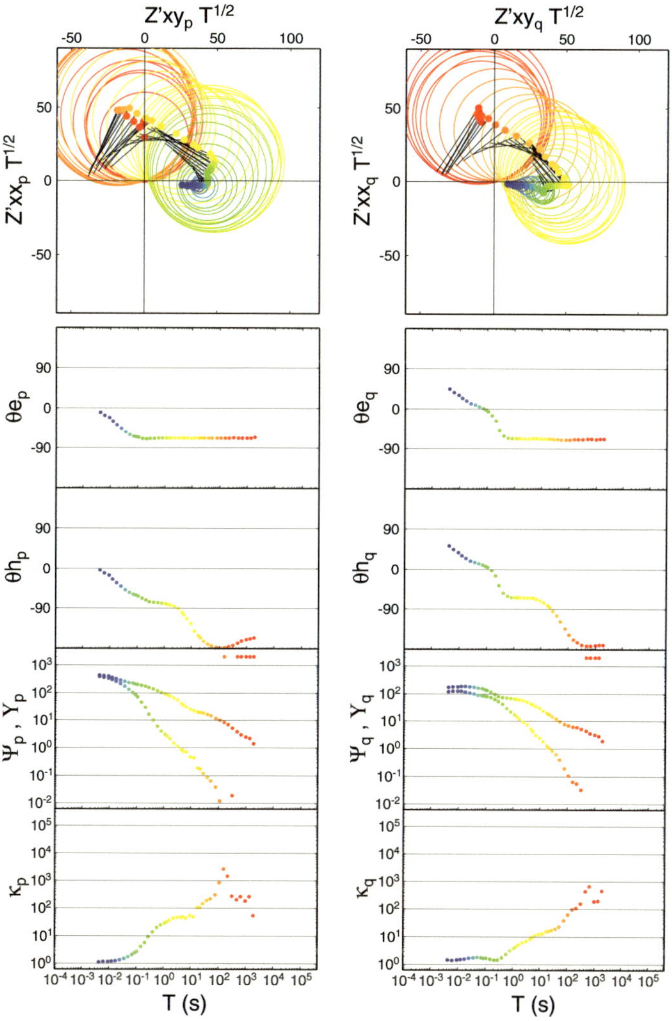

Fig. 8. The NQ142 data plotted as Mohr diagrams and analysed by SVD to give: angles θe_p, θe_q, θh_p and θh_q; the principal values Υ_p, Ψ_p, Υ_q and Ψ_q; and the condition numbers κ_p and κ_q. The variation of period with colour in the circles is the same as for the other plots. The Mohr diagrams follow the form of Fig. 2a, and have their impedance values multiplied by the square root of period (T) to make the plots more compact. Where circles enclose the origin, their lesser principal value (Ψ) is given an artificially high value to flag this circumstance; however condition numbers are computed nevertheless.

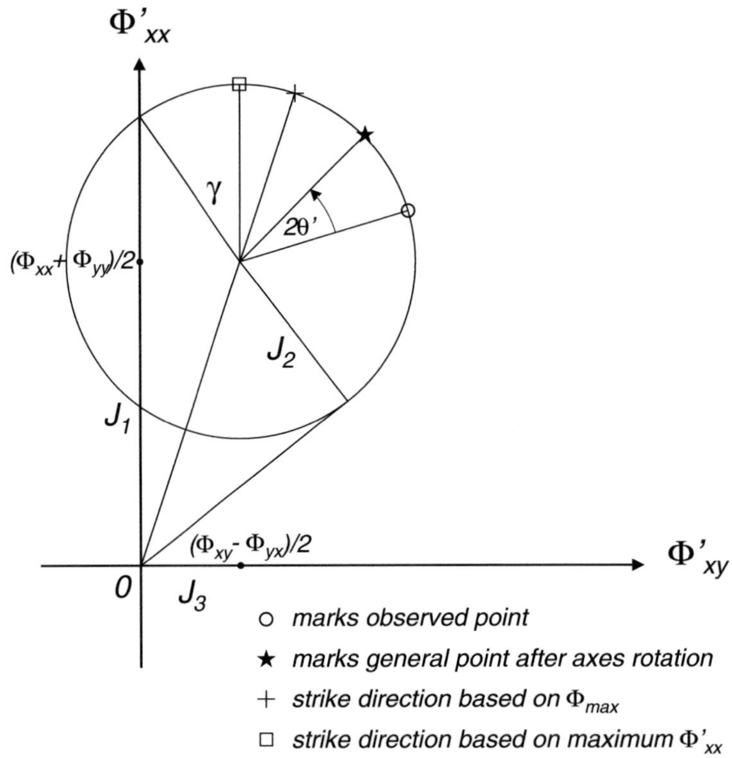

Fig. 9. Diagram showing the Mohr representation of the general phase tensor. Axes for Φ'_{yy} and Φ'_{yx} could be added (following Fig. 2a). The principal values (equivalent to Υ_p and Ψ_p in Fig. 6) are given by $(J_1{}^2 + J_3{}^2)^{1/2} \pm J_2$, and are denoted Φ_{max} and Φ_{min} by Caldwell et al. (2004).

The angle γ, which is a measure of 3D effects, is the seventh invariant of Weaver et al. (2000), described in Section 4.

The regional 2D strike direction chosen on phase considerations, as advocated by Bahr (1988) for example, is given by the point of maximum Φ'_{xx}, which is the highest point of the circle as drawn. However when the regional structure is recognised as being 3D, the best estimate of geologic strike is recommended by Caldwell et al. (2004) as the orientation of the principal axes of the phase tensor, shown in Fig. 9 as "strike direction based on Φ_{max}".

Note that a Mohr diagram for the phase tensor shows the variation of phase-tensor values as the observing axes are rotated, but that these values are generally not those of the usual MT Z_{xy} phase.

Also note that a Z_{xy} phase going "out of quadrant" does not imply that a principal value of $\mathbf{\Phi}$ is negative. In fact det$\mathbf{\Phi}$ is expected to be never negative, consistent with the principal values of $\mathbf{\Phi}$ (Φ_{max} and Φ_{min}) being never negative.

Examples of Mohr diagrams for phase tensors, which illustrate some of these points, are given in Lilley & Weaver (2010).

8.2 Mohr diagram for the 2D phase tensor

For the 2D case, the Mohr diagram reduces to a circle with centre on the Φ'_{xx} axis. The TE and TM phase values are then the arctangents of the intercepts which, from the analysis already given, are eigenvalues of the phase tensor. The two eigenvectors are orthogonal.

8.3 Mohr diagram for the 1D phase tensor

For the 1D case, the Mohr diagram reduces to a point on the Φ'_{xx} axis, marking the tangent of the phase value of the 1D MT response.

9. Invariants for the general phase tensor

To characterize the phase tensor just three invariants of rotation are needed, as pointed out by Caldwell et al. (2004), together with an angle defining the direction of the original observing axes. Options for invariants are evident from an inspection of Fig. 9, and include those adopted in Section 4 for Fig. 3. Weaver et al. (2003) chose J_1, J_2 and J_3 as shown in Fig. 9, which neatly summarize 1D, 2D and 3D characteristics, respectively.

10. Some complications illustrated by Mohr diagrams

10.1 Conditions for Z_{xy} and Z_{yx} phases out of quadrant

For simple cases of induction in 1D layered media, Z_{xy} phases will be in the first quadrant, and Z_{yx} phases in the third quadrant. However, it is not uncommon with complicated geologic structure to find phases which are out of these expected quadrants, for some orientation of the measuring axes. Experimental or computational error may be invoked in explanation, when in fact a common cause is simply distortion. Such distortion may be understood by reference to Fig. 3. Clearly once, due say to distortion, angle μ_p (or μ_q) increases to the point where the in-phase (or quadrature) circle first touches and then crosses the vertical Z'_{xx} axis, for an appropriate direction of the observing axes the phase observed of Z_{xy} will be "out of quadrant".

Adopting, in the rotated frame, the common definition for phase ϕ'_{xy} of

$$\phi'_{xy} = \arctan(Z'_{xy_q}/Z'_{xy_p}) \tag{45}$$

where the signs of numerator and denominator are taken into account, then ϕ'_{xy} is in the first quadrant when both Z'_{xy_p} and Z'_{xy_q} are positive.

The condition for the phase of Z'_{xy} to go out of quadrant is that one or both of its in-phase and quadrature parts should be negative. On the Mohr circle diagram, this condition is equivalent to either or both of the in-phase and quadrature circles for the data crossing the vertical axis. Phases greater than 90° are then possible for Z'_{xy}. If the circles however remain to the right of their vertical axes, the Z'_{xy} phase will be in quadrant for any orientation of the measuring axes.

With reference to Fig. 3, for the crossing of the vertical axis to occur, it can be seen that $\lambda_{p,q}$ and $\mu_{p,q}$ must together be greater than 90°, i.e.

$$\lambda_{p,q} + |\mu_{p,q}| > 90^\circ \tag{46}$$

With reference to the Z'_{yy_p} and Z'_{yx_p} axes also included in Fig. 2a, consequences for the phases of other elements are also clear. The symmetry of the figure is such that if the circle crosses the $Z'_{xx p}$ axis to the left, it will also cross the Z'_{yy_p} axis to the right. Thus if phases out of quadrant are possible for Z'_{xy}, so are they also possible for Z'_{yx}. However, the latter will occur for rotations 90° different from the former.

Similar considerations will give rules regarding the quadrants to be expected for Z'_{xx} and Z'_{yy} phases. An analysis of Fig. 2a shows that these phases will generally change quadrant with a rotation of the observing axes. An exception, when they would not do so, would be if the circle in Fig. 2a did not cross the horizontal Z'_{xy_p} axis, but stayed completely above (or below) this axis (and the circle for the quadrature part of the tensor behaved similarly).

10.2 Strike coordinates from phase considerations

The situation of Z_{xy} phase out of quadrant (for example for the part of the circle to the left of the vertical axis in Fig. 5c) causes difficulty in the symmetric-antisymmetric and the symmetric-2D partitions of Sections 5.1 and 5.2. First, if the circle in Fig. 5c is simply moved down so that its centre lies on the horizontal axis, the left-hand intercept with the horizontal axis is less than zero. The circle has enclosed the origin, and a negative value is obtained for the lesser of TE and TM, contrary to expectation (see Section 13.1).

Similarly there are methods which seek, as geologic strike coordinates, those given by maximum and minimum Z'_{xy} phase values when the axes are rotated. These methods will find erratic phase behaviour when either or both of the in-phase and quadrature circles cross the vertical axis as in Fig. 5c.

However Z'_{xy} phase will usually be at or near a maximum (or minimum) at the right-hand side of a circle which on its left-hand side crosses the vertical axis. Thus sensible results for maximum or minimum phase can be expected, if based only on a determination from the right-hand side.

11. Geologic interpretation of angles found by SVD, when condition numbers are high

Singular value decomposition of observed magnetotelluric tensors, taking the in-phase and quadrature parts separately, often produces directions which are frequency independent below a certain frequency which is found to be common to both in-phase and quadrature parts. This section examines how such directions may be related to the angles which arise in basic distortion models of MT data. It is found that while MT data are expected in general to be frequency dependent, there are particular limiting cases which are constant with frequency. Understanding such observed data may be useful in interpretation.

11.1 The Smith (1995) model for local distortion

This section applies the Smith (1995) description of local "static" distortion, which follows Bahr (1988) and Groom & Bailey (1989). In the absence of local distortion the measured magnetic field $\mathbf{H}$ is related to the regional electric field $\mathbf{E^r}$ by the tensor $\mathbf{Z^r}$:

$$\mathbf{E^r} = \mathbf{Z^r}\mathbf{H} \tag{47}$$

and with distortion $\mathbf{d}$ present, affecting the measured electric field $\mathbf{E^m}$ only,

$$\mathbf{E^m} = \mathbf{dE^r} \tag{48}$$

and

$$\mathbf{E^m} = \mathbf{Z^m H} = \mathbf{dZ^r H} \tag{49}$$

For the case of 2D regional structure, with measuring axes aligned with regional strike,

$$\tilde{\mathbf{Z}}^{\mathbf{r}} = [0, Z^r_{12}; Z^r_{21}, 0] \tag{50}$$

and so

$$\tilde{\mathbf{E}}^{\mathbf{m}} = \mathbf{d}[0, Z^r_{12}; Z^r_{21}, 0]\mathbf{H} \tag{51}$$

where the overscore ˜ indicates observing axes aligned with regional strike.

If the observation axes are north-south and east-west, say, and need to be rotated angle θ for alignment with regional strike, then Equation 51 (without distortion) gives

$$\mathbf{R}(\theta) \begin{bmatrix} E^m_x \\ E^m_y \end{bmatrix} = \begin{bmatrix} 0 & Z^r_{12} \\ Z^r_{21} & 0 \end{bmatrix} \mathbf{R}(\theta) \begin{bmatrix} H_x \\ H_y \end{bmatrix} \tag{52}$$

and with distortion

$$\mathbf{R}(\theta) \begin{bmatrix} E^m_x \\ E^m_y \end{bmatrix} = \mathbf{d} \begin{bmatrix} 0 & Z^r_{12} \\ Z^r_{21} & 0 \end{bmatrix} \mathbf{R}(\theta) \begin{bmatrix} H_x \\ H_y \end{bmatrix} \tag{53}$$

In co-ordinates aligned with the surficial geology the distortion amounts to scaling the electric fields by different amounts, g_1 and g_2, in directions parallel and perpendicular to the surficial strike (Zhang et al., 1987). Distortion matrix $\mathbf{d}$ then has the form

$$\begin{bmatrix} d_{11} & d_{12} \\ d_{12} & d_{22} \end{bmatrix} = \mathbf{R}(-\alpha_s) \begin{bmatrix} g_1 & 0 \\ 0 & g_2 \end{bmatrix} \mathbf{R}(\alpha_s) \tag{54}$$

where due to the restricted distortion model the d_{21} element has the value d_{12}, g_1 and g_2 are real constants (frequency independent), and α_s is the angle from the regional strike coordinates to the strike coordinates of the surficial structure causing the distortion. The decomposition of $\mathbf{d}$ in Equation 54 is displayed in Fig. 10.

Substituting the expression for $\mathbf{d}$ from Equation 54 into Equation 53 gives

$$\mathbf{E^m} = \mathbf{R}(-\theta - \alpha_s) \begin{bmatrix} g_1 Z^r_{21} \sin\alpha_s & g_1 Z^r_{12} \cos\alpha_s \\ g_2 Z^r_{21} \cos\alpha_s & -g_2 Z^r_{12} \sin\alpha_s \end{bmatrix} \mathbf{R}(\theta)\mathbf{H} \tag{55}$$

At this stage the in-phase and quadrature parts can be considered separately. Avoiding the encumbrance of additional notation, now assume that just the in-phase part of the measured E field is being considered, giving information on just the in-phase parts of the regional impedance tensor $\mathbf{Z^r}$.

Then following the form of Equation 32, SVD can be carried out on the central tensor in Equation 55, expanding it as

$$\begin{bmatrix} g_1 Z^r_{21} \sin\alpha_s & g_1 Z^r_{12} \cos\alpha_s \\ g_2 Z^r_{21} \cos\alpha_s & -g_2 Z^r_{12} \sin\alpha_s \end{bmatrix} = \mathbf{R}(-\eta_e) \begin{bmatrix} 0 & \mathrm{Y} \\ -\Psi & 0 \end{bmatrix} \mathbf{R}(\eta_h) \tag{56}$$

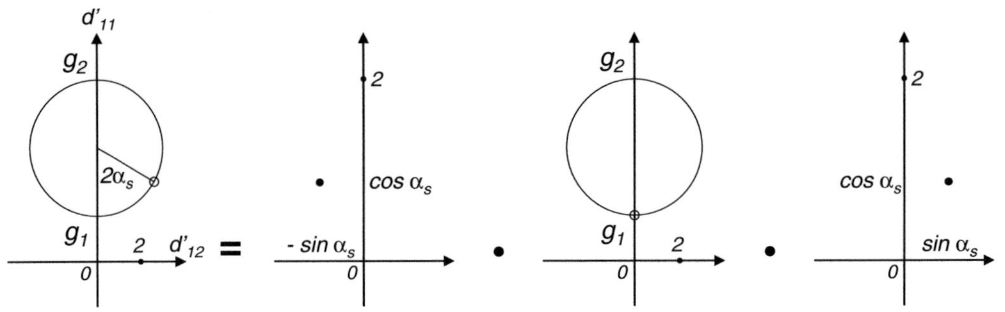

Fig. 10. Mohr diagrams for the decomposition in Equation 54. On the left the diagram shows the distortion matrix **d**. It is positive definite, due to the assumption of strictly 2D surficial anisotropy, and so has orthogonal eigenvectors. The radius of the circle is $(g_2 - g_1)/2$. The numerical matrix taken for the example is [3.5, 2.6; 2.6, 6.5], and the directions and magnitudes of its eigenvectors are shown by rotating the radial arm until it is parallel to the vertical axis. The eigenvalues then are $g_1 = 2$ and $g_2 = 8$. The angle $2\alpha_s$ on the figure is 60°, indicating that a rotation of axes anticlockwise through 30° is necessary to change the matrix to [2, 0: 0, 8]. On the right is shown the matrix decomposition as in Equation 54. The three diagrams represent the three matrices on the right-hand side of that equation, in turn. The first Mohr diagram represents a rotation by angle $(-\alpha_s)$. The second Mohr diagram is symmetric, representing a positive definite matrix according to the scaling factors g_1 and g_2. The third matrix is like the first, representing a rotation by angle $(+\alpha_s)$. Note the changes of scale between the diagrams.

where the angles η_e and η_h, and the principal values Y and Ψ are given by

$$\eta_e + \eta_h = \arctan\left[\frac{-g_2 Z^r_{12}\sin\alpha_s - g_1 Z^r_{21}\sin\alpha_s}{g_1 Z^r_{12}\cos\alpha_s + g_2 Z^r_{21}\cos\alpha_s}\right] \tag{57}$$

$$\eta_e - \eta_h = \arctan\left[\frac{-g_2 Z^r_{12}\sin\alpha_s + g_1 Z^r_{21}\sin\alpha_s}{g_1 Z^r_{12}\cos\alpha_s - g_2 Z^r_{21}\cos\alpha_s}\right] \tag{58}$$

$$\mathrm{Y} + \Psi = [(g_1 Z^r_{21}\sin\alpha_s - g_2 Z^r_{12}\sin\alpha_s)^2 + (g_2 Z^r_{21}\cos\alpha_s - g_1 Z^r_{12}\cos\alpha_s)^2]^{1/2} \tag{59}$$

and

$$\mathrm{Y} - \Psi = [(g_1 Z^r_{21}\sin\alpha_s + g_2 Z^r_{12}\sin\alpha_s)^2 + (g_1 Z^r_{12}\cos\alpha_s + g_2 Z^r_{21}\cos\alpha_s)^2]^{1/2} \tag{60}$$

Replacing into Equation 55 the matrix in its expanded form as in Equation 56, and combining adjoining rotation matrices, gives

$$\mathbf{E^m} = \mathbf{R}(-\theta - \alpha_s - \eta_e)\begin{bmatrix} 0 & \mathrm{Y} \\ -\Psi & 0 \end{bmatrix}\mathbf{R}(\theta + \eta_h)\mathbf{H} \tag{61}$$

Thus in the SVD of an MT tensor resulting from 2D surficial distortion of a regional 2D structure, the regional strike θ and the various distortion quantities g_1, g_2 and α_s occur in a way which does not allow their straightforward individual solution, because the SVD produces values for $(\theta + \alpha_s + \eta_e)$ and $(\theta + \eta_h)$. It is of interest however to examine two limiting cases, as in the following sections.

11.2 First limiting case

Now consider, in Equations 57, 58, 59 and 60, that $g_2 \gg g_1$, so that terms in g_1 are ignored with respect to terms in g_2. The results are then obtained that

$$Y = g_2[(Z_{12}^r \sin \alpha_s)^2 + (Z_{21}^r \cos \alpha_s)^2]^{1/2} \tag{62}$$

and

$$\Psi = 0 \tag{63}$$

For the angles η_e and η_h the limiting case gives

$$\eta_e = 0 \tag{64}$$

and

$$\eta_h = -\arctan(Z_{12}^r \tan \alpha_s / Z_{21}^r) \tag{65}$$

While in Equation 65 for η_h the values Z_{12}^r, Z_{21}^r and α_s still occur together, for η_e there is the simple result of zero. Fed back into Equation 61 it is seen that SVD of an MT tensor now gives, in the equivalent of θe in Equation 31, a result for $(\theta + \alpha_s)$: that is, the direction of surficial 2D strike.

Further, note that if $Z_{12}^r / Z_{21}^r = -1$, then $\eta_h = \alpha_s$, and the magnetic axes are also aligned with (or normal to) the surficial strike. This case can be seen to be that where the regional 2D anisotropy is out-weighed by the surficial anisotropy, so that the former approximates a 1D case.

11.3 Second limiting case

A second limiting case might be gross inequality in the TE and TM components of the regional 2D impedance. Consider the case where $|Z_{12}^r| \gg |Z_{21}^r|$, so that terms involving the latter may be ignored with respect to terms involving the former. Then Equations 57, 58, 59 and 60 give

$$Y = Z_{12}^r[(g_2 \sin \alpha_s)^2 + (g_1 \cos \alpha_s)^2]^{1/2} \tag{66}$$

and

$$\Psi = 0 \tag{67}$$

For the angles η_e and η_h this limiting case gives

$$\eta_e = -\arctan(g_2 \tan \alpha_s / g_1) \tag{68}$$

and

$$\eta_h = 0 \tag{69}$$

Fed back into Equation 61, it is seen that SVD of an MT tensor for the case where $\eta_h = 0$ now gives, in the equivalent of θh in Equation 31, a result for θ: that is, the direction of regional 2D strike.

12. Discussion and example

The two limiting cases, where realised, give angles of interest. Also, it should be noted that both cases correspond to the MT tensor being singular, or nearly so.

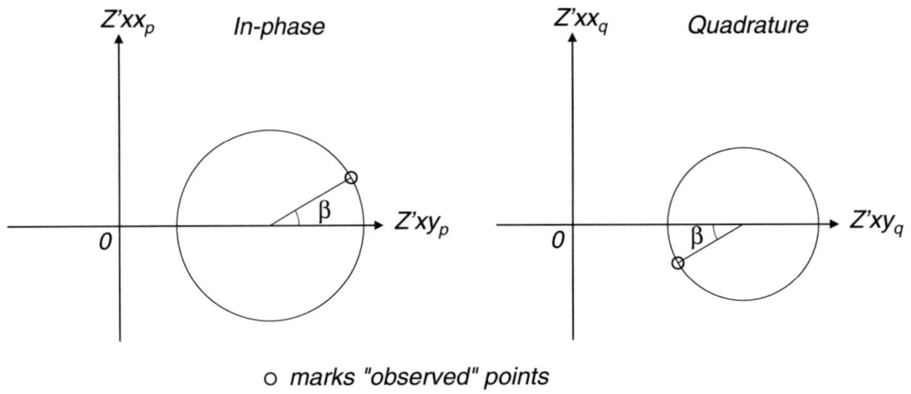

Fig. 11. Hypothetical example of a 2D case with the in-phase and quadrature radial arms anti-parallel. Such cases appear to be prohibited in nature, but inadvertently can be posed as numerical examples.

The example in Fig. 8, for frequencies less that 10 Hz, gives $\theta e_{p,q}$ values which are fixed at $-67°$, while $\theta h_{p,q}$ values continue to vary with period. The model of Section 11.1 is therefore a good candidate for the interpretation of the data. Applying the results of Section 11.2, surficial strike is determined at $-67°$ ($\pm 90°$ allowing for the usual ambiguity).

However the observation that there is a critical frequency, say f_c, only below which $\theta e_{p,q}$ values are frequency independent (and in the case of Fig. 8 are stable at $-67°$), implies a contradiction to the initial distortion model in Section 11.1, which was frequency independent, generally. The existence of such a "critical frequency" suggests distortion which is "at a distance", rather than immediately local.

13. Some remaining problems

13.1 *TE* and *TM* modes both positive?

It is well known that for the 2D case the *TM* mode is always positive (Weidelt & Kaikkonen, 1994), but proof has not been achieved for the *TE* case. It can be seen from Fig. 2b that were the *TE* case to be negative in either in-phase or quadrature part (or both) then the appropriate circles would enclose their origin of axes. Thus a general proof that circles cannot enclose their origins (discussed in Section 13.3 below) would also prove that the *TE* mode can never be negative.

13.2 Radial arms anti-parallel

The 2D example given in Fig. 2b, where the in-phase and quadrature radial arms are parallel, represents the common case observed without exception in the experience of the present author. However, in principle the case shown in Fig. 11 where the in-phase and quadrature radial arms are anti-parallel would also be two dimensional. If there is a theoretical reason why cases with radial arms anti-parallel as in Fig. 11 do not occur in practice it would be an advance to have it clarified.

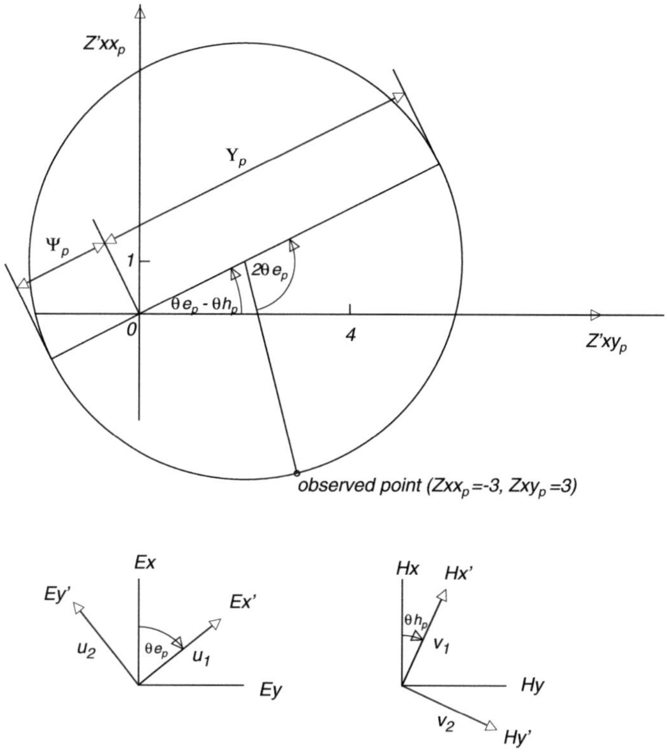

Fig. 12. The Mohr diagram for the matrix in Equation 70. The circle encloses the origin of axes because the matrix has a negative determinant. The values of θe_p, θh_p, Υ_p and Ψ_p are evident as $51°, 25°, 6.3$ and -1.8 (Ψ_p being of negative sign when the direction of Υ_p from the origin defines positive).

13.3 Circles capturing the origin: a note on negative determinants

It is also observed that circles do not capture or enclose their origins of axes, except for reasons of obvious error, or usual errors associated with singularity. The author suggests that this observed behaviour corresponds to the prohibition of components of negative resistivity in the electrical conductivity structure. Again a proof regarding this possibility would be welcome.

Negative determinants (no matter how they arise) need care in their analysis, as the following example illustrates. If, in contrast to the example in Section 6, the determinant (say of $\mathbf{Z}_p$) is negative, then the matrices $\mathbf{Z}_p\mathbf{Z}_p^{\mathbf{T}}$ and $\mathbf{Z}_p^{\mathbf{T}}\mathbf{Z}_p$ will each have a negative eigenvalue. However in a conventional SVD these negative eigenvalues are taken as positive singular values, and to allow for this action the sign is also changed of the eigenvector of $\mathbf{U}$ which they multiply.

To demonstrate this point, consider the hypothetical matrix

$$\mathbf{Z}_p = [-3, 3; -1, 5] \tag{70}$$

which has a negative determinant. The Mohr diagram representation of the matrix is shown in Fig. 12.

By Equations 33, 34, 35 and 36, the values of θe_p, θh_p, Υ_p and Ψ_p are evaluated as $51.26°, 24.70°, 6.36$ and -1.88 respectively. As before, by multiplying out the right-hand side, it can be checked that these values do indeed satisfy Equation 32. Also these values may be read off Fig. 12 as $51°, 25°, 6.3$ and -1.8, but now difficulties arising from the negative determinant are evident. Taking the direction of Υ_p from the origin to define positive, the value of Ψ_p is seen to be negative.

Put into a standard computing routine, the SVD returned for the matrix in Equation 70 is

$$\mathbf{Z}_p = \begin{bmatrix} .6257 & .7800 \\ .7800 & -.6257 \end{bmatrix} \begin{bmatrix} 6.3592 & 0 \\ 0 & 1.8870 \end{bmatrix} \begin{bmatrix} -.4179 & .9085 \\ -.9085 & -.4179 \end{bmatrix} \tag{71}$$

which may also be obtained formally by determining the eigenvalues and eigenvectors of $\mathbf{Z}_p^{\mathbf{T}}\mathbf{Z}_p$. To give it the form of Equation 32, Equation 71 may be expressed

$$\mathbf{Z}_p = \begin{bmatrix} .6257 & .7800 \\ .7800 & -.6257 \end{bmatrix} \begin{bmatrix} 0 & 6.3592 \\ -1.8870 & 0 \end{bmatrix} \begin{bmatrix} .9085 & .4179 \\ -.4179 & .9085 \end{bmatrix} \tag{72}$$

and then written

$$\mathbf{Z}_p = \begin{bmatrix} \cos 51.26° & \sin 51.26° \\ \sin 51.26° & -\cos 51.26° \end{bmatrix} \begin{bmatrix} 0 & 6.3592 \\ -1.8870 & 0 \end{bmatrix} \begin{bmatrix} \cos 24.70° & \sin 24.70° \\ -\sin 24.70° & \cos 24.70° \end{bmatrix} \tag{73}$$

The rows of the third matrix $[\cos 24.70°, \sin 24.70°; -\sin 24.70°, \cos 24.70°]$ define unit vectors v_1 and v_2 for H'_x, H'_y as shown in Fig. 12. As for Fig. 6, these indicate a rotation of the H_x, H_y observing axes.

However, the columns of the first matrix $[\cos 51.26°, \sin 51.26°; \sin 51.26°, -\cos 51.26°]$ define unit vectors u_1 and u_2 also as shown in Fig. 12. These unit vectors do not now represent just a simple rotation of the E_x, E_y axes, but also the reflection of one of them. The negative value of Ψ_p given by Equations 35 and 36 has been cast by the SVD analysis as a positive singular value in the second matrix in Equation 71. To compensate for this convention, the direction of one of the axes (originating as the direction of an eigenvector of $\mathbf{U}$ in Equation 29) has been reversed. Any model of distortion rotating the electric fields is thus contravened. An extra physical phenomenon is required to explain the reflection of one of the rotated electric field axes. Pending such an explanation, a negative principal value (-1.8870 in the present example) must be regarded with great caution.

In normal SVD formalism the change of sign of an eigenvector, as demonstrated in the above example, may occur with little comment. In the present case however it has had the profound effect of destroying, after rotation, the "right-handedness" of the observing axes of the MT data.

14. Conclusion

Basic 2 x 2 matrices arise commonly in linear algebra, and Mohr diagrams have wide application in displaying their properties. Complementing the usual algebraic approach, they may help the student understand especially anomalous data.

Such Mohr diagrams have proved useful in checking MT data for basic errors arising in the manipulation of time-series; in checking hypothetical examples for realistic form; and in checking for errors which are demonstrated by negative determinants.

Invariants of an observed tensor under axes rotation are evident from inspection of such Mohr diagrams. The invariants may be used as indicators of the geologic dimensionality of the tensor; they may also be useful in the inversion and interpretation of MT data. Mohr diagrams of the MT phase tensor also show invariants which measure geologic dimensionality, and indicate regional geologic strike.

Attention has been given to monitoring the singularity of an MT tensor with a condition number, as tensors which approach singularity are common. Carrying out SVD on observed data with high condition numbers can indicate a direction of surficial 2D strike, for the simple case of the surficial distortion of a regional 2D MT response. In some circumstances the 2D regional strike is found in this straightforward way.

Particularly noteworthy is the observation of a critical frequency below which an observed MT tensor becomes sufficiently ill-conditioned for the surficial strike to become evident. The example in Fig. 8 shows this behaviour to apply at frequencies below 10 Hz. That a frequency-independent model produces a frequency-dependent result in this way requires care in interpretation.

Remaining problems for which proofs would augment the subject are the evident prohibition of negative TE values, the evident non-observation of antiparallel radial arms, and the evident non-observation of MT tensor data with negative determinant values (corresponding to Mohr circles prohibited from enclosing their origins of axes).

15. Acknowledgements

The author thanks Peter Milligan, Chris Phillips and John Weaver for many valuable discussions over recent years on the subject matter of this chapter. Data for the NQ142 example were provided by Geoscience Australia.

16. References

Bahr, K. (1988). Interpretation of the magnetotelluric impedance tensor: regional induction and local telluric distortion, *J. Geophys.* 62: 119 – 127.

Berdichevsky, M. N. & Dmitriev, V. I. (2008). *Models and Methods of Magnetotellurics*, Springer.

Bibby, H. M., Caldwell, T. G. & Brown, C. (2005). Determinable and non-determinable parameters of galvanic distortion in magnetotellurics, *Geophys. J. Int.* 163: 915 – 930.

Caldwell, T. G., Bibby, H. M. & Brown, C. (2004). The magnetotelluric phase tensor, *Geophys. J. Int.* 158: 457 – 469.

Eggers, D. E. (1982). An eigenstate formulation of the magnetotelluric impedance tensor, *Geophysics* 47: 1204 – 1214.

Groom, R. W. & Bailey, R. C. (1989). Decomposition of magnetotelluric impedance tensors in the presence of local three-dimensional galvanic distortion, *J. Geophys. Res.* 94: 1913 – 1925.

Gubbins, D. & Herrero-Bervera, E. (eds) (2007). *Encyclopedia of Geomagnetism and Paleomagnetism*, Encyclopedia of Earth Sciences, Springer.

Hobbs, B. A. (1992). Terminology and symbols for use in studies of electromagnetic induction in the Earth, *Surv. Geophys.* 13: 489 – 515.

LaTorraca, G. A., Madden, T. R. & Korringa, J. (1986). An analysis of the magnetotelluric impedance for three-dimensional conductivity structures, *Geophysics* 51: 1819 – 1829.

Lilley, F. E. M. (1993). Magnetotelluric analysis using Mohr circles, *Geophysics* 58: 1498 – 1506.

Lilley, F. E. M. (1998). Magnetotelluric tensor decomposition: Part I, Theory for a basic procedure, *Geophysics* 63: 1885 – 1897.

Lilley, F. E. M. & Weaver, J. T. (2010). Phases greater than 90° in MT data: Analysis using dimensionality tools, *J. App. Geophys.* 70: 9 – 16.

Marti, A., Queralt, P., Jones, A. G. & Ledo, J. (2005). Improving Bahr's invariant parameters using the WAL approach, *Geophys. J. Int.* 163: 38 – 41.

Press, W. H., Teukolsky, S. A., Vetterling, W. T. & Flannery, B. P. (1989). *Numerical Recipes, The Art of Scientific Computing*, Cambridge Univ. Press.

Simpson, F. & Bahr, K. (2005). *Practical Magnetotellurics*, Cambridge Univ. Press.

Smith, J. T. (1995). Understanding telluric distortion matrices, *Geophys. J. Int.* 122: 219 – 226.

Strang, G. (2005). *Linear Algebra and its Applications, 4th Edition*, Brooks–Cole.

Stratton, J. A. (1941). *Electromagnetic Theory*, McGraw–Hill.

Szarka, L. & Menvielle, M. (1997). Analysis of rotational invariants of the magnetotelluric impedance tensor, *Geophys. J. Int.* 129: 133 – 142.

Weaver, J. T. (1994). *Mathematical Methods for Geo-electromagnetic Induction*, Wiley.

Weaver, J. T., Agarwal, A. K. & Lilley, F. E. M. (2000). Characterization of the magnetotelluric tensor in terms of its invariants, *Geophys. J. Int.* 141: 321 – 336.

Weaver, J. T., Agarwal, A. K. & Lilley, F. E. M. (2003). The relationship between the magnetotelluric tensor invariants and the phase tensor of Caldwell, Bibby and Brown, *in* J. Macnae & G. Liu (eds), *Three-Dimensional Electromagnetics III*, number 43 in *Paper*, ASEG, pp. 1 – 8.

Weaver, J. T., Agarwal, A. K. & Lilley, F. E. M. (2006). The relationship between the magnetotelluric tensor invariants and the phase tensor of Caldwell, Bibby and Brown, *Explor. Geophys.* 37: 261 – 267.

Weidelt, P. & Kaikkonen, P. (1994). Local 1-D interpretation of magnetotelluric B-polarization impedances, *Geophys. J. Int.* 117: 733 – 748.

Yee, E. & Paulson, K. V. (1987). The canonical decomposition and its relationship to other forms of magnetotelluric impedance tensor analysis, *J. Geophys.* 61: 173 – 189.

Zhang, P., Roberts, R. G. & Pedersen, L. B. (1987). Magnetotelluric strike rules, *Geophysics* 52: 267 – 278.

5

Heat and SO_2 Emission Rates at Active Volcanoes – The Case Study of Masaya, Nicaragua

Letizia Spampinato and Giuseppe Salerno
Istituto Nazionale di Geofisica e Vulcanologia, Osservatorio Etneo, sezione di Catania, Catania, Italy

1. Introduction

The necessity of understanding volcanic phenomena so as to assist hazard assessment and risk management, has led to development of a number of techniques for the tracking of volcanic events so as to support forecasting efforts. Since 1980s scientific community has progressively drifted research and surveillance at active volcanoes by integrated approach. Nowadays, volcano observatories over the world record and integrate real or near-real time data for monitoring and understanding volcano behaviour. Among the geophysical, geochemical, and volcanological parameters, the tracking of temperature changes at several volcanic features (e.g. open-vent systems, eruptive vents, fumaroles) and variations in sulphur dioxide flux and concentration at volcanic plumes are key factors for studying and monitoring active volcanoes.

Temperature is one of the first parameters that have been considered in understanding the nature of volcanoes and their eruptions. Thermal anomalies have proved to be precursors of a number of eruptive events (e.g. Andronico et al., 2005; Dean et al., 2004; Dehn et al., 2002), and once an eruption begins, temperature plays a major role in lava flow emplacement and lava field development (e.g. Ball et al., 2008; Calvari et al., 2010; Lodato et al., 2007). At active volcanoes, temperature has been measured by direct and indirect methodologies (**Fig. 1a, c**). Direct measurements represent the traditional thermal monitoring carried out at fumaroles, hot springs, molten lava bodies, and crater lakes, using thermocouples (e.g. Aiuppa et al., 2006; Corsaro & Miraglia, 2005). Indirect measurements, also known as thermal remote sensing, can be performed by satellite, ground, and airborne surveys (e.g. Calvari et al., 2006; Spampinato et al., 2011; Wright et al., 2010). Owing to the danger of most kinds of eruption, and the need of monitoring inaccessible areas on volcanoes (e.g. Wright & Pilger, 2008), indirect measurements are especially attractive. Among them, thermal imagery is one of the most widespread and results from the capability to detect the infrared radiation emitted from the surface of hot bodies, and to provide the radiometric map of heat distribution of the body's surface (Spampinato et al., 2011). This has been of primary importance for capturing the evolution of thermal anomalies, which shed light on magma movements at shallow depths (e.g. Calvari et al., 2005). While magma is rising, hot gases separate from the melt and escape either directly from the main conduits, or indirectly by leaking through fumaroles, fractures, and faults, or by dissolving within crater lakes and hot spring waters, resulting in variations in their temperature and chemical composition. At the surface, these phenomena are also associated with radiative heat fluxes, which can be detected by infrared thermal detectors. The application of thermal imaging to volcanology was largely performed using satellite surveys (e.g. Harris et al., 2011; Vicari et al., 2008), but in the last decade there has been increasing application of compact (hand-held) thermal imagers used from the air or ground (Spampinato et al., 2011).

Fig. 1. Different modes for temperature and volcanic gas sampling. Conventional in situ measurements of (a) the temperature of Hawaiian pāhoehoe lava flow fields (photo by P. Mouginis-Mark, volcano.oregonstate.edu), and (b) volcanic gas from the summit fumarole field of Kīlauea volcano in 2005. In (c) and (d) ground-based thermal imagery of the Laguna Caliente crater lake (Poás volcano, Costa Rica; 2009) and UV-DOAS measurements of the Santiago crater (Masaya volcano, Nicaragua; 2009) volcanic plume, respectively.

Volcanic degassing plays a key role in magma transport and style, and timing of volcanic eruptions observed at the Earth's surface (e.g. Carroll & Holloway, 1994; Gilbert & Sparks, 1998; Huppert & Woods, 2002; Sparks, 2003). The assessment of volcanic gas composition and flux has become a standard procedure for volcanic monitoring and eruption forecasting, since degassing regimes are fundamentally linked to volcanic processes (e.g. Aiuppa et al., 2007, 2010; Edmonds, 2008; Noguchi & Kamiya, 1963; Oppenheimer, 2003; Sutton et al., 2001). Magma contains dissolved gases that are released into the atmosphere during both quiescent and eruptive degassing phases (e.g. Oppenheimer, 2003). At high pressures, deep beneath the Earth's surface, gases are dissolved in magma; however as soon as magma rises toward the surface, where pressures are lower, gases start to exsolve according to the solubility-pressure relationship of each species, as well as compositional and diffusional constraints (e.g. Carroll & Holloway, 1994; Carroll & Webster, 1994; Oppenheimer, 2003; Spilliaert et al., 2006; Villemant & Boudon, 1999). The abundance and final gas phase composition of the emitted plume depends on magma composition(s), volatile fugacities, crystallisation, and on the dynamics of magma degassing, including kinetic effects (e.g. Giggenbach, 1996; Oppenheimer, 2003; Symonds et al., 1994, 2011). However, at the surface, the composition and flux of volcanic gases may change with time, reflecting variations in the magmatic feeding system of the volcano. Hence, by studying and tracking this variability a number of parameters, such as magma residing depths and the amount of degassing magma bodies can be determined (Allard, 1997; Steffke et al., 2010).

Among the volcanic gas species, sulphur dioxide (SO_2) is one of the most well investigated in remote sensing (e.g. Bluth et al., 2007; Carn et al., 2003; Galle et al., 2010; Hamilton et al., 1978; McGonigle et al., 2009; Salerno et al., 2009a; Williams-Jones, et al., 2008; Sweeney et al., 2008; Thomas & Watson, 2010). As for temperature, SO_2 concentration and emission rates can be measured using both direct sampling and non-contact remote sensing techniques (**Fig. 1b, d**; e.g. Finnegan et al., 1989; Giggenbach & Goguel, 1989; McGee & Sutton, 1994; McGonigle & Oppenheimer, 2003; Mouginis-Mark et al., 2000). The latter carried out during air- and ground-based surveys and on satellite platforms, are based on optical spectroscopy. Since the 1970s, SO_2 flux has been remotely measured using the COrrelation SPECtrometer (COSPEC; Newcomb & Millán, 1970; Stoiber & Jepsen, 1973; Stoiber et al., 1983) at several volcanoes worldwide (e.g. Caltabiano et al., 1994; Malinconico, 1979; Realmuto, 2000; Sutton et al., 2001; Williams-Jones et al., 2008). Over the last 10 years the advent of small, commercial and low cost spectrometers (Mini-DOAS, Galle et al., 2003; RMDI, Wardell et al., 2003; MUSE, Rodriguez et al., 2004; Flyspec, Horton et al., 2006; Dual-Field of View, McGonigle et al., 2009) offered a valuable replacement to the outdated COSPEC. In particular, the combination of Ultraviolet (UV) spectrometers with the Differential Optical Absorption Spectroscopy (DOAS) analytical method (Noxon, 1975; Platt, 1994; Platt & Stutz, 2008) improved significantly data collection, offering a number of advantages such as the possibility of obtaining measurements in the challenging environments typical of volcanic areas, detection of other plume species (Bobrowski et al., 2003; O'Dwyer et al., 2003; Oppenheimer et al., 2005), and collection of high-resolution SO_2 flux by permanent scanner networks (e.g. Arellano et al., 2008; Edmonds et al., 2003; Salerno et al., 2009a, 2009b).

Our intent here is to discuss findings and implications arising from the integration of thermal imaging-derived temperature and SO_2 emission rates by UV-DOAS spectroscopy collected in March 2009 at Masaya volcano, Nicaragua. Calibrated temperatures from thermal imagery can provide qualitative as well as quantitative information, fundamental insights and parameters contributing to understanding and modelling of several eruptive features. Anomalies in SO_2 emission rates have been often documented at several volcanoes prior to eruptive crisis (e.g. Casadevall et al., 1981; Daag et al., 1996; Kyle et al., 1994; Malinconico, 1979; Sutton et al., 2001; Williams-Jones et al., 2008; Young et al., 1998; Zapata et al., 1997). In syn-eruptive stages, anomalies in the SO_2 flux pattern might indicate variations in the eruptive style and regime associated with changes in the volcano shallow feeder system (e.g. Andronico et al., 2005; Delgado-Granados et al., 2001; Olmos et al., 2007; Spampinato et al., 2008a; Spilliaert et al., 2006). At open-vent systems, in non-eruptive phases, changes in SO_2 flux emission have provided information on increases or decreases of magma supply in the shallow plumbing system (Allard, 1997; Wallace & Gerlach, 1994) suggesting likely volcanic unrests or magma migration towards peripheral areas of the volcano edifice, respectively.

There is still much to explore about volcano behaviour and eruptive mechanisms, however, the combination of different types of monitoring techniques is crucial for constraining baselines for predicting phases of volcano unrests and for gaining useful insights for volcano hazard assessment.

2. Masaya volcano

Masaya is an open-vent, basaltic shield volcano (560 m a.s.l.) sited in western Nicaragua (Central America). The volcano edifice includes a 11 × 6 km-elongated caldera that formed ~2,500 yrs ago as a result of a 8 km^3-basaltic ignimbrite eruption (Williams, 1983). The caldera hosts a complex of lavas and cinder cones, with cones cut by pit craters, of which the Santiago is the presently active (e.g. Harris, 2009; Roche et al., 2001; **Fig. 2).** Over time, the Santiago pit crater has been characterised by the development of ephemeral lava lakes

Fig. 2. (a) Photograph of Masaya volcano taken from NNE (geoalba.com).
(b) Satellite image of Masaya volcano summit area (googleEarth 2011). San Pedro, Nindiri, Santiago, and Masaya craters are shown (see Harris, 2009 for more details). The Santiago crater is the currently active and the site of our investigation. The yellow and red dots indicate the sites from which thermal imagery and SO_2 amount measurements were respectively carried out in March 2009.

(Allen et al., 2002), explosive activity (e.g. Duffell et al., 2003; Perez et al., 2009; Rausch & Schmincke, 2010) effusive eruptions (e.g. Harris, 2009), intense degassing (e.g. Branan et al., 2008; Kern et al., 2009a; Williams-Jones, 2001), and phases of inner crater collapses (e.g. McBirney, 1956; Harris, 2009; Rymer, et al., 1998). The volcano activity has consisted of phases of quiescent degassing for over 150 years, punctuated by intermittent gas crises associated with high SO_2 emissions (e.g. Delmelle et al., 2002; Stix, 2007), and minor explosive phases throwing ejecta around the summit area, of which the most significant event of the last 30 years occurred in 2001 (e.g. Branan et al., 2008; Duffell et al., 2003).

The persistent loose of gas has been interpreted as the result of periodic magma convective overturn within the volcano shallow feeding system (Delmelle et al., 1999; Horrocks, 2001; Horrocks et al., 1999). It has been estimated that during the last 150 years, degassing has been supplied by ~10 km^3 of magma (e.g. Rymer et al., 1998; Stoiber et al., 1986).

The easy accessibility of Masaya summit area has made the volcano an ideal natural laboratory, where a number of different monitoring techniques, direct and indirect observations, have been carried out since the onset of the post-1993 degassing crisis (e.g. Allen et al., 2002; Galle et al., 2003; Mather et al., 2003; Martin et al., 2009; Nadeau & Williams-Jones, 2009). Tracking of Masaya's activity has been of primary importance not only for the understanding and modelling of the volcano deep processes (e.g. Stix, 2007; Williams-Jones et al., 2003), but also for the potential health hazard posed by the volcanic gas emissions (Delmelle et al., 2002). In fact, due to the low altitude of the volcano edifice, the continuous degassing from the Santiago crater represents a threat for people living close to the volcano foot (**Fig. 3**).

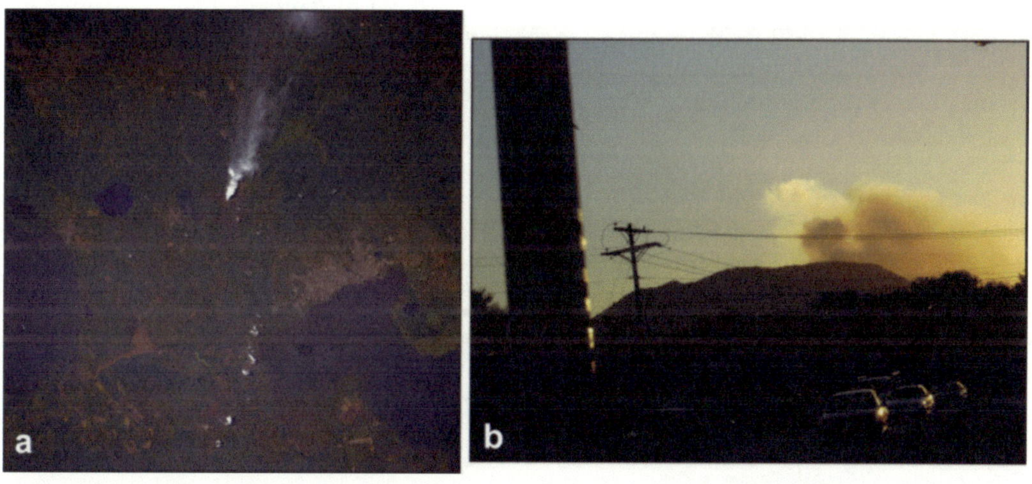

Fig. 3. (a) Satellite image showing the persistent volcanic plume from Santiago crater (zonu.com). (b) Photograph of Masaya volcano and its volcanic plume taken in March 2009 from ENE.

The persistent degassing from Santiago crater has been extensively studied by remote sensing methodologies, spanning from infrared to ultraviolet spectroscopy, carried out during ground-based surveys or by satellite platforms (e.g. Branan et al., 2008; Burton et al., 2000; Horrocks et al., 2003; Martin et al., 2010; Nadeau & Williams-Jones, 2009; Thomas &

Watson, 2010). Here we report on the Santiago's crater activity that we observed, between 20 and 24 March 2009, carrying out simultaneous volcanic plume measurements using a portable infrared imager and an ultraviolet spectrometer.

3. Infrared and ultraviolet remote sensing

Following we shortly report the main techniques for the acquisition of thermal imagery and SO_2 fluxes and amounts, and the instrumental specifications and details on the methodology of data collection during the March 2009 field campaign at Masaya volcano.

3.1 Thermal imagery

During the 20-24 March 2009, we recorded thermal imagery of the Santiago crater using a P25 FLIR (Forward Looking InfraRed Systems) portable thermal camera from the Sapper car park on the south-western crater rim. The instrument is an uncooled microbolometer with a 320 × 240 pixel array sensitive to the 7.5-13 μm wave band with a 24 × 18° field-of-view (FOV). Its quoted precision is ±2% and the thermal sensitivity is less than 273.23 K at 303.15 K. The camera is equipped with three dynamic temperature ranges 233.15 to 393.15 K, 273.15 to 773.15 K, and 623.15 to 1773.15 K, of which we used the middle one. In order to make a first-order correction for the atmospheric effects (e.g. Spampinato et al., 2011), we input in the camera internal software the measured line-of-sight from the crater bottom (~340 m; see yellow dot in **Fig. 2** for the camera site), and the daily mean temperature and relative humidity of the air (306.15 K and 38% on 20 March; 306.15 K and 32% on 21 March; 303.15 K and 40% on 22 March; 303.15 K and 42% on 23 March; and 306.15 K and 35% on 24 March). Considering the camera instrumental specifications and the path length of ~340 m, the nominal pixel size was of ~0.47 m. According to Branan et al. (2008), we used an emissivity (ε) value of the hot source of 1, and given that emissivity has non-Lambertian behaviour, we measured the inclination angle of the camera (70°) for error evaluation (e.g. Ball and Pinkerton, 2006; Spampinato et al., 2011). Images were collected every 8 seconds between 17:06:27 and 18:48:29 (here after all times are in GMT) on 20 March, 20:10:47 and 21:40:45 on 21 March, 15:53:04 and 18:22:54 on 22 March, 16:05:26 and 18:31:47 on 23 March, and 15:39:02 and 17:07:34 on 24 March. Along the five days of the survey, thermal imagery was recorded from the same identical position and viewing inclination (**Fig. 2**).

A recent account on the uncertainty in thermal imagery-derived data was provided by Spampinato et al. (2011).

3.2 UV spectroscopy

On 20, 21, 23 and 24 March 2009, we carried out SO_2 flux measurements (tonnes day^{-1}) by car-based traverses along the Llano Pacaya road (15 km downwind of the Santiago crater, see Martin et al., 2010) and along the Ticuantepe road (5 km downwind of the Santiago crater, see Martin et al., 2010). Optimal integration time for the collection of spectra in the traverse technique was 100 ms, and 50 spectra were co-added to improve the signal-to-noise ratio. Spectra were time- and position-stamped using a USB GPS receiver. In addition, between 20, 21, 22, and 23 March, we collected also SO_2 column amounts (CA, in ppm × m) using a UV spectrometer and scattered sunlight as the light source. Individual spectra were recorded from fixed position from the eastern flank of the Santiago crater, ~400 m far from the plume (see **Fig. 2** for the measurement sites) at sampling rate between 14 and 17 s.

For both kinds of measurement techniques, we used an Ocean Optics USB2000 spectrometer. The instrument comprises a 2048 pixel-detector-array and diffraction grating with 3600 grooves per mm, which combined with a 200 μm entrance slit, delivers a spectral resolution of ~0.44 nm FWHM in the 295-375 nm wavelength range. To perform SO_2 flux traverses the instrument was mounted inside a car and connected via fibre optic cable to a telescope (8 mrad FOV) oriented vertically upwards.

SO_2 CA were retrieved using the WinDoas software package (Fayt & van Roozendael, 2001) applying the standard DOAS method (Platt & Stutz, 2008). The ring spectrum (e.g. Fish & Jones, 1995; Solomon et al., 1987) was calculated from the clear sky-spectrum (spectrum collected out of the plume) following the approach of Chance (1998). Both laboratory spectra of SO_2 and O_3 (Malicet et al., 1995; Vandaele et al., 1994) and the Ring spectrum were convolved to the spectrometer's resolution. UV spectra were evaluated in the 305-316 nm window to yield the time-series of the SO_2 CA in the FOV of the spectrometer. SO_2 flux was evaluated following Stoiber et al. (1983). Wind speed was measured every 10 minutes using a portable hand-held anemometer. In the days of our observations, mean wind speed and direction were of ~5 m s^{-1} toward the SW. Error in SO_2 flux detection by UV spectroscopy depends mainly on the uncertainty in the plume-wind speed (e.g. Doukas, 2002; Mather et al., 2006). Stoiber et al. (1983) estimated uncertainty in flux calculation between 10-40%. Negligible uncertainty arises from the error in the retrieved SO_2 CA (e.g. Kern, 2009; Platt & Stutz, 2008), multiple scattering (e.g. Kern et al., 2009b; Millan, 1980), the presence of volcanic ash in the plume (Andres & Schimd, 2001), or SO_2 depletion (McGonigle et al., 2004; Nadeau and Williams-Jones, 2009; Oppenheimer et al., 1998). During our campaign, the plume always appeared to be bright and free from ash and situated below the clouds, thus we can consider the influence of multiple scattering and ash to be negligible.

4. Observations of the Santiago crater activity

Along the 5 day-observation period, the Santiago activity consisted of persistent degassing from two vents opened at the crater floor (Martin et al., 2010; Vent 1 and Vent 2 in **Fig. 4**). From the Santiago crater SW rim, from which we carried out thermal imagery (**Fig. 2**), Vent 2 was clearly visible at the naked eye, whereas Vent 1 was hidden by Vent 1 plume, and thus recognisable only by infrared optics (**Fig. 4c**). Thermal imagery showed that Vent 2 was eventually wider than Vent 1 (**Fig. 4c**). Applying an apparent temperature threshold of 300 K on thermal images and considering the nominal pixel size of ~0.47 m, we estimated an area of ~450 and ~715 m^2 for Vent 1 and Vent 2, respectively. Owning to the oblique imagery, we consider such areas as minimum estimates.

The two vents were both persistently degassing with the two plumes joining together a few seconds after the emission (**Fig. 4b**). Qualitatively, the plume seemed to be whitish and denser next to the crater floor (**Fig. 4b**) and transparent and more diluted close to crater rim (**Fig. 4a**). Plume conditions varied also according to the time of day, i.e. more transparent in the morning and more condensed in the evening (Burton et al., 2001; Martin et al., 2010; Mather et al., 2003). Along the 5-day-survey, we did not detect any explosion; however thermal images showed that the quiescent degassing observed at the crater exit, had in reality a pulsating behaviour at the vent region (**Fig. 5**).

Fig. 4. (a) Photograph of the Santiago crater rim and its volcanic plume taken from the SSE rim on 22 March 2009. (b) Photograph of the two degassing vents opened at the Santiago crater floor taken from the SW rim on 22 March 2009. (c) Zoom of the satellite image of Masaya volcano summit area (googleEarth 2011) shown in Fig. 2. The overlapped thermal image localises the position of Vent 1 and Vent 2 within the Santiago crater. The thermal image was recorded on 22 March 2009 from the SW crater rim.

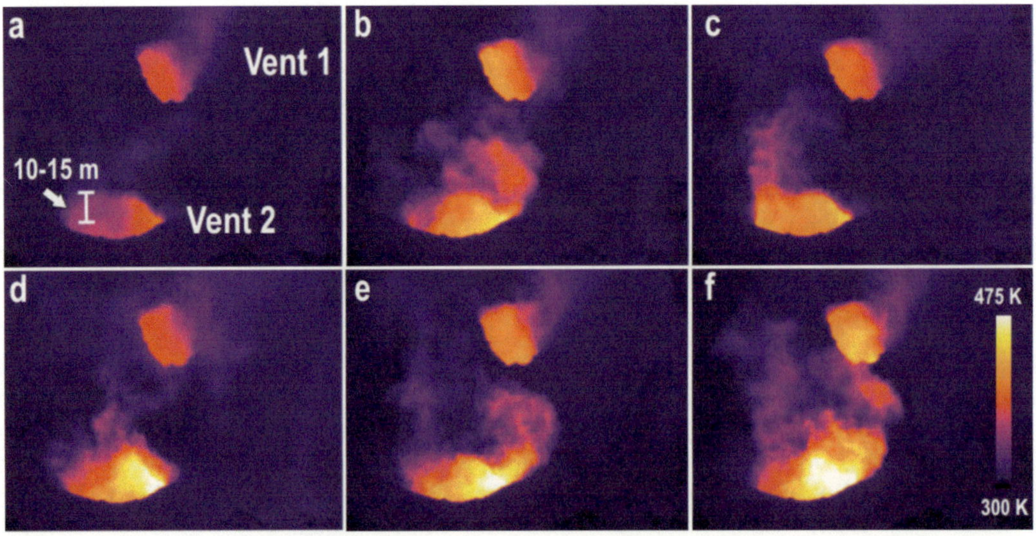

Fig. 5. (a-f) Thermal image sequence showing the degassing pulsating behaviour at both vents. The sequence was recorded on 21 March 2009 from the Santiago crater SW rim.

In addition, thermal images revealed that the magma level inside the two vents remained below both vent rims allowing estimation of the crusted crater floor thickness; that was of ~10-15 m (**Fig. 5**). Next to the degassing vents, the crater floor was characterised by talus coverage due to the collapses of the eastern crater inner walls (**Fig. 4b**).

5. Results and discussion

Following we report results of the analysis of thermal image and SO_2 CA and flux data, providing interpretation of the relationship between temperature pattern and heat flux, and SO_2 concentration and emission rates.

5.1 Thermal imaging-derived data

In figure 6, we have plotted the variability of the maximum and mean apparent temperatures (K) of Vent 1 (**Fig. 6a, b, d, f, and h**) and Vent 2 (**Fig. 6c, e, g, and i**) over time, along the five days of measurements (of which we lack Vent 2 imagery of the first day). Overall, Vent 2 plume showed somewhat higher temperatures than those of Vent 1 with peak values of ~500 K and maximum means of ~400 K, with respect to the ~460 K and ~380 K of Vent 1. However, the temperature difference between the two vents might have resulted from the viewing angle difference with which the two vents were imaged (**Fig. 5**).

In detail, Vent 1 maximum temperatures varied between 360-454 K on 20 March, 374-475 K on 21 March, 372-453 K on 22 March, 355-466 K on 23 March, and between 376-452 K on 24 March (**Fig. 6a, b, d, f, and h**). The vent mean values ranged between 338-384 K, 346-388 K, 344-384 K, 336-392 K, and 348-386 K, from 20 to 24 March, respectively (**Fig. 6a, b, d, f, and h**). Vent 2 maximum temperatures fluctuated between 385-510 K, 363-488 K, 376-500 K, and 373-504 K on 21, 22, 23, and 24 March, respectively (**Fig. 6c, e, g, and i**). Mean temperatures of Vent 2 varied between 343-405 K, 332-394 K, 340-397 K, and 336-396 K from 21 to 24 March, respectively (**Fig. 6c, e, g, and i**). Both the maximum and mean temperature trends of the two vents are characterised by the overlapping of waveforms of different amplitudes that we consider in section 5.3.

Using the estimated areas of 450 and 715 m^2 respectively for Vent 1 and Vent 2, $\varepsilon = 1$, and the most representative thermal images (i.e. those with the highest mean temperature values and the lowest standard deviations; e.g. Spampinato et al., 2008b), we have calculated magma heat loss by radiation (Q_{rad}; MW) from the two vents between 20 and 24 March 2009 (**Fig. 7**).

Figure 7 shows the variability of the daily mean Q_{rad} of the two vents (Vent 1 grey line and Vent 2 black line). As previously argued, given that the areas considered are minimum values, the Q_{rad} estimates in figure 7 correspond to minimum daily mean values. Along the days of observation, the total Q_{rad} from the two vents remained quite stable varying between 1.2 and 1.8 MW.

Note that we have considered only the radiated flux as we have assumed that the incidence of heat loss by conduction (Q_{cond}) and convection (Q_{conv}) was reduced. In particular, at Santiago crater, Q_{cond} implies heat dissipation from the walls of the conduit; however, following Giberti et al. (1992), we have assumed that after years of persistent activity the volcano shallow system is likely long-established and well insulated. Thus, we have

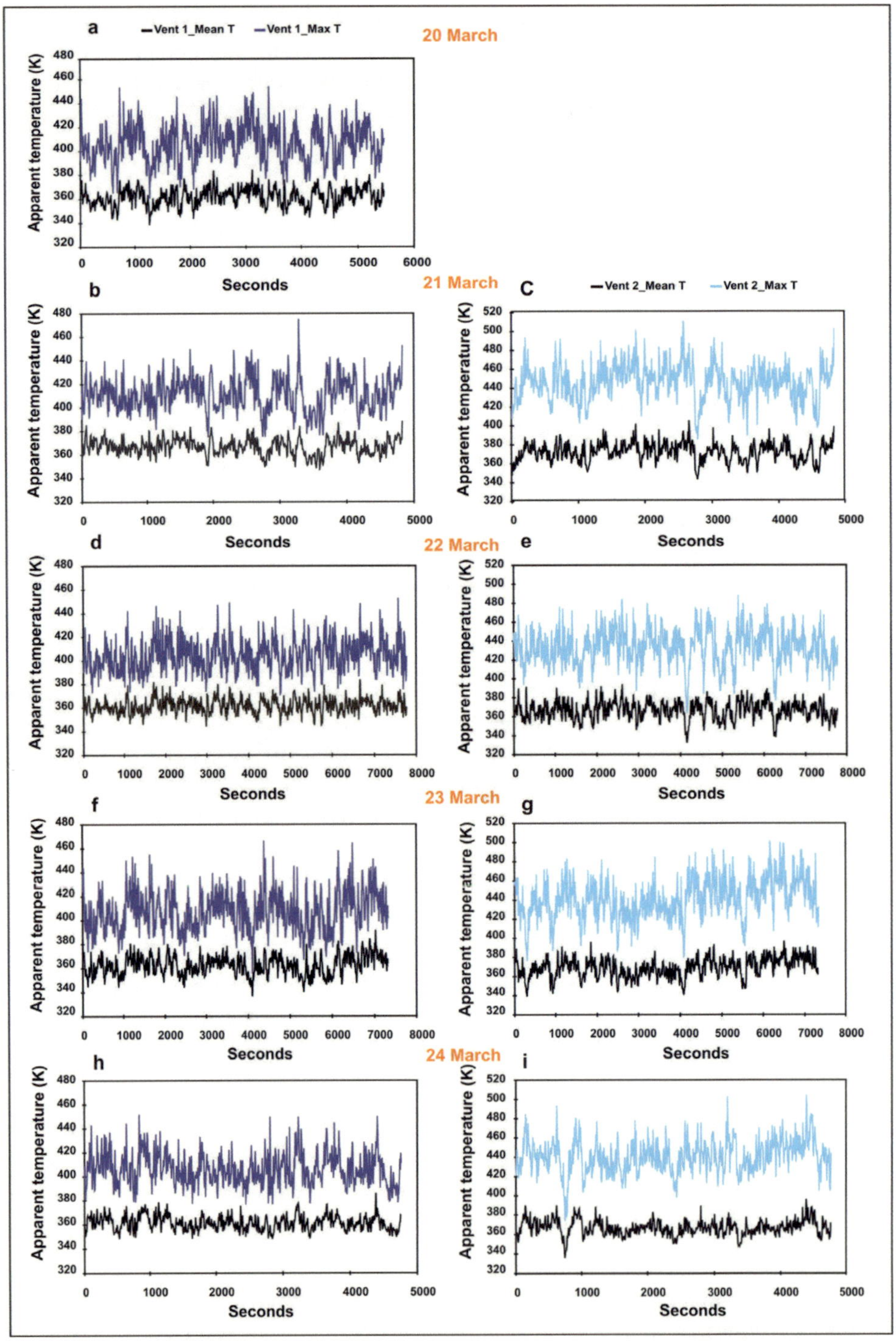

Fig. 6. Temporal variability of Vent 1 (on the left; a, b, d, f, and h) and Vent 2 (on the right; c, e, g, and i) maximum and mean apparent temperatures during the 20-24 March 2009 ground-based thermal surveys.

supposed that conduction to the country rock was irrelevant with respect to Q_{rad}. At the same manner, we have neglected the contribution of Q_{conv}. In fact, during the 5-day survey, wind conditions were quite stable (~5 m s^{-1}; free convection, e.g. Keszthelyi & Denlinger, 1996; Keszthelyi et al., 2003; Neri, 1989), and the magma level was confined at least ~15 m below the crater floor (**Fig. 5**). Hence, magma surface was not directly exposed to wind action.

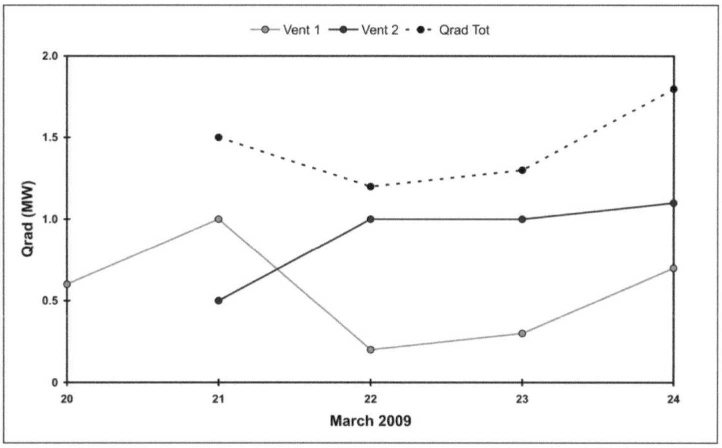

Fig. 7. Variability of Vent 1 and Vent 2 radiant heat flux from 20 to 24 March 2009. The grey line refers to Vent 1, the black line to Vent 2, and the black-dotted line to the total heat flux radiated by the two vents.

5.2 SO_2 column amounts and fluxes

Figure 8 reports the SO_2 CA collected between 20 and 23 March 2009 at the Santiago crater volcanic plume (a, b, c, and d). Given that measurements were taken out of the crater (**Fig. 2**), the amounts represent the contribution of both vents.

The highest SO_2 concentrations were detected on 20 March when the maximum CA was 5430 ppm × m (mean of 1832 ppm × m; **Fig. 8a**), with respect to the maxima of 2185, 2672, and 2590 ppm × m (means of 472, 966, and 885 ppm × m) recorded on 21, 22, and 23 March, respectively (**Fig. 8b, c, d**). As for the temperature data sets of figure 6, the SO_2 CA time-series show several high amplitude fluctuations on which higher frequency components are superimposed. In particular, on the 20 March time-series we recognised at least four main fluctuations peaking at 5430 (17:34:36), 5360 (17:56:01), 5240 (18:07:03), and 4800 ppm × m (18:21:52). In terms of maximum SO_2 concentrations, the fluctuations are characterised by a decreasing trend (**Fig. 8a**). On 21 March, we observed a more defined trend in which we clearly recognise three fluctuations of the SO_2 CA with maximum values of 1998 (20:42:51), 2186 (21:05:17), and 1583 (21:25:03) ppm × m (**Fig. 8b**). In the 22 March time-series, we distinguished four main fluctuations (**Fig. 8c**) with maximum SO_2 CA of 2058 (15:41:08), 2672 (16:26:04), 1794 (17:12:57), and 1942 ppm × m (17:43:51). Note that due to the length of the time-series, we could not determine the exact end of the last fluctuation (**Fig. 8c**). The last time-series, recorded on 23 March, displays four main SO_2 CA fluctuations with peaks of 2088 (16:42:43), 2257 (17:03:13), 2104 (17:40:54), and 2590 ppm × m (18:04:16), respectively (**Fig. 8d**). As for the maximum and mean temperature trends of figure 6, the nature of the overlapped waveforms, recognised in figure 8, are investigated in section 5.3.

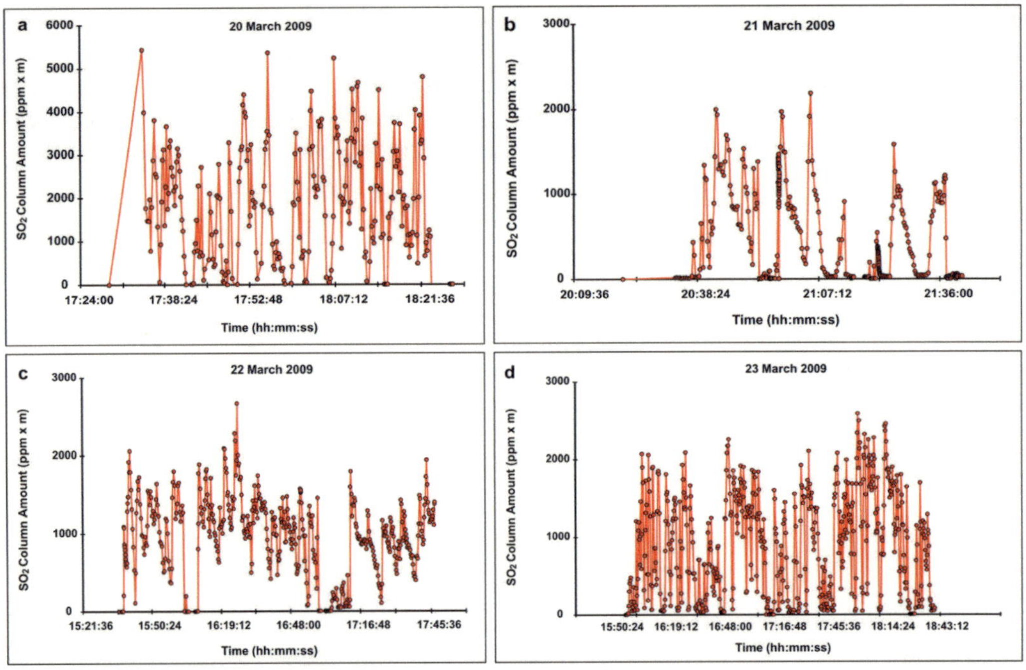

Fig. 8. Temporal variability of SO_2 CA from Vent 1 and Vent 2 between 20 and 23 March 2009. The CA were collected from fixed position and from different sites (see **Fig. 2b** for details).

During the March 2009 field campaign, we measured also the SO_2 flux by car-based traverses (**Fig. 9**; see also Martin et al., 2010). In detail, we carried out three traverses on 20 March (from 20:20:00 to 21:30:00), six on the 21st (from 16:40:00 to 18:30:00), four on the 23rd (from 20:30:00 to 22:10:00), and ten on 24 March (from 15:15:00 to 16:50:00).

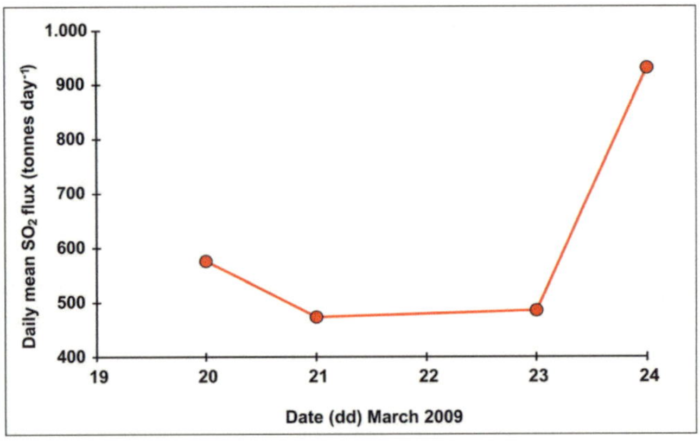

Fig. 9. Daily mean SO_2 flux measured during the 20, 21, 23, and 24 March 2009 by car-based traverses.

Overall, the daily mean SO_2 flux was characterised by an increasing trend from 20 to 24 March, when the flux reached values of 1350 and 1325 tonnes day^{-1} (**Fig. 9**). Daily mean fluxes (±1 standard deviation) were 580±180 (20 March), 470±100 (21 March), 490±170 (23 March), and 930±280 tonnes day^{-1} (24 March). The average SO_2 flux measured during the 4-day survey was of 690 tonnes day^{-1} (Martin et al., 2010).

5.3 Comparative signal processing and results

Figure 10 shows the behaviour of the mean apparent temperatures of Vent 1 and Vent 2 with respect to the pattern of the SO_2 CA from 20 to 23 March 2009. In order to make a reasonable comparison between temperatures and SO_2 CA, we have plotted the 10-point running means of both parameters. The black and grey lines refer to Vent 1 and Vent 2 mean temperatures, respectively, and the red line to the SO2 CA. In addition, moving average has allowed us to filter the very high frequency signals, which are commonly related to noise effects of variable nature such as turbulence of the volcanic plume next to the vent area and drifting of the plume within the FOV of the UV spectroscopy system.

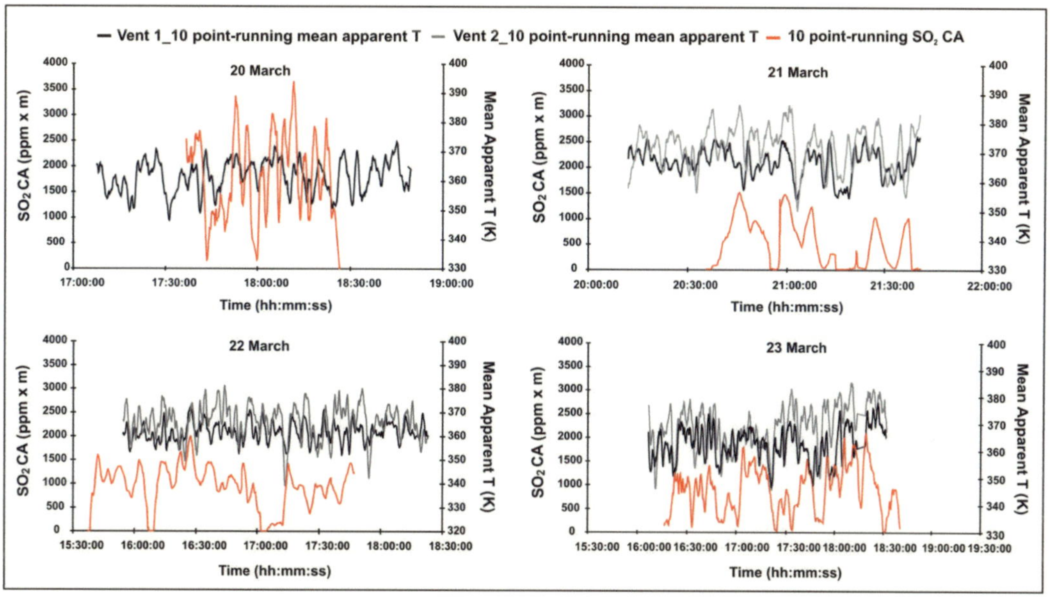

Fig. 10. Comparison between the temporal trends of Vent 1 and Vent 2-10 point running mean apparent temperatures and the 10-point running mean of the SO_2 CA.

Along the four days of measurements, temperatures and SO_2 CA are well correlated, though they show a somewhat shifting due to the different sites from which temperatures and CA were measured, i.e. the thermal camera pointed directly at the vents whereas SO_2 concentrations were taken out of the Santiago's crater rim (**Fig. 2**). Both temperatures and CA are characterised by superimposed cycles of different periods (**Fig. 10**). In order to investigate the reliability of the qualitatively observed cycles, we have carried out time-series analysis by Fast Fourier transform on both mean apparent temperatures and SO_2 CA (**Fig. 11**). Figure 11 shows the power spectra and the statistical significance calculated considering the hypothesis of a background red noise, and thus we have considered reliable only the peaks lying above the green line, which represents the 95% confidence spectrum (e.g. Spampinato et al., 2008b; Torrence & Compo, 1998).

Vent 1 shows significant periods of 1-2 min and 8 min on 20 March, 1-3 min and 15 min on 21 March, 1-3 min, 5 min, 7 min, and 13 min on 22 March, 2 min, 4 min, and 46 min on 23 March, and of 1 min, 6 min, 11 min, and 18 min on 24 March (**Fig. 11**). Vent 2 is characterised by significant peaks of 1 min and 7 min on 21 March, 1-3 min, 7 min, and 21 min on 22 March, 1 min, 7 min, and 28 min on 23 March, and of 1 min, 7 min, and 15 min on 24 March (**Fig. 11**). The SO_2 CA time-series display major peaks at 1-2 min and 4 min on the 20th, 2-3 min, 4-5 min, 7 min, and 10 min on the 21st, 1-3 min, 5 min, 8 min, and 11 min on the 22nd, and at 1-3 min and 4 min on 23 March (**Fig. 11**).

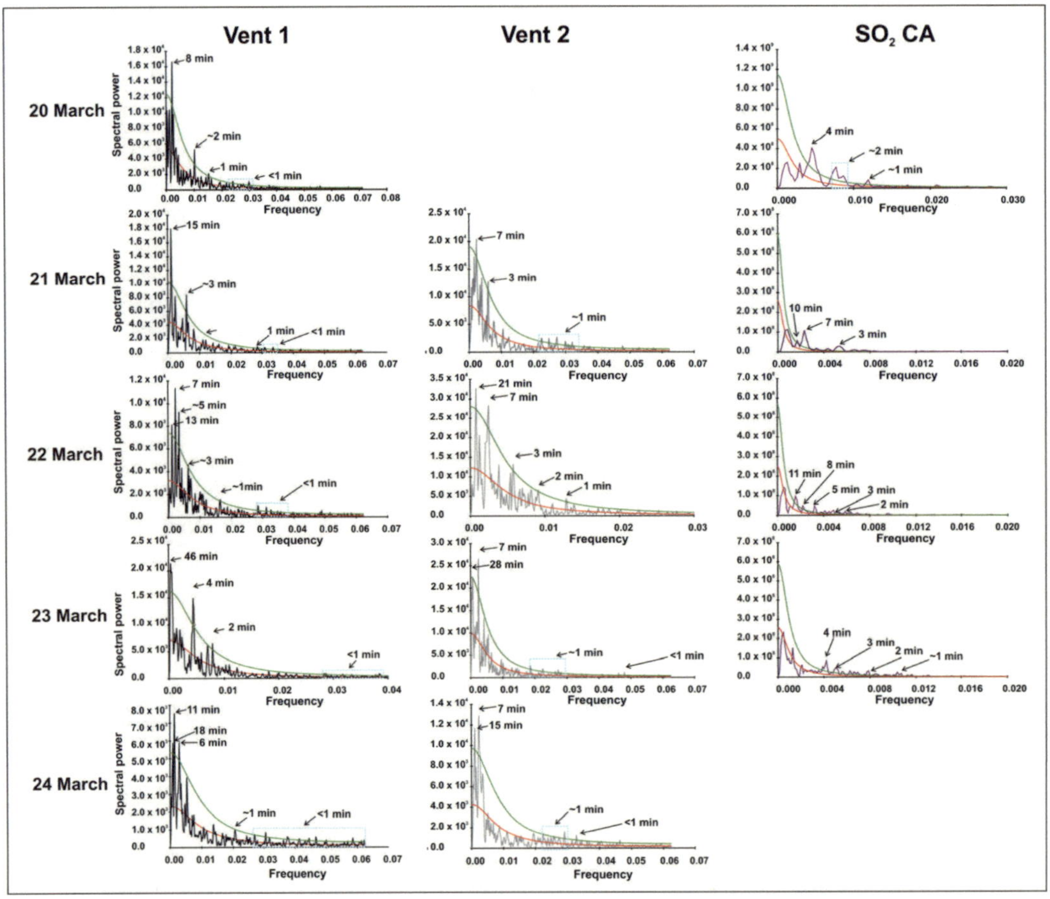

Fig. 11. Power spectra and statistical significance of Vent 1 and Vent 2 mean temperature time-series and of the SO_2 CA data sets collected between 20 and 24 March 2009. The green and red lines represent the 95% and 75% confidence spectra, respectively. In the figure, we have reported only the period of peaks above the green lines. The cyan dashed rectangles enclose low spectral power peaks with periods below 1 minute.

Observing in details the power spectrum time-series of figure 11, we detected also peaks that, though they do not overcome the green line, they are above the red lines representing the 75% confidence spectra. Most of these peaks consist of low frequency signals, between ~40 and 50 min in the temperature time-series and ~30 and 50 min in the SO_2

concentrations, which due to the reduced length of the data sets have a low spectral power **(Fig. 11)**.

5.4 Data interpretation and concluding remarks

Here we have reported on the integration of thermal imaging-derived data with both SO_2 fluxes and concentrations from the Santiago active crater of Masaya volcano in March 2009. As already reported by Martin et al. (2010), in that period the crater activity was fed by two vents opened at the crater floor. The opening and closure of vents over time (e.g. Branan et al., 2008), combined with results from structural and geophysical studies (Rymer et al., 1998; Williams-Jones et al., 2003), has suggested that the vents result from the collapses of the thin crusted roof of the volcano shallow magma accumulation zone (Martin et al., 2010). Thermal imagery collected during our campaign allowed us to infer that the magma surface within the two vents was at least ~10-15 m below the surface (**Fig. 5a**; Martin et al., 2010), suggestive of drop of the magma level over time. Magma level fluctuations have been commonly detected at several basaltic volcanoes (e.g. Stromboli, Calvari et al., 2005; Kīlauea, Tilling, 1987). In particular at lava lakes such as Erta 'Ale volcano in Ethiopia, variations in magma level within the crater have been related to magma pressures in the connected reservoir (Oppenheimer & Francis, 1997), thus to changes in the magma supply rate (e.g. Oppenheimer et al., 2004; Spampinato et al., 2008b).

Although Masaya is currently at minimum in its degassing cycle (Williams-Jones et al., 2003), during the time of our observations, the eruptive activity consisted of steady intense degassing from the two vents (the total volatile flux was of 14,000 tonnes day^{-1}; Martin et al., 2010). Except for the first day of survey, SO_2 CA recorded between 21 and 23 March were in agreement with those previously observed by Branan et al. (2008), marking the overall stable state of the volcano activity over long time-scales (Martin et al., 2010).

Peaks in brightness temperature of Vent 2, where somewhat higher with respect to Vent 1 (**Fig. 6**), likely due to vent geometry combined with the oblique imagery. However, they are comparable with temperatures recorded by Branan et al. (2008) in February 2002 and March 2003 using a thermal infrared thermometer. Whilst maximum apparent temperatures showed greater variability, mean apparent temperatures of both vents ranged between ~340 and 380 K, thus marking the quite stable background of the degassing mode.

The minimum total radiant heat power outputs estimated for the two vents did not display any remarkable variation as well, ranging from ~1.2 to 1.8 MW. The increasing trend of the radiant heat power output from 20 to 24 March can be found also in the pattern of the daily mean SO_2 fluxes, whose values pass from ~460 to 1350 tonnes day^{-1}. This suggests that, during our campaign the day-to-day variability of mean SO_2 flux might not be largely affected by wind speed uncertainty (Martin et al., 2010), as thermal imagery and SO_2 traverses were carried out from different sites and with different geometrical viewing, i.e. pointing directly the vents and crossing the plume from below. The simultaneous variations of both fluxes suggested to us that within the long-term degassing cycles (on the scale of years) of Masaya (Williams-Jones et al., 2003), there might be shorter-term sub-cycles (on the order of days) related to processes occurring within the volcano shallow feeding system (Martin et al., 2010; Nadeau & Williams-Jones, 2009; Witt et al., 2008). In detail, we believe that the increase in heat and SO_2 release might be connected to the rising of a new hot and gas-rich batch of magma from the volcano shallow reservoir feeding the persistent degassing of this volcano, through processes of magma overturning (Harris et al., 1999; Martin et al., 2010).

Figure 10 shows that both the thermal and SO_2 column amount time-series are not only correlated but they are both characterised by high and low frequency cycles, of which we have recognised periodicities on the order of minutes, of tens of minutes, and wider fluctuations of almost a hour. Owning to instrumental limitations, we could not record frequencies of tens of seconds associated with gas puffing characteristic of Santiago's degassing (Branan et al., 2008; Williams-Jones et al., 2003). Combining our observations with previous interpretations of lava lake dynamics and models (e.g. Spampinato et al., 2008b; Witham et al., 2006), we propose that cycles on the scale of minutes might relate to rates of gas bubbles/trains of bubbles bursting at the magma surface. Instead longer fluctuations in both thermal and SO_2 concentration trends might result from gradual variations in gas supply rate. However longer time-series are needed in order to better understand the meaning of these degassing cycles, especially those referring to the long fluctuations. In a site like Masaya volcano representing the ideal natural laboratory, the install of multi-parametric permanent stations will open up opportunities of long-term observations of the volcanic activity allowing refinement of models developed for open-vent volcanic systems.

6. Acknowledgements

LS and GGS thank C. Oppenheimer, GM Sawyer and RS Martin for sharing fieldwork and ideas. The field campaign in Nicaragua was part of the research project 'Understanding volcanic degassing, magma dynamics at persistently degassing basaltic volcanoes: a novel approach to linking volcanic gases and magmatic volatiles within a physical model'. University of Cambridge-University of Bristol-Natural Environment Research Council (NERC).

7. References

Aiuppa, A., Federico, C., Giudice, G., Gurrieri, S. & Valenza, M. (2006). Hydrothermal buffering of the SO_2/H_2S ratio in volcanic gases: Evidence from La Fossa Crater fumarolic field, Vulcano Island. *Geophys. Res. Lett.*, 33, L21315, doi:10.1029/2006GL027730.

Aiuppa, A., Moretti, R., Federico, C., Giudice, G., Gurrieri, S., Liuzzo, M., Papale, P., Shinohara, H. & Valenza, M. (2007). Forecasting Etna eruptions by real-time observation of volcanic gas composition. *Geology*, 35(12), pp. 1115-1118.

Aiuppa, A., Cannata, A., Cannavò, F., Di Grazia, G., Ferruccio, F., Giudice, G., Gurrieri, S., Liuzzo, M., Mattia, M., Montalto, P., Patane, D. & Puglisi, G. (2010). Patterns in the recent 2007–2008 activity of Mount Etna volcano investigated by integrated geophysical and geochemical observations. *Geochem. Geophys. Geosyst.*, 11, Q09008, doi:10.1029/2010GC003168.

Allard, P. (1997). Endogenous magma degassing and storage at Mount Etna. *Geophys. Res. Lett.*, 24(17), pp. 2219-2222.

Allen, A.G., Oppenheimer, C., Ferm, M., Baxter, P.J., Horrocks, L.A., Galle, B., McGonigle, A.J.S. & Duffell, H.J. (2002). Primary sulphate aerosol and associated emissions from Masaya volcano, Nicaragua. *J. Geophys. Res.*, 107(D23), 4682.

Andres, R.J. & Schimd, J.W. (2001). The effects of volcanic ash on COSPEC measurements. *J. Volcanol. Geotherm. Res.*, 108, pp. 237-244.

Andronico, D., Branca, S., Calvari, S., Burton, M.R., Caltabiano, T., Corsaro, R.A., Del Carlo, P., Garfì, G., Lodato, L., Miraglia, L., Muré, F., Neri, M., Pecora, E., Pompilio, M., Salerno, G. & Spampinato, L. (2005). A multi-disciplinary study of the 2002-03 Etna eruption: Insights into a complex plumbing system. *Bull. Volcanol.*, 67, pp. 314-330.

Arellano, S.R., Hall, M., Samaniego, P., Le Pennec, J.-L., Ruiz, A., Molina, I. & Yepes, H. (2008). Degassing patterns of Tungurahua volcano (Ecuador) during the 1999–2006 eruptive period, inferred from remote spectroscopic measurements of SO_2 emissions. *J. Volcanol. Geother. Res.*, 176, pp. 151–162.

Ball, M. & Pinkerton, H. (2006). Factors controlling the accuracy of thermal imaging cameras. *J. Geophys. Res.*, 111, B11203, doi: 10.1029/2005JB003829.

Ball, M., Pinkerton, H. & Harris, A.J.L. (2008). Surface cooling, advection and the development of different surface textures on active lavas on Kilauea, Hawai`i. *J. Volcanol. Geotherm. Res.*, 173, pp. 148-156.

Bluth, G.J.S., Shannon, J.M., Watson, I.M., Prata, A.J. & Realmuto, V.J. (2007). Development of an ultra-violet digital camera for volcanic SO_2 imaging, *J. Volcanol. Geotherm. Res.*, 161, pp. 47–56.

Bobrowski, N., Hönninger, G., Galle, B. & Platt, U. (2003). Detection of bromine monoxide in a volcanic plume. *Nature*, 423, pp. 273–276.

Branan, Y.K., Harris, A., Watson, I.M., Phillips, J.C., Horton, K., Williams-Jones, G. & Garbeil, H. (2008). Investigation of at-vent dynamics and dilution using thermal infrared radiometers at Masaya volcano, Nicaragua. *J. Volcanol. Geotherm. Res.*, 169, pp. 34-47.

Burton, M.R., Oppenheimer, C., Horrocks, L.A. & Francis, P.W. (2000). Remote sensing of CO_2 and H_2O emission rates from Masaya volcano, Nicaragua. *Geology*, 28, pp. 915-918.

Burton, M.R., Oppenheimer, C., Horrocks, L.A. & Francis, P.W. (2001). Diurnal changes in volcanic plume chemistry observed by lunar and solar occultation spectroscopy. *Geophys. Res. Lett.*, 28(5), pp. 843-846.

Caltabiano, T., Romano, R. & Budetta, G. (1994). SO_2 flux measurements at Mount Etna, Sicily. *J. Geophys. Res.*, 99(D6), pp. 12,809-12,819.

Calvari, S., Spampinato, L., Lodato, L., Harris, A.J.L., Patrick, M., Dehn, J., Burton, M.R. & Andronico, D. (2005). Chronology and complex volcanic processes during the 2002-2003 flank eruption at Stromboli volcano (Italy) reconstructed from direct observations and surveys with a handheld thermal camera. *J. Geophys. Res.*, 110, B02201, doi:10.1029/2004JB003129.

Calvari, S., Spampinato, L. & Lodato, L. (2006). The 5 April 2003 vulcanian paroxysmal explosion at Stromboli volcano (Italy) from field observations and thermal data. *J. Volcanol. Geotherm. Res.*, 149, pp. 160-175.

Calvari, S., Lodato, L., Steffke, A., Cristaldi, A., Harris, A.J.L., Spampinato, L. & Boschi, E. (2010). The 2007 Stromboli eruption: Event chronology and effusion rates using thermal infrared data. *J. Geophys. Res.*, 115, B04201, doi:10.1029/2009JB006478.

Carn, S.A., Krueger, A.J., Bluth, G.J.S., Schaefer, S.J., Krotkov, N.A., Watson, I.M. & Datta, S. (2003). Volcanic eruption detection by the Total Ozone Mapping Spectrometer (TOMS) instruments: a 22-year record of sulfur dioxide and ash emissions. In: Oppenheimer, C., Pyle, D.M. & Barclay, J. (Eds.), *Volcanic Degassing*. Geological Society, London, Sp. Pub., 213, pp. 177-202.

Carroll, M.R. & Holloway, J.R. (1994). Volatiles in magmas: Mineralogical Society of America *Reviews in Mineralogy*, vol. 30.

Carroll, M.R. & Webster, J.D. (1994). Solubilities of sulfur, noble gases, nitrogen, chlorine and fluorine in magmas. In: Carroll, M.R., Halloway, J.R. (Eds.), Volatiles in Magmas, *Rev. in Mineralogy*, vol. 30, pp. 231-279.

Casadevall, T.J., Johnston, D.A., Harris, D.M., Rose, W.I., Malinconico, L.L., Stoiber, R.E., Bornhorst, T.J., Williams, S.N., Woodruff, L. & Thompson, J.M. (1981). SO_2 emission rates at Mount St. Helens from March 29 through December, 1980. In: Lipman, P.W. & Mullineaux, D.L. (Eds.), The 1980 eruptions of Mount St. Helens, Washington. *U.S. Geol. Surv., Prof. Pap.*, vol. 1250, pp. 193-200.

Chance, K. (1998). Analysis of BrO measurements from the global ozone monitoring experiment. *Geophys. Res. Lett.*, 25, pp. 3335–3338.

Corsaro, R.A. & Miraglia, L. (2005). Dynamics of 2004-2005 Mt. Etna effusive eruption as inferred from petrologic monitoring. *Geophys. Res. Lett.*, 32, L13302, doi:10.1029/2005GL022347.

Daag, A.S., Tubianosa, B.S. et al. (1996). Monitoring sulphur dioxide emission at Mount Pinatubo. In: Newhall, C.G. & Punongbayan, R.S. (Eds.), Fire and mud: eruptions and lahars of Mount Pinatubo Philippines, *Philippine Institute of Volcanology and Seismology, Quezon City, University of Washington Press*, Seattle, pp. 409-434.

Dean, K.G., Dehn, J., Papp, K.R., Smith, S., Izbekov, P., Peterson, R., Kearney, C. & Steffke, A. (2004). Integrated satellite observations of the 2001 eruption of Mt. Cleveland, Alaska. *J. Volcanol. Geotherm. Res.*, 135, pp. 51-73.

Dehn, J., Dean, K.G., Engle, K. & Izbekov, P. (2002). Thermal precursors in satellite images of the 1999 eruption of Shishaldin Volcano. *Bull. Volcanol.*, 64(8), pp. 525–534.

Delgado-Granados, H., Cárdenas González, L. & Piedad Sánchez, N. (2001). Sulfur dioxide emissions from Popocatépetl volcano (Mexico): case study of a high-emission rate, passively degassing erupting volcano. *J. Volcanol. Geotherm. Res.*, 108, pp. 107-120.

Delmelle, P., Baxter, P., Baulieu, A., Burton, M., Francis, P., Garcia-Alavarez, J., Horrocks, L., Navarro, M., Oppenheimer, C., Rothery, D., Rymer, H., St-Amand, K., Stix, J., Strauch, W. & Williams-Jones, G. (1999). Origin, effects of Masaya volcano's continued unrest probed in Nicaragua. *EOS*, Transactions, Am. Geophys. Union. 80, pp. 575-581.

Delmelle, P., Stix, J., Baxter, P.J., Garcia-Alvarez, J. & Barquero, J. (2002). Atmospheric dispersion, environmental effects and potential health hazard assoviated with the low-altitude gas plume of Masaya volcano, Nicaragua. *Bull. Volcanol.*, 64, pp. 423-434.

Doukas, M.P. (2002). A new method for GPS-based wind speed determinations during airborne volcanic plume measurements. *U.S. Geol. Surv.*, Open-File Rep. 02-395, pp. 1-13.

Duffell, H.J., Oppenheimer, C., Pyle, D.M., Galle, B., McGonigle, A.J.S. & Burton, M.R. (2003). Changes in gas composition prior to a minor explosive eruption at Masaya volcano, Nicaragua. *J. Volcanol. Geotherm. Res.*, 126, pp. 327-339.

Edmonds, M., Herd, R.A., Galle, B. & Oppenheimer, C. (2003). Automated, high time-resolution measurements of SO_2 flux at Soufrière Hills Volcano, Montserrat, West Indies. *Bull. Volcanol.*, 65, pp. 578-586.

Edmonds, M. (2008). New geochemical insights into volcanic degassing. *Phil. Trans. R. Soc.*, 366, pp. 4559-4579.

Fayt, C. & van Roozendael, M. (2001). WinDOAS 2.1-Software User Manual. Belgisch Instituut voor Ruimte-Aeronomie Institut D'Aéronomie Spatiale de Belgique, Brussels, Belgium.

Finnegan, D.L., Kotra, J.P., Hermann, D.M. & Zoeller, W.H. (1989). The use of 7LiOH-impregnated filters for the collection of acidic gases and analysis by instrumental neutron activation analysis. *Bull. Volcanol.*, 51, pp. 83-87.

Fish, D.J. & Jones, R.L. (1995). Rotational Raman scattering and the ring effect in zenith-sky spectra. *Geophys. Res. Lett.*, 22(7), pp. 811-814.

Galle, B., Oppenheimer, C.M., Geyer, A., McGonigle, A. & Edmonds, M. (2003). A mini-DOAS spectrometer applied in remote sensing of volcanic SO_2 emissions. *J. Volcanol. Geotherm. Res.*, 119, pp. 241-254.

Galle, B., Johansson, M., Rivera, C., Zhang, Y., Kihlman, M., Kern, C., Lehmann, T., Platt, U., Arellano, S. & Hidalgo, S. (2010). Network for Observation of Volcanic and Atmospheric Change (NOVAC) A global network for volcanic gas monitoring: Network layout and instrument description. *J. Geophys. Res.*, 115, D05304, doi:10.1029/2009JD011823.

Giberti, G., Jaupart, C. & Sartoris, G. (1992). Steady-state operation of Stromboli volcano, Italy: Constraints on the feeding system. *Bull. Volcanol.*, 54, pp. 535-541.

Giggenbach, W.F. (1996). Chemical composition of volcanic gases. In: Scarpa, R., Tilling, R.I. (Eds.), Monitoring and mitigation of volcanic hazards, *Berlin-Heidelberg, Springer Verlag*, pp. 221-256.

Giggenbach, W.F. & Goguel, R.L. (1989). Collection and analysis of geothermal and volcanic water and gas discharges. *Department of Scientific and Industrial Research*, New Zealand, report CD2401, 81.

Gilbert, J.S. & Sparks, R.S.J. (1998). The Physics of Explosive Volcanic Eruptions. *Geol. Soc. Sp. Pub.*, vol. 145.

Hamilton, P.M., Varey, R.H. & Millán, M.M. (1978). Remote sensing of sulphur dioxide. *Atmospheric Envir.*, 12, pp. 127-133.

Harris, A.J.L., Flynn, L.P., Rothery, D.A., Oppenheimer, C. & Sherman, S.B. (1999). Max flux measurements at active lava lakes: Implications for magma recycling. *J. Geophys. Res.*, 104, pp. 7117-7136.

Harris, A.J.L. (2009). The pit-craters and pit-crater-filling lavas of Masaya volcano. *Bull. Volcanol.*, 71, pp. 541-558.

Harris, A.J.L., Steffke, A., Calvari, S. & Spampinato, L. (2011). Thirty years of satellite-derived lava discharge rates at Etna: Implications for steady volumetric output. *J. Geophys. Res.*, 116, B08204, doi:10.1029/2011JB008237.

Horrocks, L.A., Burton, M., Francis, P. & Oppenheimer, C. (1999). Stable gas plume composition measured by OP-FTIR spectroscopy at Masaya volcano, Nicaragua, 1998-1999. *Geophys. Res. Lett.*, 26, pp. 3497-3500.

Horrocks, L.A. (2001). Infrared spectroscopy of volcanic gases at Masaya, Nicaragua. *PhD thesis*, Open University, UK.

Horrocks, L.A., Oppenheimer, C., Burton, M.R. & Duffell, H.J. (2003). Compositional variation in tropospheric volcanic gas plumes: evidence from ground-based remote sensing. In: Oppenheimer, C., Pyle, D.M. & Barclay, J. (Eds.), *Volcanic Degassing*, Geological Society, London. 149-168. ISBN: 978-1-86239-136-9.

Horton, K.A., Williams-Jones, G., Garbeil, H., Elias, T., Sutton, A.J., Mouginis-Mark, P., Porter, J.N. & Clegg, S. (2006). Real-time measurement of volcanic SO_2 emissions: validation of a new UV correlation spectrometer (FLYSPEC). *Bull. Volcanol.*, 68, pp. 323-327.

Huppert, H.E. & Woods, A.W. (2002). The role of volatiles in magma chamber dynamics. *Nature*, vol. 420, pp. 493-495.

Kern, C. (2009). Spectroscopic measurements of volcanic gas emissions in the ultra-violet wavelength region. *Ph.D. thesis*, Univ. of Heidelberg, Heidelberg, Germany.

Kern, C., Sihler, H., Vogel, L., Rivera, C., Herrera, M. & Platt, U. (2009a). Halogen oxide measurements at Masaya Volcano, Nicaragua using active path differential optical absorption spectroscopy. *Bull. Volcanol.*, 71, pp. 659-670.

Kern, C., Deutschmann, T., Vogel, A., Wöhrbach, M., Wagner, T., & Platt, U. (2009b). Radiative transfer corrections for accurate spectroscopic measurements of volcanic gas emissions, *Bull. Volcanol.*, doi:10.1007/s00445-009-0313-7.

Keszthelyi, L. & Denlinger, R. (1996). The initial cooling of pahoehoe flow lobes. *Bull. Volcanol.*, 58, pp. 5-18.

Keszthelyi, L., Harris, A.J.L. & Dehn, J. (2003). Observations of the effect of wind on the cooling of active lava flows. *Geophys. Res. Lett.*, 30(19), doi:10.1029/2003GL017994.

Kyle, P.R., Sybeldon, L.M., McIntosh, W.C., Meeker, K. & Symonds, R. (1994). Sulfur dioxide emission rates from Mount Erebus, Antarctica. In: Kyle, P. (Ed.), Volcanological and environmental studies of Mount Erebus, Antarctica, Washington D.C., AGU, *Antarctic Research Series*, vol. 66, pp. 69-82.

Lodato, L., Spampinato, L., Harris, A.J.L., Calvari, S., Dehn, J. & Patrick, M. (2007). The morphology and evolution of the Stromboli 2002-2003 lava flow field: An example of basaltic flow field emplaced on a steep slope. *Bull. Volcanol.*, 69, pp. 661-679.

Malicet, J., Daumont, D., Charbonnier, J., Parisse, C., Chakir, A. & Brion, J. (1995). Ozone UV spectroscopy. II. Absorption cross-sections and temperature dependence. J. *Atmos. Chem.*, 21, pp. 263. doi:10.1007/BF00696757.

Malinconico, L.L. (1979). Fluctuations in SO_2 emission during recent eruptions of Etna. *Nature*, 278, pp. 43-45.

Manatt, S.L. & Lane, A.L. (1993). A compilation of the absorption cross-section of SO_2 from 106 to 403 nm. *J. Quant. Spectrosc. Radiat. Trasfer.*, 50(3), pp. 267-276.

Martin, R.S., Mather, T.A., Pyle, D.M., Power, M., Tsanev, V.I., Oppenheimer, C., Allen, A.G., Horwell, C.J. & Ward, E.P.W. (2009). Size distributions of fine silicate and other particles in Masaya's volcanic plume. *J. Geophys. Res.*, 114, D09217.

Martin, R.S., Sawyer, G.M., Spampinato, L., Salerno, G.G., Ramirez, C., Ilyinskaya, E., Witt, M.L.I., Mather, T.A., Watson, I.M., Phillips, J.C. & Oppenheimer, C. (2010). A total volatile inventory for Masaya Volcano, Nicaragua. *J. Geophys. Res.*, 115, B09215, doi: 10.1029/2010JB007480.

Mather, T.A., Allen, A.G., Oppenheimer, C., Pyle, D.M. & McGonigle, A.J.S. (2003). Size-resolved characterisation of soluble ions in the particles in the tropospheric plume of Masaya volcano, Nicaragua: origins and plume processing. *J. Atmosph. Chem.*, 46, pp. 207-237.

Mather, T.A., Pyle, D.M., Tsanev, V.I., McGonigle, A.J.S., Oppenheimer, C., & Allen, A.G. (2006). A reassessment of current volcanic emissions from the Central American arc with specific examples from Nicaragua. J. *Volcanol. Geotherm. Res.*, 149, pp. 297- 311

McBirney, A.R. (1956). The Nicaraguan volcano Masaya and its caldera. *Trans. Amer. Geophys. Union*, 37, pp. 83-96.

McGee, K.A. & Sutton, J.A. (1994). Eruptive activity at Mount St Helens, Washington, USA, 1984-1988: a gas geochemistry perspective. *Bull. Volcanol.*, 56, pp. 435-446.

McGonigle, A.J.S. & Oppenheimer, C. (2003). Optical sensing of volcanic gas and aerosol emissions. In: Oppenheimer, C., Pyle, D.M. & Barclay, J. (Eds.), *Volcanic Degassing*, Geological Society, London. 149-168. ISBN: 978-1-86239-136-9.

McGonigle, A.J.S., Oppenheimer, C., Hayes, A.R., Galle, B., Edmonds, M., Caltabiano, T., Salerno, G., Burton, M. & Mather, T.A. (2003). Sulphur dioxide fluxes from Mt. Etna, Vulcano and Stromboli measured with an automated scanning ultraviolet spectrometer. J. *Geophys. Res.*, 108(B9), 2455. doi:10.1029/2002JB002261.

McGonigle, A.J.S., Delmelle, P., Oppenheimer, C., Tsanev, V.I., Delfosse, T., Williams-Jones, G., Horton, K. & Mather, T.A. (2004). SO_2 depletion in tropospheric volcanic plumes. *Geophys. Res. Lett.*, 31, L13201, doi: 10.1029/2004GL019990.

McGonigle, A.J.S., Aiuppa, A., Ripepe, M., Kantzas, E.P. & Tamburello, G. (2009). Spectroscopic capture of 1 Hz volcanic SO_2 fluxes and integration with volcano geophysical data. *Geophys. Res. Lett.*, 36, L21309, doi:10.1029/2009gl040494.

Millan, M.M. (1980). Remote sensing of air pollutants. A study of some atmospheric scattering effects. *Atmos. Env.*, 14, pp. 1241-1253.

Mouginis-Mark, P.J., Crisp, J.A. & Fink, J.H. (2000). In: Mouginis-Mark, P.J., Crisp, J.A. & Fink, J.H. (Eds.), Remote Sensing of Active Volcanism, AGU, *Geophysical Monograph*, 116, pp. 1-7.

Nadeau, P. & Williams-Jones, G. (2009). Apparent downwind depletion of volcanic SO_2 flux-lessons from Masaya Volcano, Nicaragua. *Bull. Volcanol.*, 71, pp. 389-400.

Neri, A. (1989). A local heat transfer analysis of lava cooling in the atmosphere: application to thermal diffusion-dominated lava flows. *J. Volcanol. Geotherm. Res.*, 81, pp. 215-243.

Newcomb, G. & Millán, M.M. (1970). Theory, applications and results of the long-line correlation spectrometer. *IEEE Trans. Geosci. Electron.*, 8, pp. 149-157.

Noguchi, K. & Kamiya, H. (1963). Prediction of volcanic eruption by measuring the chemical composition and amounts of gases. *Bull. Volcanol.*, 26, pp. 367-378.

Noxon, J.F. (1975). Nitrogen dioxide in stratosphere and troposphere measured by round-based absorption spectroscopy. *Science*, 189, pp. 547-549.

O'Dwyer, M., McGonigle, A.J.S., Padgett, M.J., Oppenheimer, C. & Inguaggiato, S. (2003). Real time measurements of volcanic H_2S/SO_2 ratios by UV spectroscopy. *Geophys. Res. Lett.*, 30, 12, doi:10.1029/2003GL017246.

Olmos, R., Barrancos, J., Rivera, C., Barahona, F., López, D.L., Henriquez, B., Hernández, A., Benitez, E., Hernández, P.A., Pérez, N.M. & Galle, B. (2007). Anomalous emissions of SO_2 during the recent eruption of Santa Ana Volcano, El Salvador, Central America. *Pure Appl. Geophys.*, 164, pp. 2489–2506.

Oppenheimer, C. & Francis, P. (1997). Remote sensing of heat, lava and fumarole emissions from Erta 'Ale volcano, Ethiopia. *Int. J. Remote Sens.*, 18(8), pp. 1661-1692.

Oppenheimer, C., Francis, P., & Stix J. (1998). Depletion rates of sulphur dioxide in tropospheric volcanic plumes. *Geophys. Res. Lett.*, 25(14), pp. 2671-2674.

Oppenheimer, C. (2003). Volcanic degassing. In: Rudnick, R.L., Holland, H.D., Turekian, K.K. (Eds.), The crust, Treatise on geochemistry, *Oxford, Elsevier-Pergamon*, vol. 3, pp. 123-166.

Oppenheimer, C., McGonigle, A.J.S., Allard, P., Wooster, M.J. & Tsanev, V. (2004). Sulfur, heat, and magma budget of Erta 'Ale lava lake, Ethiopia. *Geology*, 32(6), 509-512.

Oppenheimer, C., Kyle, P.R., Tsanev, V.I., McGonigle, A.J.S., Mather, T.A. & Sweeney, D. (2005). Mt. Erebus, the largest point source of NO_2 in Antarctica. *Atmos. Environ.*, 39, pp. 6000–6006.

Perez, M., Freundt, A., Kutterolf, S. & Schminke, H.-U. (2009). The Masaya tripe layer: A 2100 year old basaltic multiepisodic Plinian eruption from the Masaya Caldera Complex (Nicaragua). *J. Volcanol. Geotherm. Res.*, 179(3-4), pp. 191-205.

Platt, U. (1994). Differential optical absorption spectroscopy (DOAS). In: Sigrist, M.W. (Ed.), Air monitoring by spectroscopic techniques, *Chemical Analysis Series*, vol. 127, Wiley, J., Chicherster, UK.

Platt, U. & Stutz, J. (2008). Differential Optical Absorption Spectroscopy principles and applications, Series: *Physics of Earth and Space Environment*, Springer.

Rausch, J. & Schmincke, H.-U. (2010). Nejapa Tephra: The youngest (c. 1 ka BP) highly explosive hydroclastic eruption in western Managua (Nicaragua). *J. Volcanol. Geotherm. Res.*, 192, pp. 159-177.

Realmuto, V.J. (2000). The potential use of earth observing system data to monitor the passive emission of sulfur dioxide from volcanoes. In: Mouginis-Mark, P.J., Crisp, J.A. & Fink, J.H. (Eds.), Remote sensing of active volcanism, *Geophys. Monogr. AGU*, Washington, D.C., vol. 116, pp. 101-115.

Roche, O., van Wyk de Vries, B. & Druitt, T.H. (2001). Sub-surface structures and collapse mechanisms of summit pit craters. *J. Volcanol. Geotherm. Res.*, 105, pp. 1-18.

Rodríguez, L.A., Branan, Y.K., Watson, I.M., Bluth, G.J.S., Rose, W.I., Chigna, G., Matías, O., Carn, S.A. & Fischer, T. (2004). SO_2 emissions to the atmosphere from active volcanoes in Guatemala and El Salvador, 1999-2002. *J. Volcanol. Geotherm. Res.*, 138, pp. 325-344.

Rymer, H., van Wyk de Vries, B., Stix, J. & Williams-Jones, G. (1998). Pit crater structure and processes governing persistent activity at Masaya volcano, Nicaragua. *Bull. Volcanol.*, 59, pp. 345-355.

Salerno, G.G., Burton, M.R., Oppenheimer, C., Caltabiano, T., Randazzo, D., Bruno, N. & Longo, V. (2009a). Three-years of SO_2 flux measurements of Mt. Etna using an automated UV scanner array: Comparison with conventional traverses and uncertainties in flux retrieval. *J. Volcanol. Geotherm. Res.*, 183, pp. 76-83.

Salerno, G.G., Burton, M.R., Oppenheimer, C., Caltabiano, T., Tsanev, V.I. & Bruno, N. (2009b). Novel retrieval of volcanic SO_2 abundance from ultraviolet spectra. *J. Volcanol. Geotherm. Res.*, 181, pp. 141-153.

Solomon, S., Schmeltekopf, A.L. & Sanders, R.W. (1987). On the interpretation of zenith sky absorption measurements. *J. Geophys. Res.*, 92, pp. 8311-319.

Spampinato, L., Calvari, S., Oppenheimer, C. & Lodato, L. (2008a). Shallow magma transport for the 2002-3 Mt. Etna eruption inferred from thermal infrared surveys. *J. Volcanol. Geotherm. Res.*, 177, pp. 301-312.

Spampinato, L., Oppenheimer, C., Calvari, S., Cannata, A. & Montalto, P. (2008b). Lava lake surface characterization by thermal imaging: Erta 'Ale volcano (Ethiopia). *Geochem. Geophys. Geosyst.*, 9, Q12008. doi:10.1029/2008GC002164.

Spampinato, L., Calvari, S., Oppenheimer, C. & Boschi, E. (2011). Volcano surveillance using infrared cameras. *Earth Sci. Rev.*, 106, pp. 63-91.

Sparks, R.S.J. (2003). Dynamics of magma degassing. In: Oppenheimer, C., Pyle, D.M., Barclay, J. (Eds.), Volcanic degassing, *Geol. Soc. Lond., Sp. Pub.*, vol. 213, pp. 5-22.

Spilliaert, N., Allard, P., Metrich, N. & Sobolev, A.V. (2006). Melt inclusion record of the conditions of ascent, degassing, and extrusion of volatile-rich alkali basalt during the powerful 2002 flank eruption of Mount Etna (Italy). *J. Geophys. Res.*, 111, B04203, doi:10.1029/2005jb003934.

Steffke, A.M., Harris, A.J.L., Burton, M., Caltabiano, T. & Salerno G.G. (2010). Coupled use of COSPEC and satellite measurements to define the volumetric balance during effusive eruptions at Mt. Etna, Italy. *J. Volcanol. Geotherm. Res.*, 205, pp. 47-53.

Stix, J. (2007). Stability and instability of quiescently degassing active volcanoes: The case of Masaya, Nicaragua. *Geology*, 35(6), pp. 535-538.

Stoiber, R.E. & Jepsen, A. (1973). Sulphur dioxide contribution to the atmosphere by volcanoes. *Science*, 182, pp. 577-578.

Stoiber, R.E., Malinconico, L.L. & Williams, S.N. (1983). Use of the correlation spectrometer at volcanoes. In: Tazieff, H. & Sabroux, J.-C. (Eds.), Forecasting volcanic events, *Elsevier*, Amsterdam, pp. 425-444.

Stoiber, R., Williams, S. & Huebert, B. (1986). Sulfur and halogen gases at Masaya caldera complex, Nicaragua: total flux and variations with time, *J. Geophys. Res.*, 91(B12), pp. 12,215-12,231.

Sutton, A.J., Elias, T., Gerlach, T.M. & Stokes, J.B. (2001). Implications for eruptive processes as indicated by sulfur dioxide emissions from Kīlauea Volcano, Hawai`i, 1979-1997. *J. Volcanol. Geotherm. Res.*, 108(1-4), pp. 283-302.

Symonds, R.B., Rose, W.I., Bluth, G.J.S. & Gerlach, T.M. (1994). Volcanic gas studies methods, results, and applications. *Rev. Mineral.*, 30, pp. 1–66.

Symonds, R.B., Gerlach, T.M. & Reed, M.H. (2011). Magmatic gas scrubbing: implications for Volcano monitoring. *J. Volcanol. Geothermal Res.*, 108, pp. 303–341.

Sweeney, D., Kyle, P.R. & Oppenheimer C. (2008). Sulfur dioxide emissions and degassing behavior of Erebus volcano, Antarctica. *J. Volcanol. Geother. Res.*, 177(3), pp. 725-733.

Thomas, H.E. & Watson, I.M. (2010). Observations of volcanic emissions from space: current and future perspectives. *Nat. Hazards*, 54, pp. 323-354.

Tilling, R.I. (1987). Fluctuations in surface height of active lava lakes during 1972-1974 Mauna Ulu eruption, Kīlauea volcano, Hawai'i. *J. Geophys. Res.*, 92, 13721-13730.

Torrence, C. & Compo, G.P. (1998). A practical guide to wavelet analysis. *Bull. Am. Meteorol. Soc.*, 79, pp. 61–78.

Vandaele, A.C., Simon, P.C., Guilmot, J.M., Carleer, M. & Colin, R. (1994). SO_2 absorption cross section measurement in the UV using a Fourier transform spectrometer. *J. Geophys. Res.*, 99, pp. 25,599-25,605.

Vicari, A., Ciraudo, A., Del Negro, C., Herault, A. & Fortuna, L. (2008). Lava flow simulations using discharge rates from thermal infrared satellite imagery during the 2006 Etna eruption. *Nat. Hazards*, doi:10.1007/s11069-008-9306-7.

Villemant, B. & Boudon, G. (1999). H_2O and halogen (F, Cl, Br) behavior during shallow magma degassing processes. *Earth Planet. Sci. Lett.*, 168, pp. 271-286.

Wallace, P.J. & Gerlach, T.M. (1994). Magmatic vapour source for sulphur dioxide released during volcanic eruptions: evidence from Mount Pinatubo. *Science*, 265, pp. 497-499.

Wardell, L.J., Kyle, P.R. & Campbell, A.R. (2003). Carbon dioxide emissions from fumarolic ice towers, Mount Erebus volcano, Antarctica. In: Oppenheimer, C., Pyle, D.M. & Barclay, J. (Eds.), Volcanic degassing, *Geol. Soc. Lond., Sp. Pap.*, vol. 213, pp. 231-246.

Williams, S.N. (1983). Geology and eruptive mechanisms of Masaya caldera complex. *PhD thesis*, Dartmouth College, Hanover, N.H..

Williams-Jones, G. (2001). Integrated geophysical studies at Masaya volcano, Nicaragua, *Ph.D. thesis*, Open Univ., UK.

Williams-Jones, G., Rymer, H. & Rothery, D.A. (2003). Gravity changes and passive SO_2 degassing at the Masaya caldera complex, Nicaragua. *J. Volcanol. Geotherm. Res.*, 123(1-2), pp. 37-160.

Williams-Jones, G., Stix, J. & Hickson, C. (2008). The COSPEC Cookbook: making SO_2 measurements at active volcanoes. IAVCEI, *Methods in Volcanology*, vol. 1.

Witham, F., Woods, A.W. & Gladstone, C. (2006). An analogue experimental model of depth fluctuations in lava lakes. *Bull. Volcanol.*, 69, pp. 51-56.

Witt, M.L.I., Mather, T.A., Pyle, D.M., Aiuppa, S., Bagnato, E. & Tsanev, V.I. (2008). Mercury and halogens emissions from Masaya and Telica volcanoes, Nicaragua. *J. Geophys. Res.*, 113, B06203, doi: 10.1029/2007JB005401.

Wright, R. & Pilger, E. (2008). Satellite observations reveal little inter-annual variability in the radiant flux from the Mount Erebus lava lake. *J. Volcanol. Geotherm. Res.*, 177, pp. 687-694.

Wright, R., Garbeil, H. & Davies, A.G. (2010). Cooling rate of some active lavas determined using an orbital imaging spectrometer. J. Geophys. Res., 115, B06205, doi:10.1029/2009JB006536.

Young, S.R., Sparks, R.S.J., Aspinall, W.P., Lynch, L.L., Miller, A.D., Robertson, R.E.A. & Shepherd, J.B. (1998). Overview of the eruption of Soufriere Hills volcano, Montserrat, 18 July 1995 to December 1997. *Geophys. Res. Lett.*, 25(18), pp. 3389-3392.

Zapata, J.A., Calvache, M.L. et al. (1997). SO_2 fluxes from Galeras Volcano, Colombia, 1989-1995: Progressive degassing and conduit obstruction of a Decade Volcano. *J. Volcanol. Geotherm. Res.*, 77, pp. 195-208.

6

What Caused the Ice Ages?

Willy Woelfli[1] and Walter Baltensperger[2]
[1]*Institute for Particle Physics, ETHZ Hönggerberg, Zürich*
[2]*Centro Brasileiro de Pesquisas Físicas, Urca, Rio de Janeiro,*
[1]*Switzerland*
[2]*Brazil*

1. Introduction

In this didactic paper basic features of the Ice Ages will be illustrated using well known data. The empirical facts are confronted with two theories.

The usual attempt to explain the Ice Ages is based on two phenomena. First, the other planets of the solar system induce small and slow variations of Earth's orbit. They were introduced by Milankovich (Milankovitch, 1941). They lead to variations of the insolation on Earth. Secondly, various feedbacks in Earth's climate system are considered. An example of a positive feedback: cold brings snow, which reflects solar light efficiently, so that the temperature decreases. The question is, whether the feedbacks are large enough to produce the Ice Ages. The small quasi periodic changes of Earth's orbit follow directly from the laws of mechanics and gravity; they are not questioned. Milankovitch cycles in the climate should belong to any climate model. The problem is the size of the feedbacks, which at present produce our regular climate. This model of the Ice Ages, often simply called "the Milankovitch theory", is used by practically all members of the scientific community for the interpretation of the data.

The second theory, which has been developed by us, involves a very unusual astronomical action on Earth's climate system during the last 3 million years. The revealing observation is an asymmetry with respect to the present North Pole: when the glaciation was at its maximum, it reached the region of New York City, while East Siberia remained ice free. Evidently, the North Pole was in Greenland. Therefore, at the end of the Ice Ages it shifted geographically to its present position in the Arctic Sea. This involves a motion of the globe, not of the rotation axis. We claim that there are circumstances in which this is compatible with laws of nature and observations. A planetary object with a mass at least 1/10 of Earth's mass must pass very close to Earth. We shall call this planet "Z". The tidal forces induce a 1 per mil stretching deformation of Earth's globe. While this relaxes in a time of order one year the pole moves. Since Z has disappeared, it must have disintegrated and the fractions evaporated. By necessity, Z was hot, liquid and radiant. It moved in an extremely eccentric orbit. Its evaporation created a disk shaped cloud of ions around the Sun.

A disk shaped cloud has been postulated by R.A. Muller and G.J. MacDonald (Muller& MacDonald, 1995; 1997), since Earth's orbit would be in the cloud with a Milankovitch period of 100 000 years. This is the dominant climate cycle. During cold periods Earth's orbit was in the cloud. This is when fine grained inclusions were deposited into the ice. The cloud has its proper dynamics. It builds up to a density at which inelastic particle collisions induce its collapse. The resulting near-periodic time dependence resembles that of Dansgaard-Oeschger (Dansgaard, 1993) climate events. When the plane of Earth's orbit was outside the cloud, the Earth received additional scattered radiation. This occurred during the last interglacial, about 120 000 years ago, and again during most of the last 10 000 years. In this theory with a shift of the poles, the Ice Ages were a singular and now definitely finished period of Earth's history.

In short, this is the contents of this chapter. New is only the interpretation of "reverse Dansgaard–Oeschger events" in subsection 4.4.

At the present time, the study of the Ice Ages is a detective story. This chapter describes two possible culprits. One is the system itself, which produced the Ice Ages. Somewhat like the weather, which creates a storm on one day and sunshine on the next. The other possible culprit is an exceedingly rare external object: a hot planet. A good detective keeps more than one hypothesis in his mind.

2. Basic features of the Ice Ages

2.1 When did the Ice Ages begin?

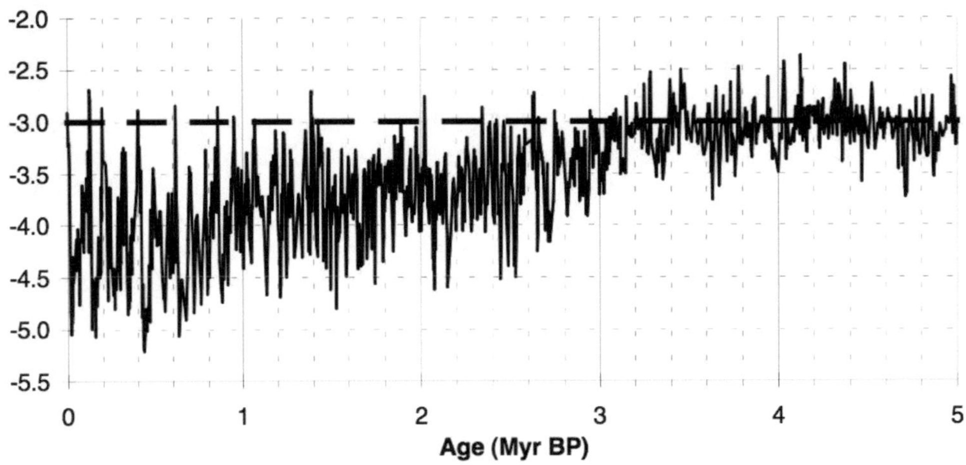

fig 1. Marine sediments contain information of the past. The ^{18}O to ^{16}O isotope ratio was measured on organic matter (Benthic foraminifera) as function of depth of the sediment. The diagram shows $\delta\,^{18}O$ for 5 million years. $\delta\,^{18}O$ is 1000 times the difference of the ^{18}O to ^{16}O ratio for the sample and a standard divided by the standard value. Lower $\delta\,^{18}O$ corresponds to colder climate. Measurements from the Atlantic Ocean east of Africa. (site 659: 18° 05′ N, 21° 02′ W). The horizontal dashed line corresponds to the present value. Data from (Clemens&Tiedemann, 1997).

We see in fig 1 that the Ice Ages began about 3.2 million years ago. Before that time the temperature was similar to that of our time. The mean temperature of the Ice Ages was low, but the temperature variations were large. In fig 1 with the scale of million years the end of the Ice Ages and the present situation are not visible.

2.2 When did the Ice Ages end

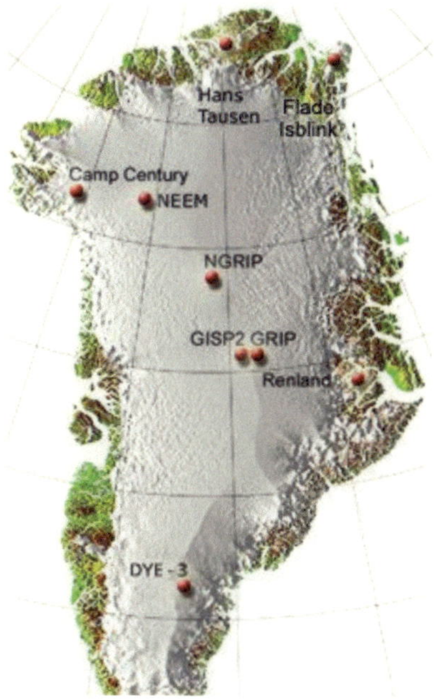

fig 2. Location of the GISP2 and NGRIP stations in central Greenland.

The ^{18}O to ^{16}O isotope ratio can also be measured as a function of depth in an ice core. Fig 2 shows the position of the GISP2 research station. Variations of the isotope ratio of Oxygen in Greenland arise from the following process. Water evaporates from tropical seas. On the way to arctic regions it rains. In the formation of ice particles or water droplets there is a slight preference for the heavier atom. This is actually a quantum effect, due to the fact that lighter atoms have a higher energy in a localized state. This selection is more effective at lower temperatures. Therefore, the lower the temperature between the tropical sea and the arctic position of GISP2, the more deficient in the heavier atom will the isotope ratio be. This indicates a temperature variation of the hemisphere. In fig 3 the red curve shows this as temperature at the site in Greenland for the last 20 000 years.

10 000 years ago the Ice Ages with its cold periods and violent climate changes stopped. This is the end of the geological period called Pleistocene and the beginning of the Holocene, which in its usual definition includes the present time. Most of the Holocene was warmer than at present. Its temperature variations of up to two degrees were much smaller than those of the Ice Ages. Still, they could be a respectable threat to civilization. The last 200 years were calm, so that man made changes (subsection 2.4) became a serious consideration.

2.3 Are we at present between two cold periods or did the Ice Ages end definitely?

This is the basic question of this paper. Are we in a warm peak like those of Fig 12 or in a period as before 3.2 Myr in fig 1 ?

The dominant theory assumes that the Ice Ages with all its variations were a product of the system. Small changes of Earth's orbit due to the other planets lead to small changes in solar

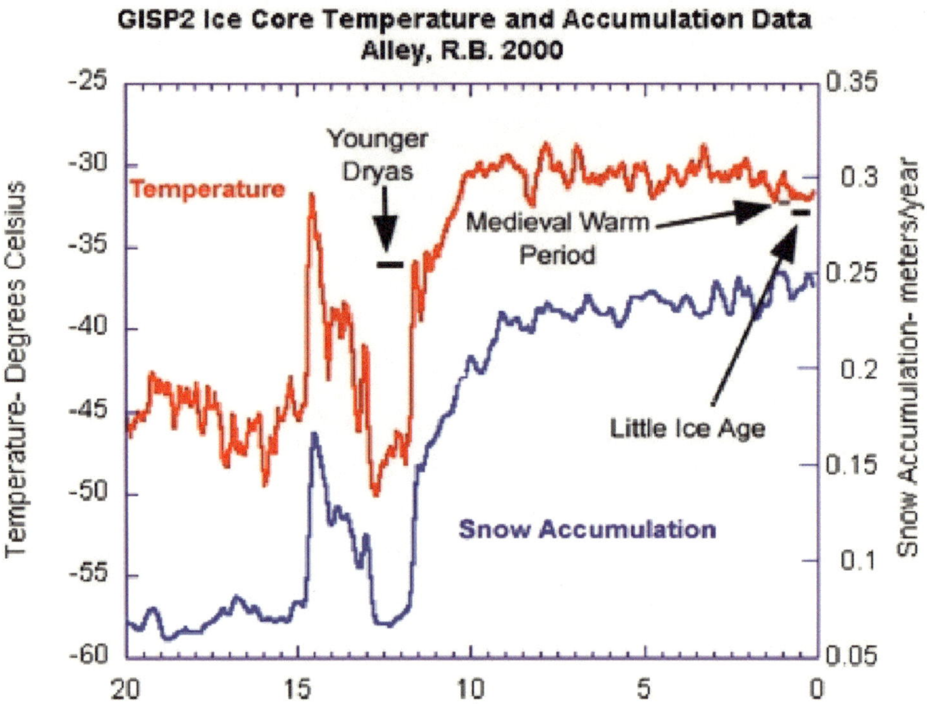

fig 3. Temperature at the site GISP2 in Greenland as inferred from the ^{18}O to ^{16}O ratio in an ice core for the last 20 kyr. Also shown (blue) is the yearly snow accumulation, which increases with temperature. Data from (Alley, 2000).

radiation on Earth, as first calculated by Milankovitch (Milankovitch, 1941). In many research projects Milankovitch cycles have been identified in the empirical data. This is common to all theories, since the variations of the solar radiation and some feedback undoubtedly exist. The distinction is quantitative: the main stream theory assumes the feedbacks to be sufficiently large to produce non linear effects and thereby the Ice Age mode of the climate system with cold periods and strong variations.

In the theory with a pole shift the Ice Ages were produced by an exceptional astronomic object, a hot evaporating planet Z, which no more exists. In this model the exceptional situation is gone and will not come back. This theory is actually not a wide open field, but a narrow window. It describes a barely possible sequence of events.

The mere fact that the Ice Ages had a beginning and apparently an end may favor the second theory, but it does not contradict the first. As an example, you may drive a car on a rainy day. The windscreen wiper works smoothly. Suddenly it starts jumping noisily over the window. The system went into a different mode. This may reverse after a while. In this case the noisy mode may start again. This simple mechanical system can work in two modes.

For much older geological times (about 600 to 750 million years ago) there are evidences for a "Snowball Earth" (Snowball, Bibl.). Then the continents were joined together. When the continent is one block, then it surrounds the rotation axis in the stable motion of the rotating globe. The causes of the "Snowball Earth" may be quite different from those of the Ice Ages.

2.4 How important are the man made climate changes?

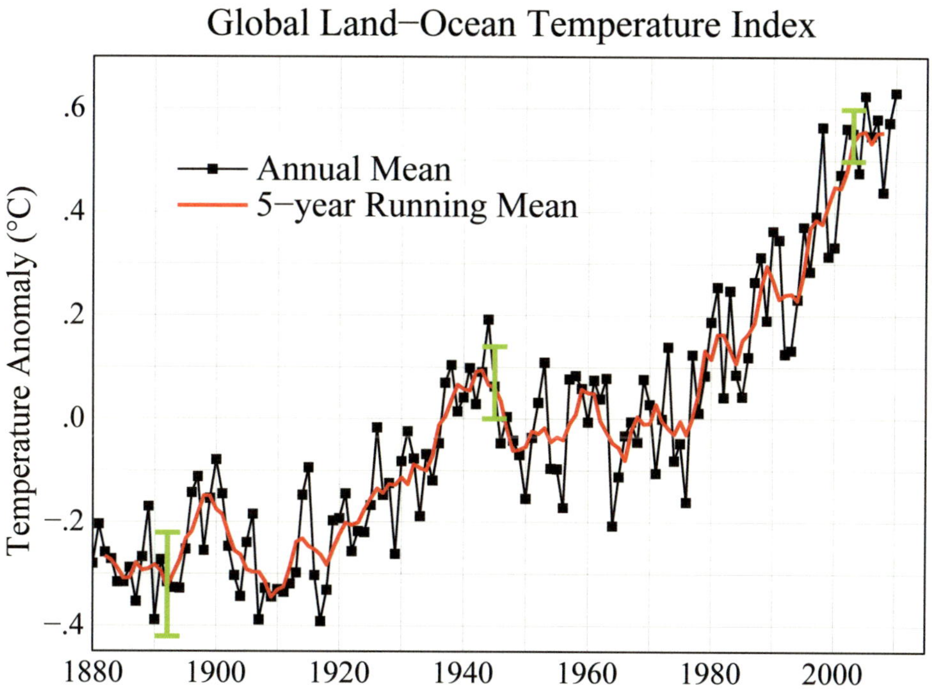

fig 4. Global averages of measured temperatures since 1880. Green intervals are estimated uncertainties. Figure from NASA.

Fig 4 shows instrumental mean global temperatures since 1880. The temperature increase of about 1 degree centigrade is most likely due to greenhouse gases, notably CO_2; it is human made. Since it modifies the environment to which plants, animals and ourselves are adjusted, we must consider it as bad. Fig 3 shows that the variations of the temperature of the last 10 000 years were larger, and in the Ice Ages they were so enormous that the mere survival of a species must have been an achievement.

The development of the human species in the last 2 to 3 million years was simultaneous with the Ice Ages; this is certainly not a mere coincidence. In the dramatic climate changes the innate reactions of the individuals may often have been inadequate. The large brain of the human species can carry a complex language. Thus humans dispose of a new type of information: orally transmitted experience. This was an advantage. It also created new problems: the language had to learned, so that the time of childhood was increased, the brain is energy consuming, and - last not least - there are two decision makers in one brain: the instinctive and the rational.

The climate changes of the Ice Ages inhibited organized civilizations; therefore we probably tend to underestimate the capabilities of the humans of those times. Even 8 000 years ago, when organized civilizations arouse, the climate showed considerable variations (Fig 3). The quiet period of the last 300 years may have helped the astounding development of our modern world.

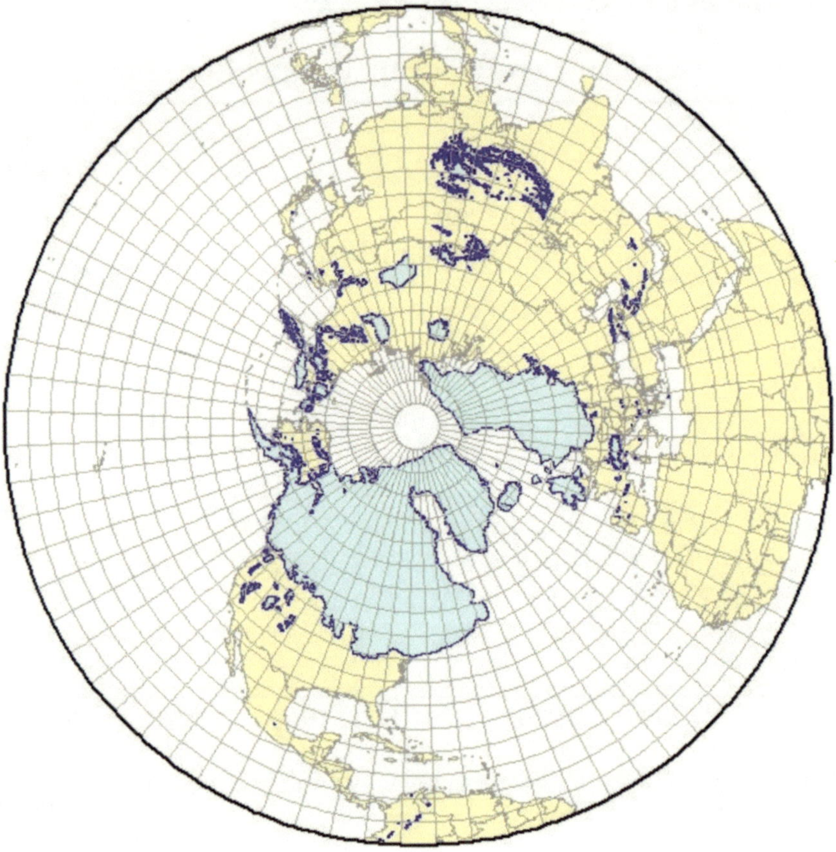

fig 5. The largest glaciation (about 20 000 years ago) was not symmetric with respect to the present position of the North Pole in the Arctic Sea. fig by J, Ehlers. Data from (Ehlers& Gibbard, 2004)

What value should we give to the measured mean temperature changes of fig 4? They are now larger than intrinsic fluctuations, which reversed the trend between 1940 and 1980. Clearly, we should avoid modifying the global climate, which is the basis of life of all plants and animals and ourselves. However, the answer depends on the theory of the Ice Ages. If the system itself produced the Ice Ages, then the most urgent questions are: Do we understand how the Ice Ages were switched on and off? Do the greenhouse gases increase or lower the chances that the system drops again into its Ice Age mode?

In the theory with a pole shift there is no danger of a new Ice Era, since astronomical observations show that the causes (a hot planet or its fractions and a disk shaped cloud of ions around the Sun) have disappeared. Therefore, the planetary system is now clean, and it is up to us that we do not perturb our climate. In this view, at least after the Little Ice Age, i.e. about 300 years ago, a new, clean period had started. In the Holocene, the warmer temperatures and their variations were produced by the cloud from the fractions of Z. At present this cloud does not exist any more. This geological period deserves to have a new name, such as the sometimes used "Anthropocene".

2.5 Why was the glaciation asymmetric with respect to the present position of the North Pole?

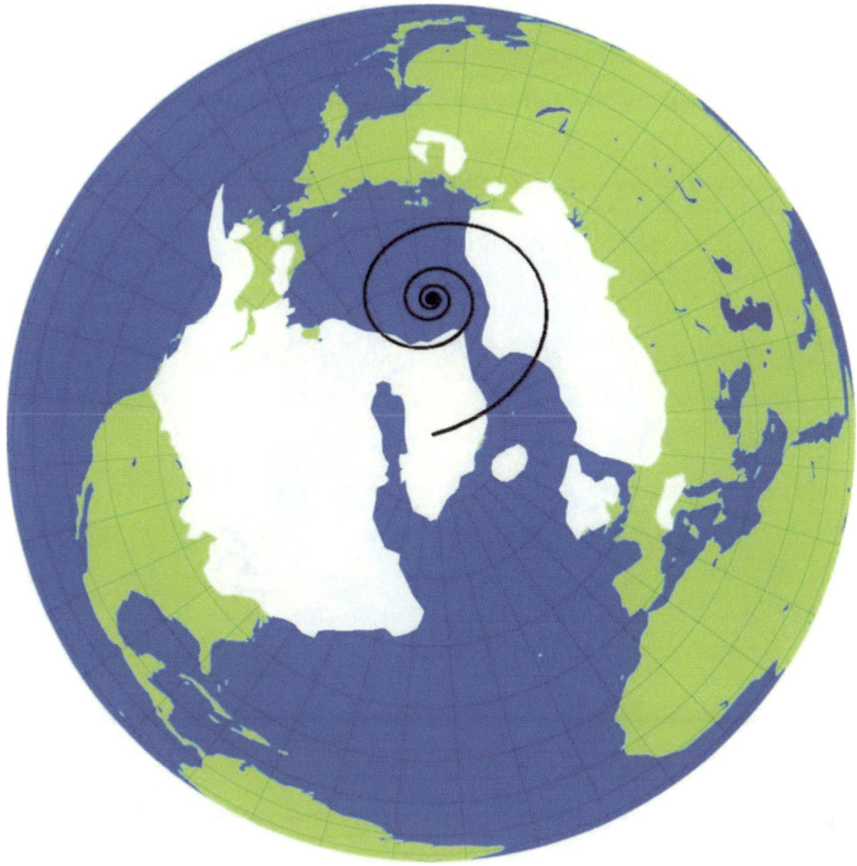

fig 6. Geographic shift of the North Pole: path from Greenland, the assumed center of the largest glaciation, to the present position in the Arctic Sea. One turn of the spiral takes about 400 days (Chandler period). fig from Woelfli et al. (2002)

The maximum ice distribution on land is clearly visible from erosion and moraines. It occurred about 20 000 years ago. The ice reached New York City, while East Siberia remained ice free. In fig 5 the center of the ice distribution is obviously not the present North Pole in the Arctic Sea. The glaciation suggests a position in Greenland, about 17° away from the present North Pole. The best suggested position depends on the influence attributed to the Atlantic Ocean.

Remains of mammoths have been found in East Siberia in regions with a high latitude, where at present these herbivores could not exist. In East Siberia herds of mammoths grazed within the arctic circle, even on islands in the Arctic Sea, which were connected with the mainland during the glacial periods. These facts were first reported in the Soviet literature and are confirmed by current investigations (Bocherens, 2003; Orlova et al., 2001; Schirrmeister et al., 2002). The data are complex, since the temperature varied between stadials and interstadials, but consistently this area was not ice covered. The yearly insolation decreases with increasing latitude. The present distribution of the flora on the globe suggests that in arctic regions the yearly insolation is insufficient for steppe plants.

While the North Pole changed from land to sea, the South Pole remained within the Antarctic continent. Correspondingly, the climate was less affected on the southern hemisphere.

These facts are explained by the shift of the poles, which is the basic assumption of the alternative theory.

The dominant theory discusses changes of the water currents in the oceans. In particular, during cold periods the Golf Stream is supposed to sink into the deep Atlantic and revert its direction before it reaches Northern Europe (fig 7). At present the importance of water streams as compared to air currents is questioned (Seager, 2006). The absence of glaciation in East Siberia is attributed to cold and dry air with the corresponding lack of snow (fig 3).

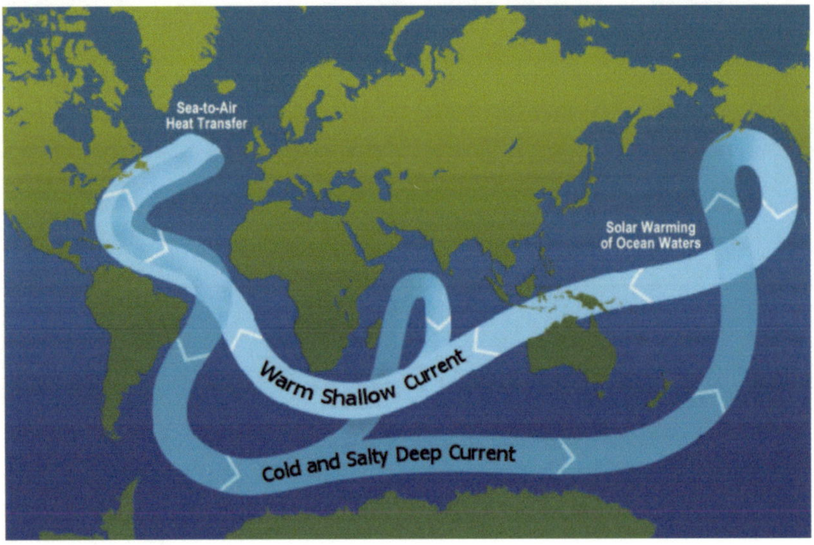

fig 7. Simplified picture of the ocean streams.

3. The rapid geographic shift of the poles

The observation that during the ice ages the North Pole was in Greenland was the motivation for postulating planet Z. At the the end of the Pleistocene the Earth turned relative to the fixed rotation axis. We claim, that a close passage of an object with about 1/10 the mass of the Earth could do the trick. Fig. 6 shows a calculated geographic path of the North Pole. One turn of the spiral takes about 400 days.

This event of the close passage of Z was complex. Z was scattered by a few degrees and, aided by the pressure of the hot interior, it decayed into several parts. The evaporation rate from the fractions is vastly increased, since their escape velocity is reduced. At present they do not exist any more. However, during their evaporation a dense disk shaped cloud around the Sun made an angle of a few degrees with Earth's orbit. Then most of the time the Earth was outside this cloud and received an increased amount of radiation due to scattered light from the cloud.

3.1 What is the physics of a rapid geographic pole shift?

This subsection is destined to those who want to know the mechanics that we used to describe the geographic motion of the pole. The inertial tensor of a rigid body with mass density $\rho(\vec{r})$ in a coordinate system fixed to this body and with the origin at the center of gravity is

$$I_{jk} = \int d^3r\, \rho(\vec{r})(r_j^2\delta_{jk} - r_j r_k) \tag{1}$$

The tensor I enters into the equation of motion for the rotation of the solid. This is the Euler equation, which determines the motion of the angular velocity vector $\vec{\omega}$ in coordinates fixed to the body:

$$\frac{dI\vec{\omega}}{dt} = [I\vec{\omega}, \vec{\omega}] \tag{2}$$

where the bracket signifies the vector product. The equation expresses the conservation of angular momentum in the moving coordinates.

We introduce a (dominant) relaxation time τ for global deformations. Let $I_0[\vec{\omega}]$ be the inertial tensor for the equilibrium shape of the globe with a rotation vector $\vec{\omega}$, i.e. for a globe with an increased radius at the corresponding equator. The inertial tensor $I(t)$ relaxes in direction to the one of the equilibrium shape $I_0[\omega(t)]$:

$$\frac{dI(t)}{dt} = -\frac{I(t) - I_0[\omega(t)]}{\tau} \tag{3}$$

The spiral of fig 6 is a numerical solution of the Euler and relaxation equations with τ =1 000 days. The Euler equation is valid for rigid solids only, but since $\vec{\omega}(t)$ varies much more rapidly than $I(t)$, it should hold approximately.

As initial condition, $I(0)$ was calculated for a globe with the increased radius at the equator and in addition a stretching deformation of one per mil in a direction 30° from the rotation axis. A real symmetric tensor can be represented by three axis. In the initial situation the longest axis of $I(0)$ makes an angle with the rotation axis. Then the rotation axis will precess around the main axis of I. This motion is similar, although much larger, than the observed "Chandler wobble". This is a minute geographic motion of the position of the rotation axis on the globe, in which the axis circles irregularly with a Chandler period of about 400 days. If the global deformation relaxes in a time, which is neither very short nor very long compared to this period, the geographic path of the rotation axis will be a spiral that ends at a different place than the starting point. This is the mechanics of a rapid geographic polar shift

3.2 Why was a polar shift considered impossible?

Already in the 19th century geologists wondered about the asymmetry of the glaciations as shown in fig 5. Then the possibility of a rapid geographic shift of the poles was studied by leading physicists including Lord Kelvin, J. C. Maxwell, G. Darwin and G.W. Schiaparelli. They concluded that a shift was impossible. A direct hit of an astronomic object of the required size would liberate an amount of energy incompatible with the continuation of life on Earth. A deformation of Earth, as we assume it, would relax too rapidly to allow the necessary shift. At that time, condensed matter was considered to be either solid or liquid, and in both cases the relaxation time would not be in the range of the Chandler period. Later, F. Klein and A. Sommerfeld (Klein & Sommerfeld,1910) in their compendium on the theory of the gyroscope remark, that there are also substances such as tar. The true relaxation of a deformed Earth to a new equilibrium shape is a complex phenomenon. We simply assume a relaxation time in the good range. Then the two formulas (2) and (3) indicate the essence of the phenomenon. That a rapid pole shift is possible has been pointed out by T. Gold (Gold,1955).

3.3 How could the Earth become stretched?

When a mass is not far from Earth, it accelerates the Earth as a whole, but the side near the mass more and the opposite side less. Thus the Earth is stretched by the tidal force. When the distance between Earth's center and the mass is large compared to the radii of these objects, the tidal force varies with the inverse third power of the distance. If the Moon were 10 times nearer, i.e. at 40 000 km, its tidal force would be 1 000 times larger. For the required pole shift this would not be sufficient. A 10 times bigger mass has to pass at half this distance to produce the 1 per mil stretching. When this happened, it produced a cataclysmic earthquake. Heavy animals could hardly survive.

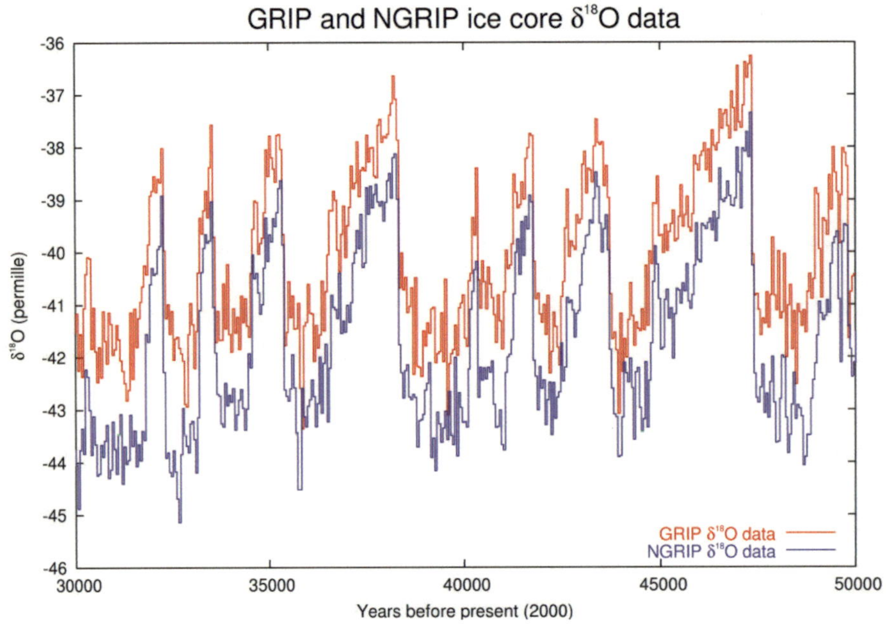

fig 8. Dansgaard-Oeschger Oxygen isotope variations in ice cores of two sites in Greenland

3.4 How could this passing mass disappear?

Earth created a tidal force on Z . Could this disintegrate this Mars sized object? In 1993 the comet Shoemaker-Levy 9 passed near Jupiter and broke into 21 fragments. This event was studied with a computer model by Erik Asphaug and Willy Benz (Asphaug & Benz,1996). Their fig 13 indicates that a Mars sized pile of stones (and therefore probably also a fluid sphere) passing near the Earth would not disintegrate into several fractions. However, if it were hot inside, the resulting pressure would tend to make it explode when deformed.

The evaporation from a planet is determined by gravity. The escape velocity of a fraction is smaller. This highly increases the evaporation rate. As the object loses mass, even molecules

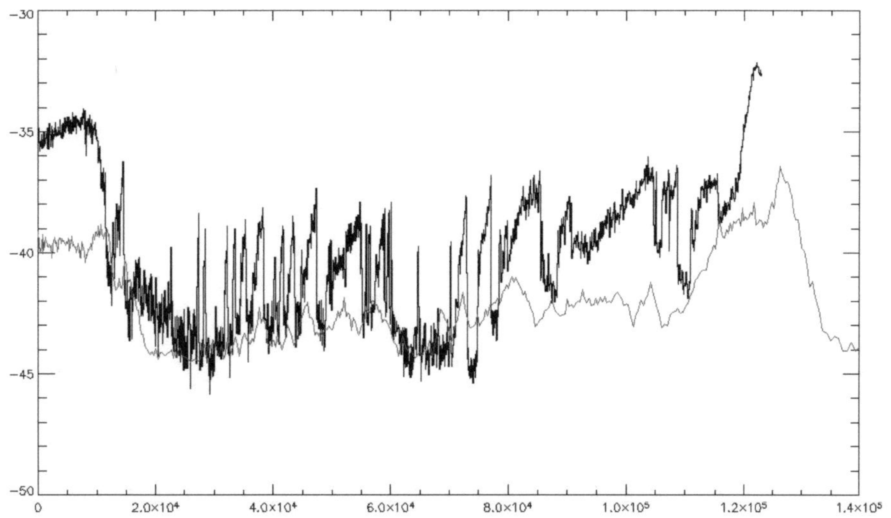

fig 9. Temperatures from Oxygen isotopes in Greenland (black) and Antartica (blue) between present and 140 000 years before present.

and clusters escape. Therefore, a fraction of a Mars sized object could evaporate within the 10 000 years of the Holocene.

The cloud produced by the evaporation of the fractions of Z will influence the climate on Earth during the Holocene. The close passage of Z near Earth scatters the fractions by a few degrees. In the case of the comet Shoemaker-Levy 9 the 21 fractions moved one behind the other practically on the same orbit. This indicates the possibility that during the Holocene the cloud continued to be disk shaped and furthermore that it made an angle with the plane of Earth's orbit.

3.5 How could this planet Z be hot?

Z must be hot, since otherwise it cannot disappear. Its orbit has to be extremely eccentric, so that it passes close to the Sun. Near the Sun it is heated inside by tidal work and outside by solar radiation. In our examples we often used an eccentricity $\epsilon = 0.973$. Then the ratio of the distance to the Sun at aphelion to that at parhelion becomes $(1+\epsilon)/(1-\epsilon) = 73$. During the Ice Ages Z was liquid and shining. It evaporated particles. Due to tidal work the parhelion distance slowly diminished. The cloud of evaporated particles became denser. Therefore, the characteristics of the Ice Age climate gradually increased. This effect is clearly visible in fig 1.

3.6 What were the chances of a close approach?

Since Z was in an extremely eccentric orbit, it passed radially through the sphere at the distance of Earth's orbit. This distance is the astronomic unit $A = 1.5 \cdot 10^8$ km. Let's suppose that the orbit of Z remained near the invariant plane, which is perpendicular to the angular momentum of the planetary system. Let's take near to mean within an angle of ± 1 degree. Then the crossings of Z at distance A are limited by a surface $S = 2\pi A \cdot 2(\pi/180)A$. For a near passage or hit the distance between the centers of Z and Earth have to come closer than $R = 20\,000$ km. The target surface is $T = \pi R^2$. Then $S/T = 4 \cdot 10^6$ is the typical number of

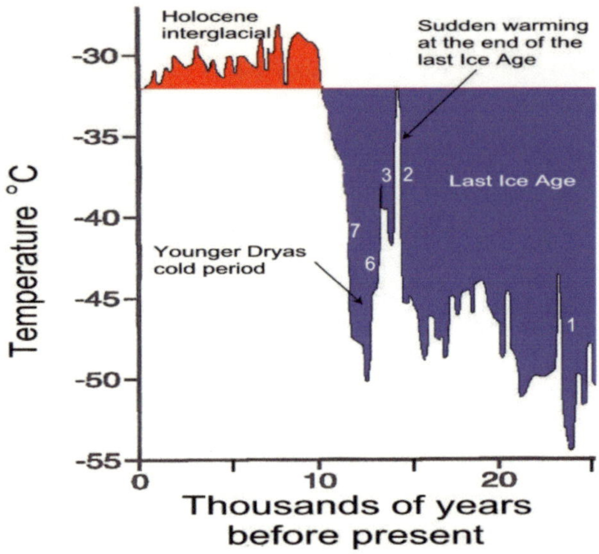

fig 10. The temperature change from Pleistocene to Holocene at GISP2. The change is representative for the northern hemisphere.

passages for one encounter. There is a lot of space in the planetary system! If the period of Z's orbit was of order 1 year, then it is expected to have spent of order 2 million years before the close encounter. This is the order of the time that the Ice Ages lasted.

An actual hit would occur at distances under $R/2$. Fortunately, a close passage between $R/2$ and R is 3 times more probable. Nevertheless, this shows that the scenario could not have happened several times while life on Earth developed. This hot planet must be something exceedingly rare and therefore unlikely to be observed.

The fact that Z is singular corresponds to a rare origin: perhaps a moon of Jupiter or even an object from interplanetary space (MOA & OGLE, 2011).

This theory with a pole shift is a narrow window. The necessary sequence of events is quite well defined. This is actually a strength of this model: if it fits the facts, then this is not adapted by hand. However, we must admit that our estimates were crude, mostly determining orders of magnitude. More detailed estimates should be made. If they differ from ours, they could close the window.

4. The disk shaped cloud around the Sun

4.1 How was this cloud created?

Z had sufficient mass to produce a 1 per mil stretching deformation on Earth during its close passage and it had to disintegrate into fractions in this event. Therefore it must have been hot inside. If its orbit was very eccentric, near the perihelion, say at about 4 million km from the Sun, it was heated inside by tidal work and on the illuminated surface by solar radiation. Evaporation was possible for particles which exceeded the escape velocity, which must have been about 5 km/s. As an example, for an ^{16}O atom the corresponding kinetic energy is 2.1

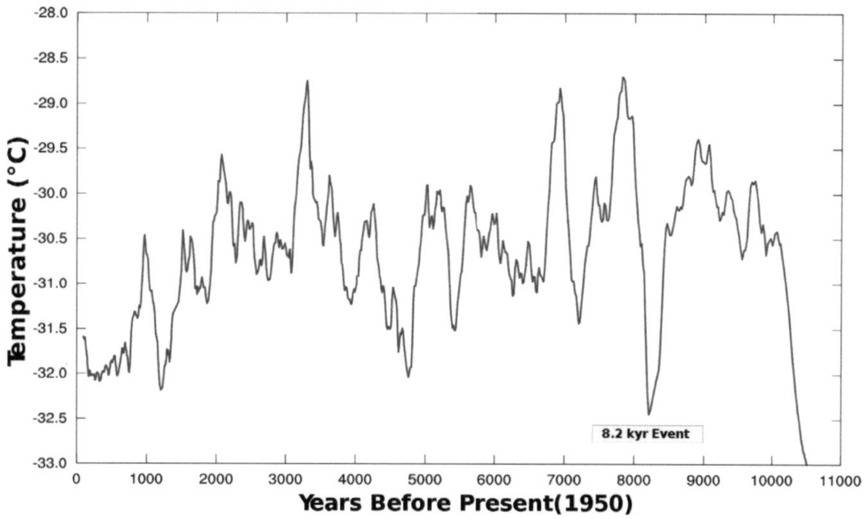

fig 11. Temperatures in the Holocene from isotope variations in ice at the Greenland station GISP2.

eV. Even for a temperature of $T = 1000°$ this is 16 times larger than the mean thermal kinetic energy $\frac{3}{2}k_BT = 0.13$ eV. Evaporation against gravity favors light particles. The evaporated elements have a strong isotope effect.

The speed of Z is much higher than thermal velocities. The emitted particles start their journey practically with the velocity of Z. They are attracted to the Sun by gravity and repelled by the light pressure, which also decreases with the inverse square of the distance. Effectively, the particles suffer a reduced gravitational attraction. If light pressure on particles were a purely geometric question, the cross section of a spherical cluster would increase with the square of the diameter and the mass with the cube. Then for a density of 1000 kg/m^3, light pressure would dominate gravity, when the diameter of the sphere is less than $1.16 \cdot 10^{-6}$ m. At present in the solar system a weak dust cloud exists, which is visible in a dark night as Zodiacal light. This dust is due to asteroids, which disintegrate near the Sun.

The cloud produced by the evaporation from Z is quite different. Scattering of light from small objects is not a geometric problem. For atoms, ions or molecules quantum phenomena rule. If the particle has a dipole transition in the main solar spectrum, light pressure dominates. For that reason molecules and atoms (with the possible exception of Helium) cannot remain in an orbit around the Sun. Actually, molecules dissociate and atoms ionize. Then excitations by the solar radiation become quite rare. Some weak light scattering by the ions will persists from resonant scattering near the high end of the solar spectrum or from forbidden transitions.

4.2 How did the cloud disappear?

When Earth's orbit is in the cloud, it receives less solar radiation. A decrease of a few percent reduces the global temperature to the level of the cold periods.

The cloud is composed of ions in orbits, which are slightly different from that of Z due to thermal velocities and the interaction with solar light. The density of ions in the cloud

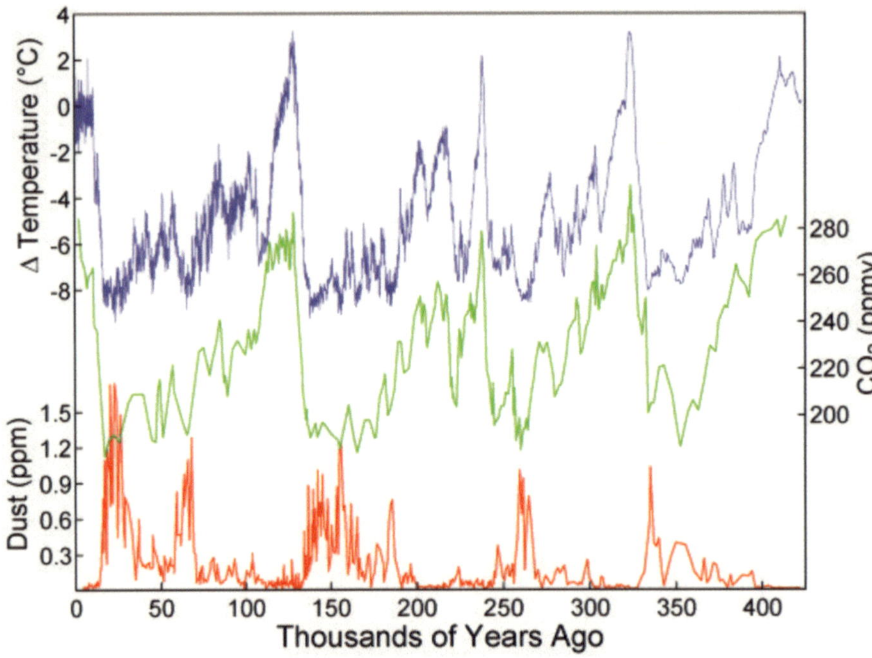

fig 12. Temperature Variations (blue), CO_2 (green) and Dust content (red) in an ice core from the Vostok station in Antarctica over the last 420 000 years. Data from (Petit et al.1999)

increases with each passage of Z near the Sun. Since the particles have planetary velocities, collisions between ions can be inelastic. They convert kinetic energy into light. The reduced orbits have a higher probability for further collisions. Light carries little momentum, so that these processes end in small orbits with the initial angular momentum (Woelfli& Baltensperger, 2007a). The Poynting-Robertson drag (Gustafson, 1994), a sort of friction with light, finally leads the particles into the Sun.

We frankly admit that this cloud of ions and electrons, this plasma, may have many properties unknown to us. Just imagine how a purely theoretical study of water vapor in Earth's atmosphere could miss the clouds and their amazing variety, which determine the weather. Possibly the plasma lowers the energy of electric fields diminishing the width of the cloud. Then this width could be smaller than what is expected from the initial thermal velocities of the particles. For us, the question whether electric discharges exist in the cloud, is open.

4.3 What caused the Dansgaard-Oeschger events?

Fig 8 shows isotope variations due to temperature changes of the northern hemisphere, which have first been described by Dansgaard and Oeschger (Dansgaard, 1993). The periods are too short to be connected with Milankovitch cycles of Earth's orbit. In the dominant model these variations are attributed to changes of water streams on the surface and depth of the oceans (fig 7).

We suggest that the gradual increase of the cloud's density followed by its collapse corresponds to the lowering of the global temperature followed by its rapid increase.

Fig 9 shows, that the Dansgaard-Oeschger temperature variations are much smaller in the Antarctic than in Greenland. This may be due to stronger feedbacks on the northern hemisphere.

4.4 Why was the Holocene mostly warmer than the present?

Various aspects of climate variability of the last 10 000 years, i.e. the Holocene, have been examined (Mayewski et al.,2004). Global temperature variations were found. Fig 11 shows data from the GISP 2 station: the temperature was mostly higher during the Holocene than at present. The main stream theory does not offer processes, that would explain the large temperature increases at the GISP2 station over the present temperature. The CO_2 content varied. However, the changes of CO_2 content followed the temperature variations and could not be the cause. Warmer oceans emitted CO_2 into the atmosphere and colder oceans absorbed it. The temperature varied too rapidly to be connected with Milankovitch changes of Earth's orbit.

In our model (Woelfli & Baltensperger,2007) a close approach of Z with Earth happened at the End of the Pleistocene. The hot planet disintegrated into several fractions, which evaporated during the Holocene. Therefore an intense disk shaped cloud existed around the Sun. Due to the scattering at the close encounter the plane of the orbits of the fractions probably made an angle of one or several degrees with the plane of Earth's orbit. Each year the Earth crossed this cloud twice, but most of the time it was outside and received scattered light. Since the outgoing radial component of the radiation flux is diminished inside the cloud, it must be larger outside. In addition the scattered light has tangential components. Therefore, outside the cloud there is additional radiation.

Thus the cloud cools the Earth in cold periods and it warms it in warm periods. In this last case, an increase in the clouds density increases the temperature gradually and the cloud's collapse diminishes it abruptly. Warm periods have reverse Dansgaard-Oeschger events with gradual warming and rapid cooling. In fig 11 this tendency is recognizable in contrast to the cold periods of fig 8 or fig 9.

4.5 Why was the last interglacial warmer than the present?

As seen in fig 12 the cold periods were separated by warm periods roughly with a 100 ka periodicity. The last interglacial began 127 kyr ago and lasted for about 9 000 years. It was warmer than the present time. Forests reached the North Cap, where now there is tundra. The sea level was 4 to 6 m higher. This indicates warmer oceans and less glaciation.

The cloud again offers an explanation. When the inclinations of Earth's orbit and of the cloud differed sufficiently so that the Earth was outside the cloud, then the Earth received solar light scattered by the cloud. Half a year one side of the Earth was illuminated and, after crossing the cloud, half a year the other side. Since the light comes from a planar cloud and Earth's obliquity to the orbit is only about 23°, this additional illumination changes essentially from one hemisphere to the other. When within the year this change happens depends on unknown details of the orbits. The additional illumination would be most effective, if each hemisphere received it during its winter.

What was the color of the scattered light ? This is not an easy question, since the particles in the cloud are ions without a dipole transition in the main solar spectrum. That may be an argument for the violet end of the solar spectrum. However, weak transitions of lower energy could also contribute.

4.6 Why were small grained inclusions in ice cores dense in cold periods?

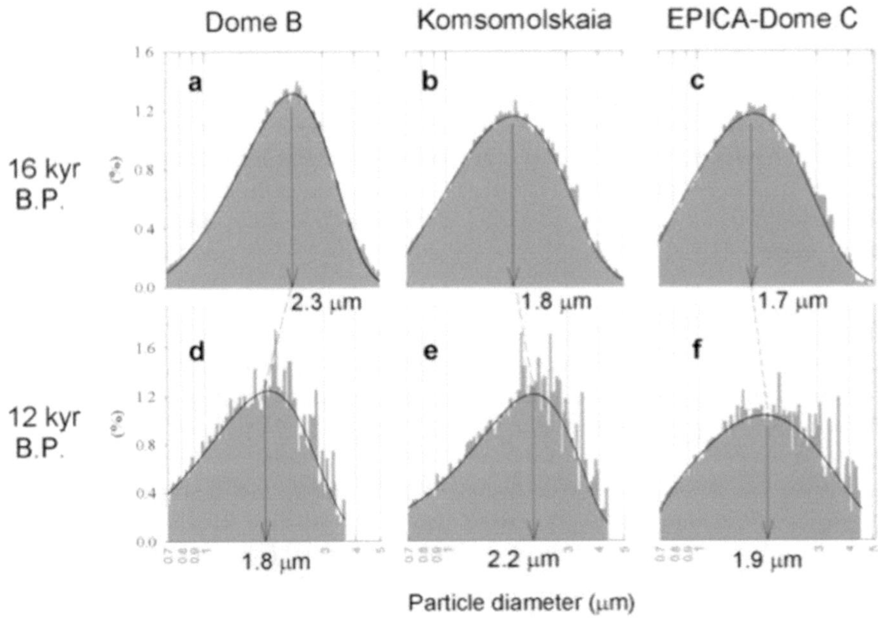

fig 13. Distribution of the small grained dust (radius $< 4\mu$m) from 3 Antarctic sites and two dates. Data from (Delmonte et al.2003).

The red curve of fig 12 represents the dust content in ice cores of the Vostok station in Antarctica. A comparison with the blue curve shows that the dust density is high during cold periods. Planet Z theory offers this interpretation: in cold periods Earth's orbit is in the cloud, so that particles enter into Earth's atmosphere, where they coagulate and are included in precipitation or sink to the ground.

The dust in the ice cores contains small (radius $< 4\mu$m) and large grains. The small grains have a bell shaped distribution as seen in fig 13. This regular distribution may result from a coagulation in the upper atmosphere (Baltensperger& Woelfli, 2009). In this case, the small grained dust would be extraterrestrial. The evaporation from Z would modify its isotope distributions. We suggest that this should be examined using the three stable isotopes of Mg. The absence of an isotope effect in Mg would be a severe difficulty for our theory. In order to determine the origin of the small grains the isotope distributions of Sr and Nd has been compared with that from samples from many regions of the globe (Delmonte et al., 2004b). An agreement was found between material from Antarctica and from Patagonia. In Table 1b of (Delmonte et al., 2004b) the samples of South America are dated; those of the Pampas have ages between 10 and 25 kyr. It was concluded that the small grains have been transported from Patagonia to Antactica (Delmonte et al., 2004b; Gaiero, 2007). However, it is also possible that the grains from the two regions have the same extraterrestrial origin.

Large dust particles in ice cores have more irregular size distributions. They have been examined carefully, since their mineralogical structure reveals their origin. The large grains from antarctic sites come from Patagonia, while those from Greenland originated in the Gobi desert (Biscaye et al., 1997). They must have been transported by storms.

5. The Milankovitch cycles

5.1 What are Milankovitch cycles?

The mass of the Sun is about 1000 times larger than the mass of all its planets. Therefore, the motion of a planet is an approximate two-body problem: planets move in Kepler orbits. Milankovitch considered the gravitational interactions between the planets and found that the parameters of the Kepler orbits suffered slow variations. Typically, the variations of an orbital parameter are dominated by one or a few frequencies. These are the Milankovitch cycles.

5.2 What produced the 100 000 year cycle of the Ice Ages?

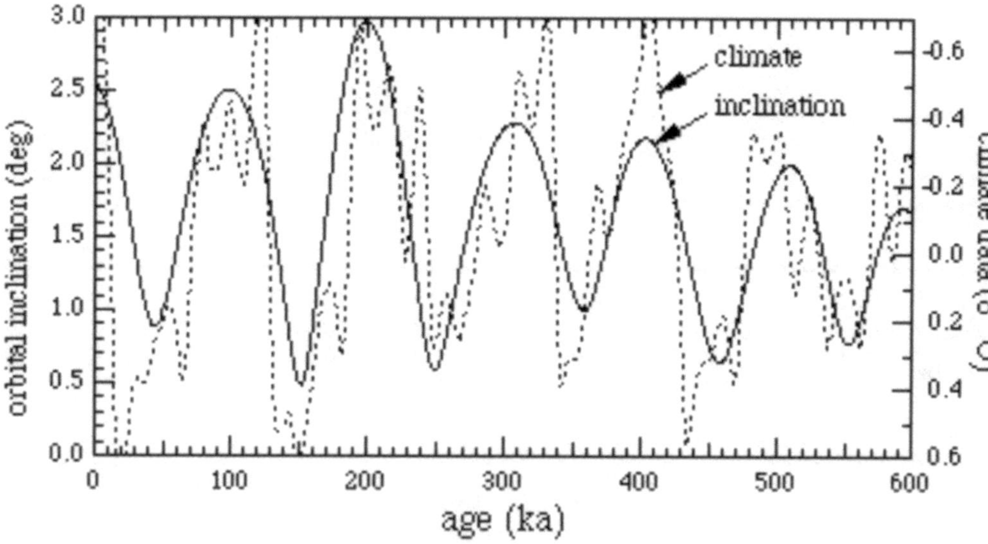

fig 14. The calculated inclination of Earth's orbit (full line) and the isotope variation $\delta^{18}O$ (dotted line) from (Imbrie et al., 1984) for 600 000 years. Figure from (Muller& MacDonald, 1995)

Figures 12, 14 and 15 show that 100 000 years is the strongest period of the Ice Ages. Earth's inclination, i.e. the angle between the plane of Earth's orbit and the invariant plane (perpendicular to the angular momentum vector of the planetary system) varies with this period. In view of this, R.A. Muller and G.J. MacDonald (Muller& MacDonald, 1995; 1997) postulated the existence of a disk shaped cloud around the Sun. Depending on the value of the inclination, the Earth would be inside the cloud and therefore in a cold period or outside in a warm period. In the planet Z theory, this cloud necessarily exists before the pole shift.

The dominant theory is based on the Milankovitch cycles. However, changes of the inclination are not relevant in this theory, since the Sun radiates with equal strength in all directions.

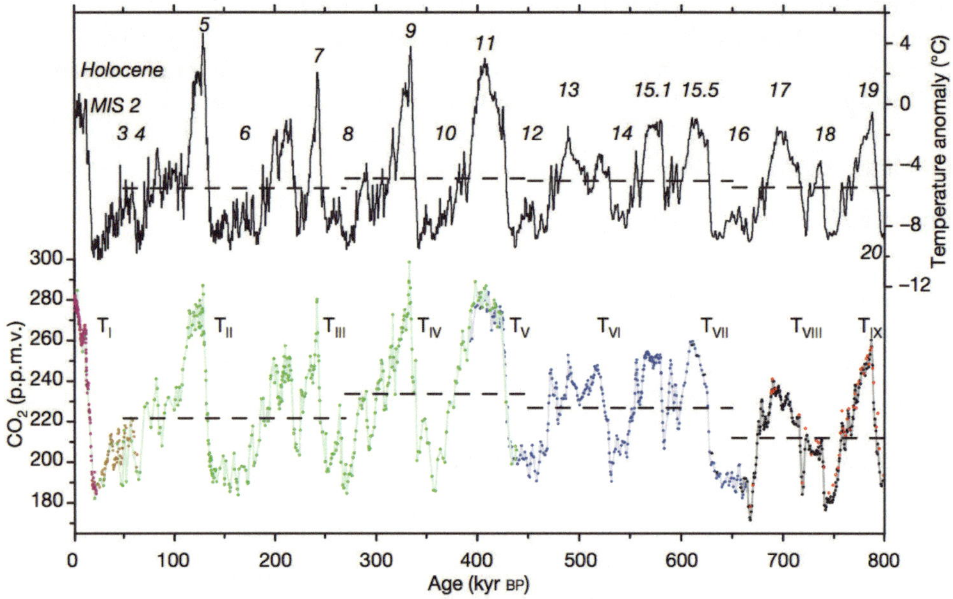

fig 15. Temperature anomaly with respect to the mean temperature of the last millennium (EPICA from antarctic Dome C) and CO_2 record from several research groups. A tendency of a sharp increase in temperature followed by a gradual decrease exists in the 100 kyr period. Figure from (Luethi & al., 2008)

Earth's orbital eccentricity has periods at 95 kyr, 105 kyr and 400 kyr. The problem is that with Earth's small eccentricity (< 0.05) the variations have little effect on the insulation. If one argues that strong feedbacks could amplify these effects, then it remains unclear, why the 400 ka period is not visible. Thus, the strongest period of the Ice Ages of the last million years remains unexplained in the dominant theory.

5.3 What determined the shape of the 100 kyr cycles?

fig 12 and 15 show that the 100 kyr cycles have a similar behavior as the Dansgaard-Oeschger events in fig 8: a rapid increase in temperature is followed by a gradual decrease. We do not know the origin of this. If in Dansgaard-Oeschger events the cloud builds up in a time of about 2 kyr and then collapses, this cannot be responsible for a 100 kyr buildup. There must be a climate relevant process on Earth or in the cloud or in their coupling, which requires 100 kyr for its formation. Does this shape simply indicate that creating ice in the cold takes more time than melting it?

5.4 Why could the 40 000 year cycle be dominant before 1 million years?

The 100 kyr cycle was only dominant in the last million years. Before that time a 40 kyr cycle appears. Since the Earth is essentially a sphere and its orbit nearly a circle, the total insulation on Earth is essentially constant. However, the axial tilt, the angle between Earth's axis and the perpendicular direction to the orbital plane, varies between 22.1° and 24.5° with a 41 kyr cycle. The larger this obliquity the more winter differs from summer on both hemispheres. Since the hemispheres differ from each other and feedbacks depend on the seasons, the obliquity affects

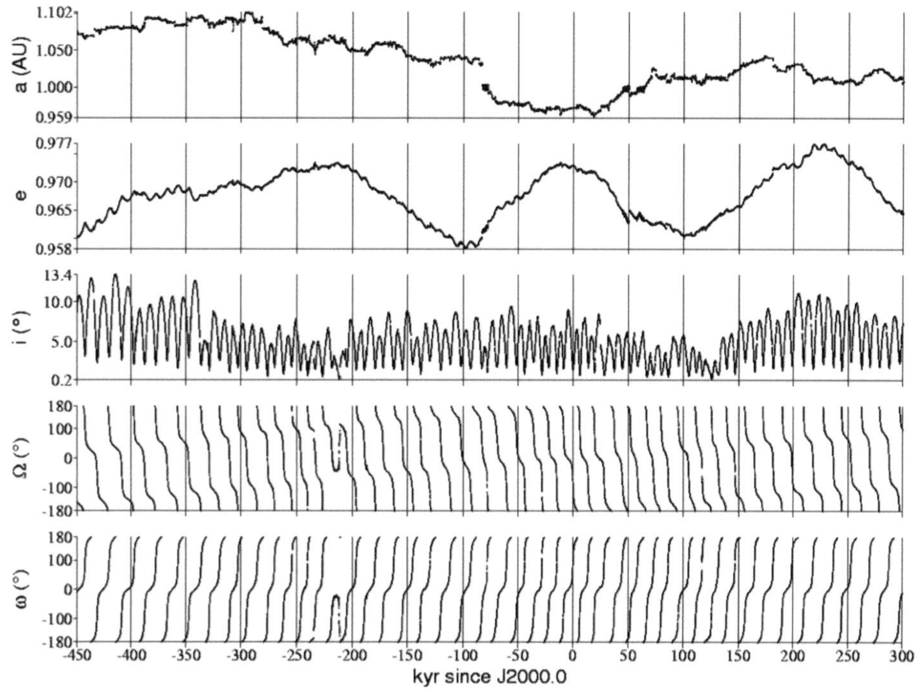

fig 16. Calculated Milankovitch variations of the orbital parameters of Z with assumed mass $0.11M_E$. (a, semi-major axis; e, eccentricity; i, inclination; Ω, longitude of the perihelion; ω, argument of the perihelion) over 750 kyr. Note the short cycles of the inclination. Details of the calculation are found in section 2 of (Nufer et al., 1999).

the mean temperature. Evidently, an explanation of the 41° cycle involves detailed knowledge of the climate system.

The disk shaped cloud due to Z explains the 100 kyr climate cycle by the inclination cycle. However, in the theory with a pole shift we only know Earth's orbit after the cataclysmic encounter. Then Earth was scattered and this could change the angle of its motion by a sizable fraction of a degree. The fact, that the calculation backward in time is made with an invalid orbit may be a reason for a failure for large times. Furthermore, the existence of Z may have an unknown influence on Earth's orbit. In the theory with a pole shift long backward calculations are doubtful.

5.5 What about Milankovitch cycles of Z ?

For Z in an eccentric orbit a calculation of the Milankovitch variations is reported: fig 16 (Nufer et al., 1999). In this example, the inclination of Z shows variations of 10° with periods less than 10 kyr. Consequently, a single particle of the cloud emitted from this Z should have similar variations. However, in the interpretation of the 100 kyr climate cycle the cloud behaves like a disk in the invariant plane. For this to be true, the particles must be coupled somehow in this plasma of ions and electrons, so that their variations average out. At present this is an open problem.

In fig 11, occasionally, the temperature drops for times of the order of a few hundred years, notably at the 8.2 kyr Event. Then Earth entered the cloud in a motion too rapid for Earth's periods. In this case evidently the cloud continued to grow, since afterwards the temperature had gradually increased, as it should in a reverse Dansgaard–Oeschger event.

6. Does this paper lead to conclusions?

This depends on the reader, especially since two very different theories are confronted.

The dominant theory is based on the Milankovitch cycles. A wealth of research has shown that Milankovitch cycles appear in the climate data. The production of these data are one of the truly great scientific achievements of our time. In our view an influence of the Milankovitch cycles on climate is undisputed. The question is, whether the Ice Ages were produced by the Milankovitch variations and Earth's climate system alone. The dominant theory assumes that this is the case. It claims that feedbacks amplify small differences of the radiative input to the extend that the climate system can enter into a different mode.

The starting point of the alternative theory with a pole shift is the asymmetry of the glaciation with respect to the present position of the North Pole, as seen in fig 5. We claim that the geographic pole shift shown in fig 6 is both necessary and possible. It requires a hot planet, which produces a disc shaped cloud of ions around the Sun.

In a way the situation resembles that of a century ago, when the energies of electrons emitted by light from a metallic surface could not be explained by Maxwell's equations. These well proven equations continued to be valid, but something else had to be introduced: quantization. Quantum theory became a vast science with many applications. It was somewhat crazy at the beginning and continues to be so at present.

We claim that the Milankovitch cycles are valid, but a hot planet Z has to be introduced. Z was probably unique in the history of the Solar system, since it involved a danger of collision with Earth, which would have stopped the development of life. The requirements on Z are stringent. This theory is not a vast field, but a narrow possible window. Although the theory involves only known scientific concepts, the uniqueness of Z makes it look somewhat crazy.

If you accept Z, you understand

- the asymmetry of the glaciation (fig 5),
- the gradual lowering of the temperature and increase of the fluctuations during the Ice Ages (fig 1),
- the fact that the Ice Ages lasted for a time of order a few million years (fig 1),
- the rapid geographic shift of the poles, a cataclysmic event, which killed heavy animals (fig 6),
- the dominant 100 kyr cycle of the ice ages (fig 14),
- the high density of small grains in ice cores during cold periods (fig 12),
- the gradual decrease and rapid increase of the temperature of Dansgaard-Oeschger events (fig 9),
- the enhanced temperatures during the Holocene (figs 10 and 11) and the last interglacial (fig 12) as compared to the present,
- the reverse Dansgaard-Oeschger events (i.e. gradual increase and rapid decrease of the temperature) in the Holocene (fig 11),

This list contains basic properties. There are may special facts that become plausible: e.g. northern Amazon was a desert (Filho et al., 2002), because before the pole shift its latitude was that of the present Sahara, or Bolivia was humid (Baker, 2001), because it was at the equator, or lake Baikal was not frozen during the whole year in cold periods (Kashiwaya, 2001), since it was about 10° further south than at present.

So far the theory of the pole shift has only been sketched. Some of our estimates were just order of magnitude considerations. Further work would be highly valuable. Research on the Ice Ages is a detective story. A scientist should be able to keep more than one theory in his mind. Also, a hypothesis must be excluded by scientific arguments only.

The concept "reverse Dansgaard-Oeschger event" visible in temperature variations of the Holoceen, fig 11, and its interpretation may be new. Otherwise, the aim of this chapter was didactic.

7. References

Alley R.B., (2000). The Younger Dryas cold interval as viewed from central Greenland. *Quaternary Science Reviews,* 19, 213–226.

Asphaug E. & Benz W. (1996). Size, density and structure of comet Shoemaker– Lewy-9 inferred from the physics of tidal breakup. *Icarus* 121, 225–248.

Baker A., Rigsby C.A., Seltzer G.O., Fritz Sh.C., Lowenstein T.K., Bacher N.P., Veliz C., (2001). Tropical climate changes at millennial and orbital time scales on the Bolivian Altiplano. *Nature* 409, 698–701.

Baltensperger W. & Woelfli W. (2009). Interpretation of the small grains in the inclusions of ice cores. URL: arxiv.org/abs/0909.5089

Biscaye P.E. et al. (1997). Asian provenance of glacial dust (stage 2) in the Greenland ice sheet project 2 ice core, Summit, Greenland. *Jour. Geophys. Res.* 102, 26 765–26 781.

Bocherens H. (2003). Isotopic biogeochemistry and the paleoecology of the mammoth steppe fauna. *Advances in Mammoth Research,* Deinsea 9, 57–76.

Clemens S.C. & Tiedemann R. (1997). Eccentricity forcing of Pliocene - Early Pleistocene climate revealed in marin oxygene-isotope record. *Nature* 385, 801–804.

Dansgaard W. et al. (1993). Evidence for general instability of past climate from a 250-kyr ice–core record. *Nature,* 364, 218–220.

Delmonte B. et al. (2004). Dust size evidence for opposite regional atmospheric circulation changes over east Antarctica during the last climatic transition. *Climate Dynamics* 23, 427–438.

Delmonte B. et al., (2004). Comparing the Epica and Vostok dust records during the last 220 000 years: stratigraphical correlation and provenance in glacial periods. *Earth Science Reviews* 6, 63–87.

Ehlers J. & Gibbard P. (2004). Quaternary glaciations – Extent and Chronology. *Elsevier.*

Filho A. C., et al. (2002). Amazonian Paleodunes Provide Evidence for Drier Climate Phases during the Late Pleistocene-Holocene. *Quaternary Research* 58, 205–209. DOI: 10.1006/qres.2002.2345.

Gaiero Diego M., (2007). Dust provenance in Antarctic ice during glacial periods: From where in southern South America? *Geophysical Res. Lett.,* 34, L17707, doi:10.1029/2007GL030520.

Gold T. (1955). Instability of the Earth's axis of rotation. *Nature* 175, 526–529.

Gustafson Bas, (1994). Physics of zodiacal dust. *Annual Rev. Earth and Planet. Science* 22, 552–554.

Imbrie J.J., (1984). *Milankovitch and Climate,* Part 1, (eds. A. L. Berger et al.), pp. 269–305.

Kashiwaya K, Ochiai S, Sakai H, Kawai T, (2001). Orbit-related long-term climate cycles revealed in a 12-Myr continental record from Lake Baikal. *Nature* 410, 71–74.

Klein F. & Sommerfeld A. (1910). Über die Theorie des Kreisels. *Teubner,* Leipzig.

Luethi D. et al. (2008). High–resolution carbon dioxide concentration record 650 000–800 000 years before present. *Nature* 453, 379–382.

Mayewski Paul A. et al. (2004). Holocene climate variability. *Quaternary Research* 62, 243-255.

Milankovitch M.M. (1941). Kanon der Erdbestrahlung und seine Anwendung auf das Eiszeitproblem. *Königliche Serbische Akademie,* Spez. Publikation 133, 1–633, Belgrad.

The Microlensing Observations in Astrophysics (MOA) Collaboration & The Optical Gravitational Lensing Experiment (OGLE) Collaboration, (2011). Unbound or distant planetary mass population detected by gravitational microlensing. *Nature* 473, 349–352.

Muller R.A. & MacDonald G.J. (1995). Glacial cycles and orbital inclination. *Nature* 377, 107–108.

Muller R.A. & MacDonald G.J. (1997). Glacial cycles and astronomical forcing. *Science* 277, 215–218.

Nufer R., Baltensperger W. & Woelfli W. (1999). Long term behaviour of a hypothetical planet in a highly eccentric orbit. URL http://xxx.lanl.gov/abs/astro-ph/9909464

Orlova L.A. et al. (2001). Chronology and environment of woolly mammoth (Mammuthus primigenius Blumenbach) extinction in northern Asia. *The World of Elephants,* International Congress, Rome, 718–721.

Petit J.R. et al. (1999). *Nature,* 399, 429–436.

Schirrmeister L. et al. (2002). Paleoenvironmental and paleoclimatic records from permafrost deposits in the Arctic region of Northern Siberia. *Quaternary International* 1998, 97–118.

Seager R. (2006). The Source of Europe's Mild Climate. *American Scientist* 94, 334–341.

Snowball bibliography, URL: www.snowballearth.org/bibliography.html

Tiedemann R., Sarntheim M. & Shackleton N.J. (1994). Astronomic time scale for the Pliocene Atlantic ^{18}O and dust flux records of Ocean Drilling site 659. *Paleoceanography* 9, 619–638.

Woelfli W.; Nufer R. & Baltensperger W. (2002). An additional planet as a model for the Pleistocene Ice Age. URL: arxiv.org/abs/physics/0204004

Woelfli W. & Baltensperger W. (2007a). Arctic East Siberia had a lower latitude in the Pleistocene. *Anais da Academia Brasileira de Ciências,* 79(2): 183–193.

Woelfli W. & Baltensperger W. (2007b). On the change of latitude of Arctic East Siberia at the end of the Pleistocene. URL: arxiv.org/abs/0704.2489

7

Origin of HED Meteorites from the Spalling of Mercury – Implications for the Formation and Composition of the Inner Planets

Anne M. Hofmeister and Robert E. Criss
Department of Earth and Planetary Sciences, Washington University in St Louis, MO, USA

1. Introduction

The bulk chemical composition of the Earth is poorly constrained because samples are limited to only the outer 10% of its radius. Meteorite data are used to represent the huge zones that we cannot sample, but for this approach to provide a realistic model of the Earth, we need to know where various meteorites originated and what processing they underwent. Currently popular models of the Solar System presume that the Sun formed first, followed by assembly of the planets from mineral dust in a disk that condensed locally from hot gas (e.g., Boss, 1998). More recent findings that comets are mixtures of phases formed at high and low temperature (Zolensky et al., 2006) came as a great surprise to planetary scientists because this contraindicates the existence of a condensation gradient as proposed by Lewis (1974). The alternative to dust condensation around a pre-formed Sun is a nebula containing dust inherited from older generations of stars. Meteorites possess small amounts of isotopically distinctive pre-solar material, proving that the nebula included some dust that predates our Sun (e.g., Bernatowicz & Zinner, 1997). Recent detection of silicate dust in molecular clouds (van Breeman et al., 2011), which are loci of star formation, suggests that pre-solar dust was abundant. Disk models also have problems conserving angular momentum (Armitage, 2011) and fail to explain the first order characteristics of the Solar System (upright axial spin and nearly circular orbits). Thus, independent evidence points to formation of planets and Sun simultaneously from a 3-dimensional dusty nebula (Hofmeister & Criss, 2012), in which case, virtually all meteoritic material has been processed. What represents the bulk Earth?

It is generally accepted that chondritic meteorites are very primitive material, and bulk compositions of the silicate Earth have been derived from various averages. Models based on enstatite chondrites (Javoy, 1995; Lodders, 2000), which have Earth-like oxygen isotopes and Fe-metal content, agree with heat flux measurements (Hofmeister & Criss, 2005). Newly measured flux of neutrinos from within the Earth (Gando et al., 2011) support the enstatite chondrite model. However, because even the most primitive meteorites show some evidence of alteration and are comprised of non-equilibrium assemblages (e.g., Brearley & Jones, 1998), bulk compositions so derived are estimates, not exact representations of the Earth. Hence, it is worthwhile to consider alternative means of inferring bulk composition.

The present paper considers how more processed meteorites (the differentiated achondrites) may be related to the Earth. We demonstrate that the largest class of achondrites (the HEDs) has been wrongly assigned to an asteroidal parent body: a planet is required. The likely source is evolving Mercury, which suffered severe impacts that ejected the majority of its mantle to space (Section 1.2). This connection permits us to decipher information about the Earth during core formation, a very early and important step in its history which is not documented in terrestrial samples, and to estimate bulk planetary composition.

1.1 Problems in assigning HED meteorites to the asteroid Vesta

Many different classes of achondrites are recognized, but most of these classes contain few meteorites (e.g., Mittlefehldt et al., 1998, summarized in Table 1). The largest single class of achondrites, the HEDs (= howardite, eucrite, and diogenite) is a distinctive petrographic family whose affinity has been confirmed by oxygen isotopes (e.g., Taylor et al., 1965). Eucrites are similar to pigeonite-plagioclase basalts and basaltic cumulates, while diogenites are similar to orthopyroxenites, and howardites are brecciated mixtures of these two types (e.g., Dymek et al., 1976; Shearer et al., 1997). It is further recognized that mesosiderites, pallasites, and IIIAB iron meteorites are geochemically similar to each other and to HEDs (e.g., Duke & Silver, 1967; Clayton & Mayeda, 1996). Section 5 provides further discussion.

Main type	Description	Type	Mineralogy	ρ, g cm^{-3}	Number	Total kg
Chondrites	unmelted stones*	ordinary	pyroxene, olivine	3.3	36326	43000
		carbonaceous	hydrated, C	2.1-3.5	1385	3281†
		enstatite	~$MgSiO_3$, albite	3.6	506	850
Achondrites	melted stones	HEDs‡	basaltic, see text	2.9-3.3	1033‡	1250‡
		Martian	pigeonite, olivine	3.2	99	100
		Lunar	anorthite, basalt	2.7, 3.8	146	60
		other§	as above, fassaite	~3	576§	4700§
Stony- irons	Metal+ silicates	mesosiderites‡	HED minerals	4.2	170‡	6300‡
		pallasites‡	olivine	4.8	89‡	13100‡
Irons	Fe with<20% Ni	non-magmatic	most have low Ni	7.5	280	150000
		magmatic IIIAB‡	most have low Ni	7.5	288‡	136000‡
		other magmatic	variable Ni	7.5	485	164000

* Most contain chondrules (spheres of once melted material, typically olivine+pyroxene+feldspar rim or of forsterite+enstatite+Fe^0), calcium-aluminum-rich inclusions, and/or pre-solar grains.
† Virtually all carbonaceous chondrite mass is in the Allende meteorite (3000 kg).
‡ Part of the HED family, sensu lato.
§ Of the remaining achondrites, 298 are C-rich ureilites but 67 aubrites (related to enstatite chondrites) have 4300 kg of the mass. Brachinites and angrites have chemical affinities to HEDs (Section 5).

Table 1. Summary of meteorite types with simple descriptions. Density from Britt and Consolmagno (2003). Number and mass from *The Meteoritical Bulletin* database (www.lpi.usra.edu/meteor/metbull.php, accessed 9/26/2011).

Sourcing the HEDs to the 3rd largest asteroid, Vesta, is based on spectral comparisons (e.g., Gaffey, 1997). Slight variations in spectra across its surface were interpreted as differing proportions of pigeonite to other phases. Detection of a basaltic mineral was taken as evidence that the asteroid had differentiated, and thus has a core (e.g., Ruzika et al., 1997).

Attribution of HED meteorites to Vesta and a few spectrally similar asteroids, despite weak evidence for orthopyroxenes or olivine (e.g., Burbine et al., 2001), sets the view that this tiny object is a major parent body. Spectral variations do not require differentiation: 1) dust on the surface could be non-uniform in grain-size, distribution, or mineral proportions, 2) Vesta could be a remnant from a larger body, or 3) Vesta could be a conglomerate, as are many meteorites. Section 2 shows that Vesta is too small to sort phases by density in its own gravitational field. Section 4 shows that grain-size variations explain Vesta's spectra.

The Vesta hypothesis is counter-evidenced by the abundance of HEDs, which outnumber all other achondrites combined (Table I). Assignment of the HEDs to Vesta further requires a few asteroids provide IIIAB irons, and most mesosiderites and pallasites, which constitute a huge proportion of the collection (Table I). It is far more likely that Vesta and similar asteroids are cousin fragments of a much larger parent body.

Dynamical arguments that were set forth against proposals of Mars as the source of SNC meteorites (e.g., Vickery & Melosh, 1983) were refuted by subsequent isotopic and chemical evidence (summarized by Grady, 2006). Recognition of lunar and Martian meteorites, coupled with Mercury being of intermediate size to these parent bodies led Love & Kiel (1995) to suggest that some Mercurian samples may reside in available collections. Although impact calculations have been extensively revised to explain the existence of lunar and Martian meteorites, the view is nevertheless still held that dynamics forbid Mercury as a meteorite source (e.g., Melosh & Tonks, 1993). Clearly, dynamical arguments can neither prove nor preclude planetary ejection, let alone establish transport. Problems with such modeling are covered in Sections 1.2 and 3.

1.1.1 Is Vesta a mesosiderite? Was Vesta highly impacted?

Low orbital eccentricity of ~0.09 for Vesta and Ceres, compared to an average of 0.26 for 25 large asteroids which is similar to those of Pallas and Juno (http://nssc.gsfc.an.gov), indicates that Vesta was not more heavily impacted. We propose that Vesta is a mesosiderite with ~20% iron, based on its density of 3.9 g cm^{-3}, and spectra (Section 4), and represents below of 1% of the mass ejected from Mercury. High iron content provides material strength. The lumpy shape resembles that of iron-rich meteorites.

1.2 Is early Mercury the source of HEDs and related meteorites?

Mercury's unique characteristics are its high orbital eccentricity (0.2056 compared to 0.0067-0.094 for the other seven planets), the large size of its Fe core compared to its mantle, and its similarity to the Moon, with craters upon craters, ejecta, impact basins, and smooth plains that are probably volcanic (e.g., Spudis & Guest, 1988). Delivery of immense kinetic energy in an impact is evidenced by the Caloris basin, and its antipodal structure.

Various hypotheses attempt to explain Mercury's special characteristics. Attribution of Mercury's Fe rich composition to high-temperature nebular condensates being located near the Sun (Lewis, 1974) cannot be the cause because condensation temperatures of Fe^0 and Mg_2SiO_4 (forsterite) only differ by a few degrees (Lodders, 2003) which precludes their sorting during condensation. Attribution of Fe-enrichment to vaporization of its surface (Cameron et al., 1988) does not address high orbital eccentricity and requires excessively high surface temperatures. It is difficult to reconcile the topography with vaporization. Loss of Mercury's mantle through catastrophic fragmentation following many giant impacts, in which most of the material was reassembled while Mercury's orbit ranged dramatically, including beyond the Earth (Wetherill, 1988) has problems as regards orbital excursions, and incorrectly assumes that the impactor plus Mercury constituted a bound state, see Section 3.

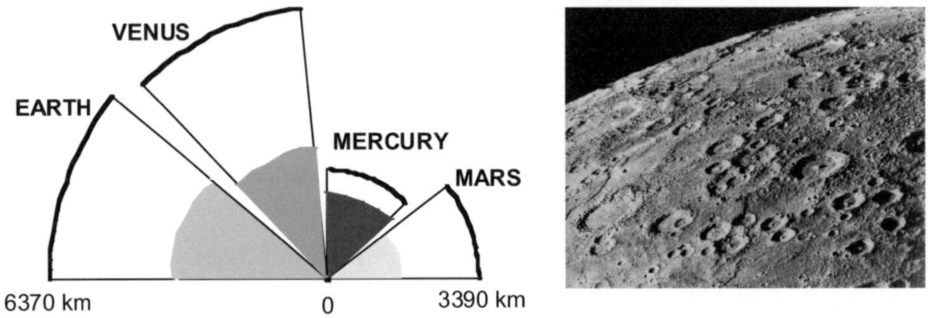

Fig. 1. Mercury's internal structure and surface features. Left = comparison of the terrestrial planets to scale. Surfaces = heavy lines. Metal cores = shaded sections. Data from Lodders & Fegely (1998). Right = photo of a limb of Mercury taken by the Mariner spacecraft. The lower edge shows about 560 km across. From http://history.nasa.gov/EP-177/ch2-2-2.html.

Hofmeister and Criss (2012) recently proposed a different history for Mercury, whereby repeated impacts stripped away the mantle through the late heavy bombardment (LHB) that occurred up to 800 Ma after formation (Gomes et al., 2005). Given the physical characteristics of Mercury, its proximity to the gravitational focus of the Solar System (the Sun), and events like the LHB we propose that the spallation of Mercury before core formation was complete provided HEDs, pallasites, mesosiderites, irons, and others. Minimal addition of kinetic energy is needed to promote ejecta to distant orbits (Section 3). We conclude that the asteroid belt is a junkyard of debris that originated from diverse regions of the Solar System over its history, particularly small bodies (e.g., various moons, Mars, and Mercury as it evolved) and outer Solar System primitive material drawn in, e.g., during the LHB. Oxygen isotope data implicate early Mercury as the parent body of the HEDs (Section 5) and lead to bulk chemical compositions for Earth and its core (Section 6).

2. Did Vesta differentiate under its own gravity?

This section compares characteristics of Vesta to those of bodies with known cores. The observed trends are explained using Stokes' law and stability arguments.

2.1 Limitation of known cores to round, large bodies

Small planets are cold relative to large, due to their high surface area to volume ratio, which provides for greater radiative cooling to space, and shorter distances over which internal heat diffuses. Hence, their potential for differentiation via convection and gravitational segregation is limited. Mass fractions of planetary cores depend on planetary radius (Fig. 2a). Trends of the rocky bodies suggest that a minimum radius of ~1200 km is required to form a core. Considering icy and rocky bodies for which the moment of inertia is known, differentiation is also associated with a minimum surface gravitational acceleration (g) of

~1.2 ms^{-2} (Fig. 2b). The trends show that an object must have a mass in excess of ~10^{22} kg to have a metallic core, which is greater than the mass of the entire asteroid belt.

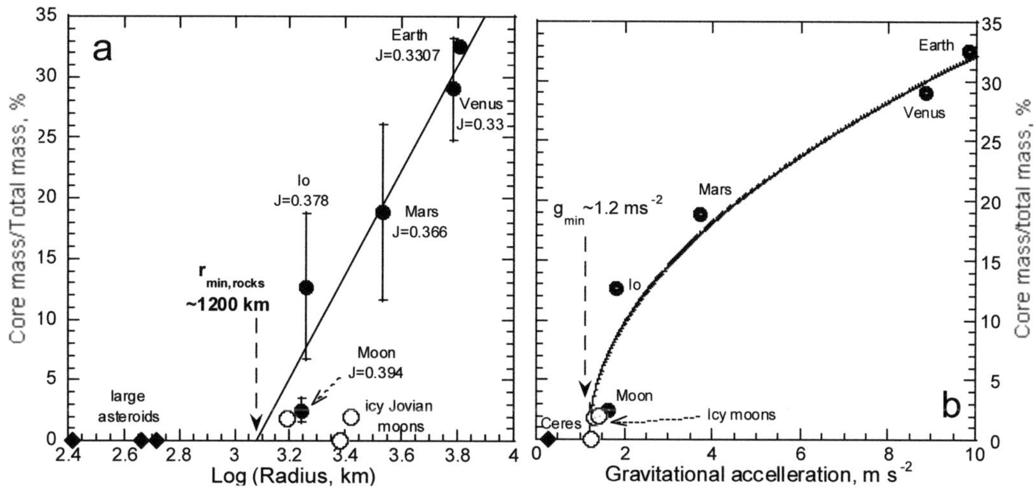

Fig. 2. Dependence of core size, expressed as the ratio of core mass to total planetary mass, on (a) the logarithm of the surface radius and (b) the surface gravitational acceleration. Diamonds = asteroids. Dots (with error bars) = rocky moons. Open circles = icy moons with highly uncertain core sizes. Mercury was not included in the linear regression, although the x-intercept of the fits would change little if it were. Measured moments of inertia ($J=I/Mr^2$ where r is radius and M is total mass) are indicated. Bodies with cores have $J<0.4$, which is the value for a sphere of constant density. Dashed vertical arrows emphasize minimum values needed for core formation. Data from Lodders & Fegley (1998).

2.2 A minimum planetary size is required for gravitational settling

Core size correlates with both g and R (Fig. 2) because the latter are related through:

$$g = 4\pi\rho G R/3 \tag{1}$$

where G is the gravitational constant, and ρ is the mean density. Forming a planetary core involves sorting of different density phases in a gravitational field. We therefore focus our discussion on gravitational acceleration. Because iron and silicate melts are immiscible (e.g., Goodrich and Bird, 1985) and density of liquids and crystals generally correlate, core formation can be viewed as sinking of Fe^0 particles or blobs within a silicate "fluid" in the hot interior of the planet. Settling is described in terms of a Stokes (terminal) velocity:

$$v = g\Delta\rho d^2/(18\eta), \tag{2}$$

where η is dynamic viscosity of the "fluid," d is particle diameter, and Δρ is the density difference between the particle and the fluid. The percentage difference between $\rho_{silicate}$ and ρ_{ice} being similar to that of ρ_{Fe} and $\rho_{silicate}$ and η being related to the density (compression) explains why one trend in surface g exists for bodies with different compositions (Fig. 2b).

The core grows at a rate comparable to Stokes' velocity. Large planets with large *g* remain hot longer, forming proportionately larger cores (Fig. 2b). For bodies with *g* exceeding ~2 $m{\cdot}s^{-2}$, core size directly depends on *g*, as indicted by Eq. 2. Because large planets retain more heat than small, viscosity decreases as planet size increases which makes the dependence of *v* on g weaker than linear. However, Eq. 2 assumes that particles sink and reach a terminal velocity within the static fluid, but does not describe whether sinking is possible. Figure 2 demonstrates that a minimum g is needed for coherent, downward motions.

Asteroids are too small to possess cores. It is immaterial whether asteroidal bodies are mixtures of iron and silicate or of silicate and ice, given similar density contrasts. Because the density contrast among various silicate mineral pairs are an order of magnitude smaller, asteroids could not have sorted olivine from plagioclase. Our conclusion that Vesta and all other asteroids could not differentiate due to internal gravitational sorting does not preclude these being differentiated material. Rather, our analysis shows that the differentiation occurred within a larger body with sufficient mass to permit density sorting.

3. Derivation of achondrites from the inner Solar System

Derivation of meteorites from Mars has been demonstrated (summarized by Grady, 2000). Mercury has g = 3.70 $m{\cdot}s^{-2}$, close to that of Mars (3.71 $m{\cdot}s^{-2}$) and an escape velocity of 4.3 $km{\cdot}s^{-1}$, low compared to Mars (5 $km{\cdot}s^{-1}$). Together these characteristics show that ejection of material from Mercury is not merely possible, but expected. Hence, objections to Mercury as a parent body rest on the notion that transport of material outward from the innermost Solar System is prohibited due to the huge gravitational influence of the Sun (e.g., Melosh & Tonks, 1993; summarized by Love & Kiel, 1995). Such difficulties have been overstated. As discussed below in general terms, classical mechanics indicates transport is feasible.

3.1 Dynamical constraints

Calculations concerning ejection of objects from planetary bodies and the consequent orbits are a variant of the three-body problem of classical mechanics. A system of three or more masses, moving under their mutual gravitational forces, cannot be solved in any general way (e.g., Goldstein, 1950; Symon, 1971). For example, the three-body problem cannot be reduced to three one-body problems. Very few special solutions exist: 1) the restricted three-body problem can be solved, wherein one mass is small and does not perturb the motions of the other two large masses (e.g., the trajectory of a rocket from Earth to the Moon); 2) motions of the planets can be accurately calculated, which represent a numerical solution obtained from specified initial conditions that holds over some period of time; and 3) a particular solution exists for three-bodies orbiting about the center of mass (discovered by Lagrange). Dynamical studies of ejection of material from planets, and the resultant trajectories, are neither general nor particular solutions to the n-body problem, but are numerical results predicated on various assumptions, whether implicitly or explicitly stated. Hence, the fate of an impact can only be deduced in general terms from simple approaches.

3.2 Outward transport of material ejected during collisions

The total kinetic and potential energy of a particle in an elliptical orbit depends on the semi-major radius but not on the eccentricity (*e*), semi-minor radius, or angular momentum (e.g., Symon, 1971). Hence, an object in a circular orbit of radius *r* from the Sun has the same total energy as an object in extremely eccentric orbits whose distance at aphelion can approach 2*r*. Due to this property, ejecta can be transported to beyond the orbit of Venus without adding any kinetic energy (Fig. 3).

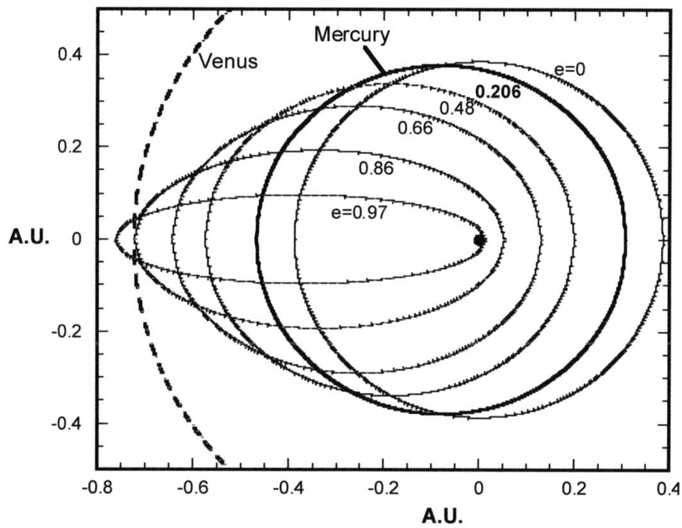

Fig. 3. Elliptical orbits for objects sharing the orbital energy of Mercury (solid curves, Mercury in bold), compared to the nearly circular orbit of Venus (dashed curve). Labels indicate the eccentricity of the orbits, which vary from a perfect circle (e = 0) to nearly unity.

3.2.1 Sling-shotting to distant reaches

For any ejecta, including those in elliptical orbits sharing the kinetic energy of Mercury's orbit, promotion to even more distant parts of the Solar System is possible. We base this remark on use of Jupiter's gravitational well to "slingshot" the famous Voyager spacecraft into the outer Solar System. If Venus were in a suitable position relative to the trajectories of ejected material in a highly eccentric orbit (e>0.86, Fig. 3), it is possible for that material with minimum kinetic energy to be promoted far beyond Venus' orbit.

3.2.2 Effect of kinetic energy beyond the minimum

If the perturbing collision adds energy to the object being ejected, far greater distances than Venus' orbit can be realized. Escape from Mercury itself requires 4.3 $km{\cdot}s^{-1}$, a mere 1% increase over existing orbital energy. For transport to infinity, the velocity needed for an object to escape the gravitational well of the Sun at Mercury's orbit is 68 $km{\cdot}s^{-1}$. Because the orbital speed of Mercury is 48 $km{\cdot}s^{-1}$, escape of ejecta from the Sun's pull to infinity requires only doubling their kinetic energy. It should be clear that energies arising from impacts were sufficient to move material from the inner to the outer Solar System.

The kinetic energy of Mercury in its orbit is ~10^6 $J{\cdot}g^{-1}$ whereas that of the asteroids is much smaller, ~10^5 $J{\cdot}g^{-1}$, because orbital velocities become progressively slower with distance from the Sun. Hence, material ejected from an inner planet, if on an appropriate trajectory, has more than enough kinetic energy for placement in the asteroid belt. Once material reaches the belt, interactions and collisions with inward bound material could reduce their kinetic energy to an amount appropriate to the orbital energy in this region.

3.2.3 Solar controls on the draw of Mercury and on impact and ejecta trajectory

Geometry, properties of central forces, and relative sizes of Mercury and the Sun suggest possible paths. Importantly, gravitational competition with the Sun severely limits Mercury's draw. Due to the central nature of this force, for a particle within a dust cloud to be incorporated in a body orbiting a larger central mass (Fig. 4a),

$$\left(r_{\text{cloud}} / r_{\text{orbit}} \right)^2 \leq M / M_{\text{central}} \qquad (3)$$

(Hofmeister and Criss, 2012). Although Mercury would actually draw in more distant material above and below its orbital plane and away from the Sun, our spherical formula suffices due to the large orbital radii and small planet mass: using the masses of the Sun and Mercury, provides a cloud size of 4.4×10^7 m, which is only 20 times Mercury's planetary radius (2.4×10^6 m) and an insignificant portion of its orbital radius (5.79×10^{10} m). Hence, a tiny spherical cloud reasonably describes the gravitational draw of Mercury. Impacts derived from cloud material are radial inward and help assemble the planet's mass. Our interest is rather collisions occurring after formation that would scour evolving Mercury.

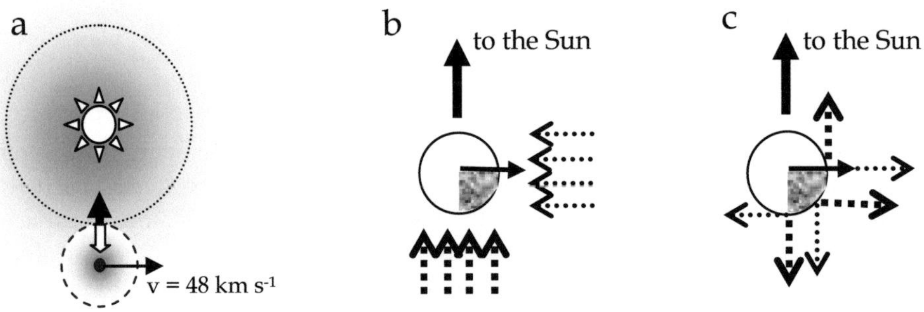

Fig. 4. Geometrical control on impacts on a planet. (a) Gravitational competition. The size of a cloud collecting unto an orbiting body (dashed circle) is determined by where the gravitational force (black arrow) of the central object (Sun) balances that (white arrow) of the orbiting protoplanet (dot). (b) Input trajectories. The planet (circle) in its orbit, intercepts material that is stationary or moving towards the planet (dotted arrows). Material focused on the Sun strikes the planet at nearly right angles (heavy dashed arrows). Shaded area = the most affected quadrant. (c) Initial output trajectories are shown for direct hits, glancing blows and strikes in the middle of the preferred quadrant.

Once Mercury was formed, two paths for impact are relevant. One is a "windshield" effect, whereby Mercury collides with debris entering its orbit (Fig. 4b). Because of the great distance to the Sun, these trajectories are approximately tangential to the orbit. The other impact path is approximately radially inward to the Sun (Fig. 4b), which is the gravitational focus of the Solar

System. Given the great distances, parallel paths reasonably approximate these impacts. Consequently, Mercury's outer leading quadrant is struck preferentially, and most of the resulting recoil paths point away from the Sun or are tangential to Mercury's orbit (Fig. 4c).

A minimum energy escape trajectory from an isolated body follows a parabolic path ($e = 1$). With increasing energy, hyperbolae are expected ($e = 2$; e.g., Symon, 1971). However, Mercury is not isolated and the Sun pulls ejecta with $v < 68$ km·s^{-1} into bound orbits (Fig. 5a). With more energy, the ellipse is larger, making excursion to the location of the asteroid belt possible for certain values of kinetic energy and initial paths. Aperiodic bounded orbits which exist for central forces (Symon, 1971) would be possible trajectory paths for intermediate energy particles, with initially tangential trajectories (Fig. 5b). This type of orbit can be open as shown, or closed, similar to the patterns of a drawing implement known as a "spirograph." Orbits of meteor strikes are presumed ellipses (e.g., Gounelle et al., 2006). To our knowledge, aperiodic bound orbits have not been considered, yet these are not only less restricted by geometry but are more likely to intersect Earth over time.

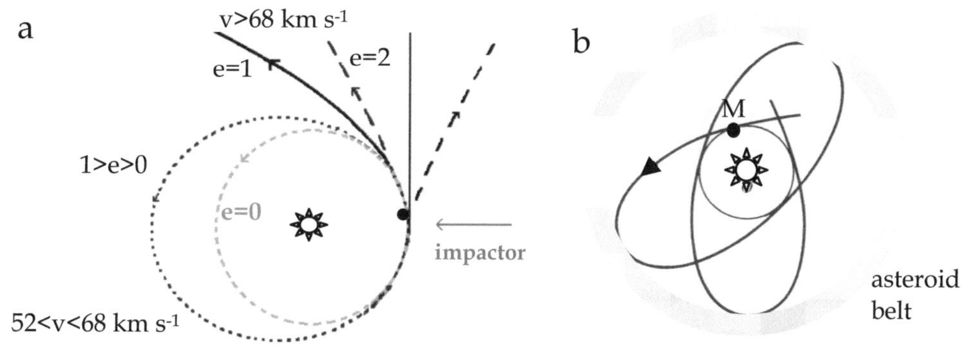

Fig. 5. Schematic of possible ejecta orbits. (a) Simple orbits showing the connection of eccentricity with velocities needed to escape the Sun's control. Only for v>68 km·s^{-1} can the ejecta escape the Solar system. (b) Aperiodic bound orbits are permissible for a central force (Symon, 1971) and could arise from initially tangential recoil paths.

3.3 Mass arguments for the asteroid belt as debris

Regular and distinct trends exist for the mass of dusty and gassy bodies that orbit the Sun with distance (Fig. 6). These trends result from 3-d accretion of the pre-solar nebula (Hofmeister & Criss, 2012), such that the dusty bodies form first, and gas is augmented onto dusty protoplanets in accord with Eq. 3. The decrease in mass for the dusty bodies going as $r^{-1.7}$ is consistent with conservation of angular momentum and energy during 3-collapse. Independent of our model, the trend for dusty primary satellites shows that objects in the asteroid belt formed in a different manner than did the planets and dwarf planets, especially insofar as the total mass in the asteroid belt falls two orders of magnitude below the trend.

3.4 Type and mass of material excavated from evolving Mercury

The innermost planet seems to have lost practically all of its mantle, as its core extends out to nearly 80% of the planet's radius, and constitutes ~⅔ of its total mass (Fig. 1). A plausible interpretation is that proto-Mercury was once at least as large as Mars (Fig. 1) and that

>3x10²³ kg mostly of silicates was blasted off this battered planet early in its history. The total mass of the asteroid belt (~3x10²¹ kg) is only 1% of Mercury's loss, and half of the mass in the belt is sequestered in one object, Ceres. Much of Mercury's lost mantle would have been absorbed within the inner Solar System, but this huge quantity of material will offset the low probabilities of ejecting material and transferring it to Earth-crossing orbits. Transport of <1% of Mercury's loss to the asteroid belt is reasonable, and, more importantly, is much more than sufficient to generate the aggregate mass of achondrite and other classes of meteorites in available collections, including the ~1000 kg of HEDs (Table 1). Other bodies will provide achondritic material, but in much smaller amounts and ejecta originate at various times. Chondritic material would represent dust originating from this region.

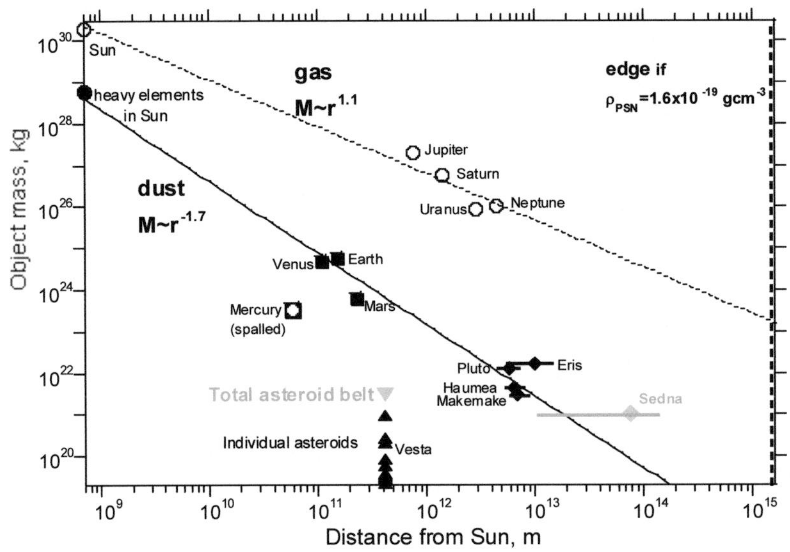

Fig. 6. Mass of large bodies orbiting the Sun vs. distance from the origin. The Sun is plotted at its body radius to use a logarithm plot, needed to represent the vastness of the Solar System. Open circles and dotted line = gas giants. Dashed line at right side = radius of the pre-solar nebula, calculated from the Sun's mass, assuming an initial density like that of molecular clouds. Filled circle = heavy elements in the Sun (Basu & Antia, 2008), taken to represent the dust component of the pre-solar nebula (PSN). Squares = inner rocky planets, except Mercury (open square) which has lost significant mass. Diamonds = dwarf planets of the outer Solar System. Line = fit to the circles, squares and diamonds. Grey diamond = Sedna. Horizontal lines describe substantially elliptical orbits. Triangles = individual asteroids (grey for the total mass). Data from Lodders & Fegely (1998) and http://pds.jpl.nasa.gov/.

The heavily cratered surface of Mercury and its location as the innermost planet provide evidence for deep excavation. Saturation of the surfaces of Callisto and Ganymede with craters implies that the Jovian moons were also extensively bombarded, due to the gravitational focus of impactors by Jupiter. By analogy, early, evolving Mercury received an intense flux of impactors, due to its proximity to the Sun.

Our understanding of impacts is skewed toward late, low energy events which do not cause great loss of planetary matter. Such minor impacts provide dust (rayed craters), breccias, shock metamorphism, large volumes of melt, and small volumes of melt ejecta (tektites and small spherules) (French, 1998). Mercury has low gravity and lacks an impeding atmosphere. Impacts into pools of melt on Mercury splattered innumerable chondrules into space. Crystallization of the melt in space would provide round rather than the aerodynamic shapes of tektites, and would have formed bodies of diverse sizes. More massive impacts would have vaporized material, which would have escaped and then condensed. The internal structure of Mercury suggests that most ejecta would be from the mantle. However, impacts add their own matter to the surface, which would also be subsequently ejected. Hence, primitive material, differentiated material, both their melts, and mixtures of all such debris are possible. Vesta could be a conglomerate of ejecta from Mercury.

3.5 Statistical arguments for planetary origins of meteorites

Asteroids are classified by remotely sensed data by several schemes. Spectra (Bus & Binzel, 2002) or spectra plus albedo (Tholen, 1989) are further used to link these types to meteorites. Broad categories of asteroids common among these schemes occur in percentages that are inconsistent with the numerical occurrence of meteorites (Fig. 7). Current collections have ~40,000 samples (Table 1) and should be statistically representative. Mass distribution gives us a different perspective and is skewed towards high iron content, due to material strength. Also, spectra of asteroids preferentially represent larger bodies.

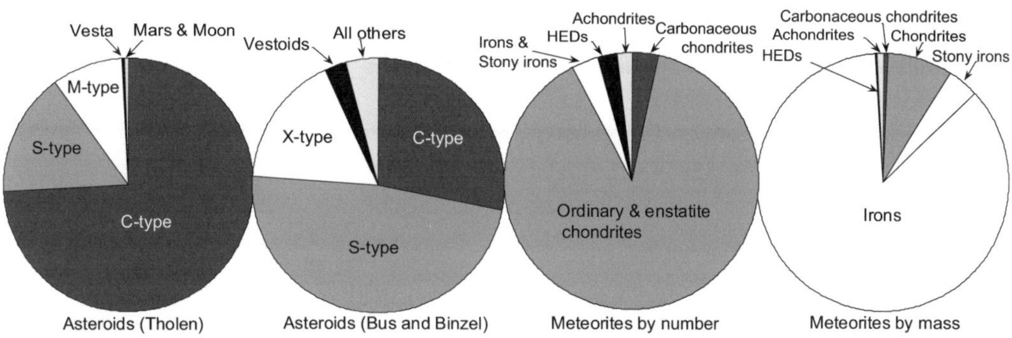

Fig. 7. Large disparity between two different asteroidal classifications (left) and meteorite types (right). Dark "C-type" asteroids represent a majority, yet are considered as the primary sources of the rare carbonaceous chondrites. Vesta is only one of >10^6 asteroids, but is presumed to be the source of the HEDs, which represent >½ of the achondrites, and are geochemically tied to ~ ⅓ of the irons, and most stony-iron meteorites. Data from Table 1.

The current linkage is statistically untenable, as is well known. Problems exist with attribution of C-type asteroids to carbonaceous chondrites. Spectra of C-types are featureless with low albedo, consistent with dusty surfaces; it may simply be that their mineralogy is equivocal. Grain-size and dust coverage affect surface spectra (Section 4).

3.6 The asteroid belt as the junkyard of the Solar System

The above dynamical, mass, and statistical considerations lead to an alternative view of the asteroid belt. Classical mechanics requires that ejecta from early Mercury predominately occupy elliptical orbits and aperiodic bound orbits. Over time, the amount of material is non-circular orbits will diminish. The inner Solar System has been swept clean. The asteroid belt has a large number of objects because the distances are large, so the clearing rate is low, and a planet did not form in this area. On this basis, a likely repository for the ejecta and leftovers of accretion is the asteroid belt. Gravitational forces (e.g. Jupiter) have shepherded errant objects into this veritable junkyard of debris. Nothing requires this zone to be the locus of accretion for all of its material.

4. Limitations of spectroscopic assignments

Spectra from remote objects are obtained under uncontrolled conditions and contain a mixture of absorption, emission, and reflection features. To relate these data to laboratory mineral spectra, we need to understand how spectra are affected by sampling conditions.

4.1 Effect of temperature on spectra

For a remotely probed surface, the most important effect of temperature (*T*), is the control it exerts on whether the object is emitting or reflecting light over any given frequency range.

Blackbody emission curves depend strongly on *T* and frequency (Fig. 8a). The 1st law of thermodynamics requires that the flux from Vesta match that received from the Sun (Fig. 8b); roughly speaking, the areas under the curves must be equal. Due to the properties of Planck curves, Vesta, which is cold (85<*T*<255 K; e.g., Lucey et al., 1998) outputs virtually all its light below 2200 cm^{-1}, and therefore is emitting in the infrared but reflects light in the near-IR to visible.

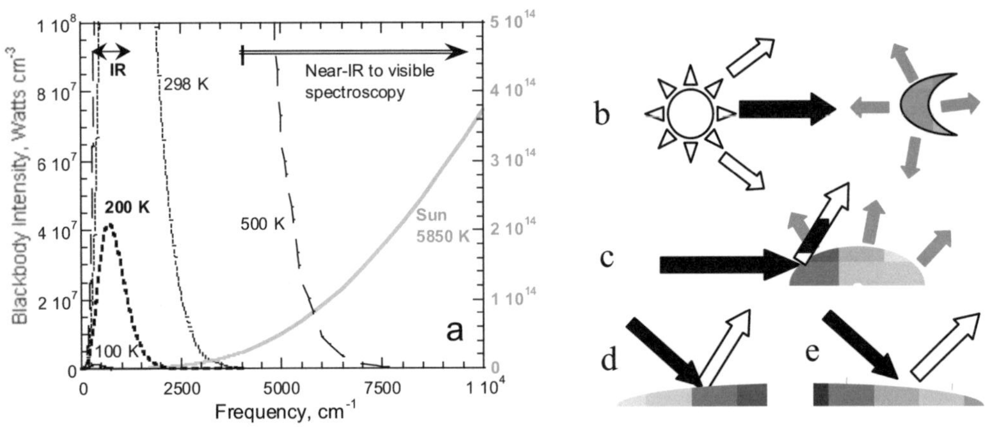

Fig. 8. Factors governing whether light is emitted or reflected. (a) Dependence of blackbody curves on temperature, as labeled. Except for the Sun at ~6000 K (grey), the left y-axis pertains. Arrows indicate spectral ranges sampled by different instruments. (b) Schematic of the first law of thermodynamics. Black arrow = light from Sun. Grey arrows = light emitted, which integrated over the area and frequency must equal the flux received.
(c) Cold objects emit at low frequency, and reflect sunlight at high frequency (speckled arrow). Both are modified by spectral properties, which depend largely on whether the surface is bare (d) or dusty (e), as sketched for the case of reflected sunlight.

4.2 Effect of grain-size on spectra

Grain-size controls whether peaks are superimposed positively or negatively on the baseline, and strengths of the features.

Light produced by a body at any frequency is the product of the blackbody function times the emissivity. Kirchhoff's law for a body in thermodynamic equilibrium requires that absorptivity ($\alpha = I_{abs}/I_0$, where I_{abs} = the intensity of light actually absorbed) equals emissivity (ε) (e.g., Bates, 1978; Brewster 1992). For an opaque material such as a metal, $1 = \alpha + r$, where $r = I_{ref}/I_0$ is reflectivity. Although asteroids are large and opaque, dust grains on their surface are partially transparent. For the part of the asteroid that is sampled, $1 = \alpha + r + t$, where $t = I_{tran}/I_0$ is transmittivity of the uppermost layer as derived by Bates (1978) for dielectrics (e.g. silicates).

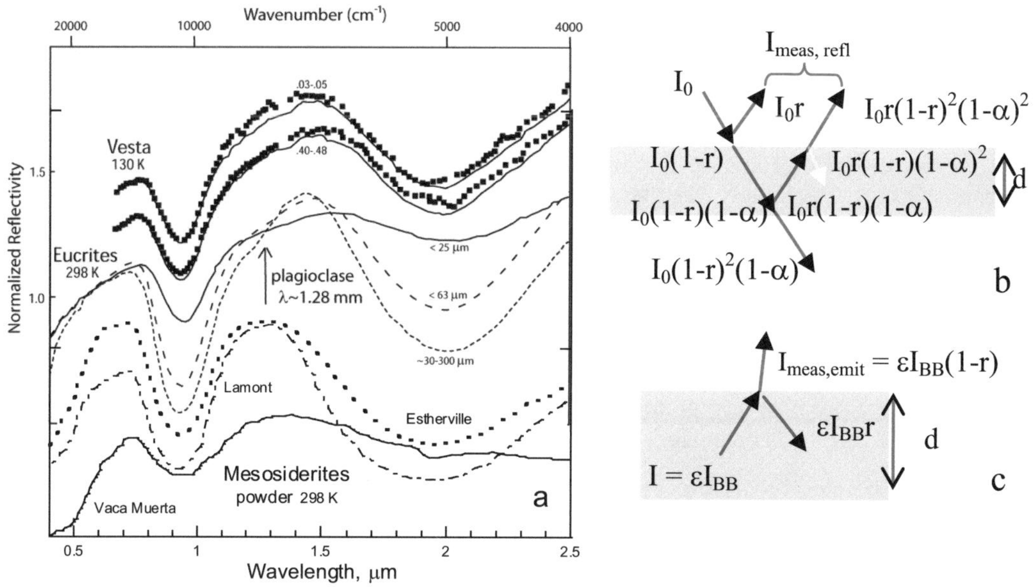

Fig. 9. Effect of grain-size. (a) Comparison of orientational differences in spectra for Vesta to the grain-size dependence of reflectance-absorbance spectra from eucrites [solid = Padvarminkai (Hiroi et al., 1995); dashed = Macibini (Burbine et al. 2001); dotted = Padvarminkai (Gaffey, 1976)] and powdered mesosiderites as labeled (Burbine et al., 2007). Vesta spectra (Gaffey, 1997) are shown for two rotational aspects (dots), with each compared to the average (lines) of 4 to 5 rotational aspects. The uncertainties are roughly 2-3 times the symbol size, and the placement of the points near 2 μm is not exact because spectra presented at different scales were merged. Spectra offset for clarity. (b) Schematic of reflections for the most intense ray paths at near-normal incidence (the angle is exaggerated and the white arrow is not labeled). The light received by a detector (I_{meas}) relative to incident intensity (I_0) is a combination of reflection and absorption, depending on thickness, *d*. (c) Schematic of emissions, where I_{BB} is blackbody intensity.

Measurements are affected by back reflections (e.g., Hofmeister et al., 2003). If the surface layer is very thick then the back reflections are not sensed. Reflectivity at high frequency is flat for a bare, large, single crystal due to Fresnel's law, ($r = [(n-1)^2+k^2]/[(n+1)^2+k^2]$ where n is the index of refraction, $k = a/4\pi\nu$ where a is the absorption coefficient, and ν = frequency.

For grain sizes of rocks, measured reflectance is reduced by back reflections (Fig. 9b). As spectra of eucrites demonstrate (Fig. 9a), larger grain size at the surface means deeper absorption features. Small grains are not detected: these contribute only to scattering.

Regarding emitted light, for a bare, thick surface or large grains, $\varepsilon = \alpha$, so emissions go as $I_{BB}(1\text{-}r)$. At frequencies where transitions occur less light is received, i.e., peaks point down as seen in the spectra of Christensen et al. (2000). (Note: 1-r has been incorrectly interpreted as emissivity for large grains). If the surface is covered with dust, this layer is heated by lattice conduction from below and emitted light comes only from this layer. Because reflection peaks are broader than absorption peaks, emissions from a dusty surface go as $\varepsilon I_{BB} = \alpha I_{BB}$ and peaks point up as seen in spectra from small grains (Low & Coleman, 1966; Bates, 1978) and references therein.

4.3 Effect of Fe^{2+} contents and grain-size on near-IR reflectance spectra of Vesta

High frequency reflection spectra (Fig. 9a) have the strong peaks of pigeonite (as in eucrites). Fe-bearing plagioclase may be present: its 8000 cm^{-1} band (Hofmeister & Rossman, 1984) is weak for Serra de Magé eucrite. No obvious peaks for orthopyroxene (dioginites) or olivine (pallasites) exists. Such may be extracted by peak fitting, but variations in grain-size of pigeonite suffice to explain the scant differences, as is clear from Fig. 9. Variation in grain-size were inferred from recent analysis of Vesta's mid-IR light-curves (Chamberlain et al., 2011).

Mesosiderite spectra are similar to eucrites, but the presence of orthopyroxene broadens the peak at 2000 nm and shifts it to 1800 nm (Fig. 9). The surface of Vesta was impacted. Rapid surface cooling would produce pigeonite as in terrestrial lavas. From Fig. 9, the surface of Vesta could be a mixture of pigeonite and metal. The proportion of pigeonite is high, based on peak depths, which is consistent with Vesta's density (Section 1.1.1).

Importantly, the depths of the peaks are connected with Fe^{2+} content of the minerals. Orthopyroxene in the HEDs has Fe/(Mg+Fe) of 0.25 which is half that of pigeonites. Therefore, pigeonite will dominate if in equal proportions. Orthopyroxene is also coarse grained: large crystals will not provide back reflections. The spectra of Vesta could contain all phases of mesosiderites, but spectra only prove that pigeonite is present and abundant.

4.4 Problems with interpretation of mid-IR emission spectra from Vesta

Spectra obtained of Vesta using the Infrared Satellite Observatory (ISO) record its emitted light. A broad peak near 450 cm^{-1} (Heras et al., 2000) indicates a blackbody temperature of 130 K, consistent with previous inferences (e.g., Lucey et al., 1998). ISO spectra in this range and below are noisy, obscuring features of minerals. Mid-IR features are weak, consistent with a dust covering (Dotto et al., 2000; Lim et al., 2005). Hence, these features are in emission. Unfortunately, spectral analyses by these authors assume that emissions are that of an opaque (metallic) body [= $I_{BB}(1\text{-}r)$] whereas presence of a dust cover requires that $I_{meas,emit} = \varepsilon I_{BB}$ for a dielectric (silicate) material (Fig. 9bc). The fits need redoing.

5. Geochemical arguments for HEDs as inner Solar System material

Meteorites display a bewildering complexity of chemical and isotopic relationships. The following discussion focuses on characteristics that reveal radial chemical gradients in the

Solar system inherited from the pre-solar nebula. An important marker is the ratio of volatile to refractory elements, which is exemplified by K/U and Rb/Ba ratios. A ratio involving Al is useful because this refractory element has a short-lived isotope. We selected Ga for normalization as it is a more volatile Group IIIA element with a similar ionic radius.

Stable isotope ratios (e.g. Si, O) have proven utility to petrogenesis. Oxygen isotopes are exceedingly important because this element is abundant and its fractionation and distribution are well-understood.

5.1 Geochemical characteristics of meteorites

Detailed descriptions of the geochemical and petrological characteristics of HED meteorites are available (see references in Section 1.1). Compared to other achondritic meteorites, the HED family includes the oldest known basalts in the Solar System, is the most volatile depleted, the most strongly reduced, and distinct in terms of isotopes (Table 2).

Ordinary chondrites, by far the most abundant stony meteorites (>90%), are widely believed to originate from the asteroid belt. This attribution is consistent with chemical uniformity of this group of meteorites. Persuasive arguments for the origination of SNC meteorites from Mars have been advanced (see Grady, 2006). In contrast, assigning the HED family to Vesta or Vestoids (e.g., Drake, 2001) contradicts physical data (Section 1), abundances (Fig. 7), and geochemical characteristics, discussed below.

	A.U.	K/U	Rb/Ba	10^4 Ga/Al	$\Delta^{17}O$	$\delta^{18}O$	$\delta^{30}Si$
R chondrite	~3	>40000	>0.35	8	2.8	4.3	
Ordinary chondrites	2-3	60000	0.6	5	0.7 to 1.3	3.7 to 5.6	-0.46
Mars, SNCs	1.52	16000	0.17	4	0.27	4.5	-0.48
Earth upper mantle	1.00	10000	0.10	2	0.00	5.7	-0.68
Moon, Lunar	1.00	3000	0.015	0.3	0.00	5.7	-0.45
Venus	0.72	7000*					
HEDs, protoMercury	0.39	3700	0.007	0.2	-0.25	3.7	-0.45
Ureilites, CAI, K-ch	~0	<1000	0.002	0.005	<-0.8	<8.1	-0.47

Table 2. Geochemical and isotopic characteristics of meteorites and rocky planets. Data Sources: BVSP (1981), Lodders and Fegley (1998), Lodders (1998), Kitts and Lodders (1998), Molini-Velsko et al. (1986), McKeegan et al., (1998); MacPherson et al. (2005), Franchi et al. (2008), Armytage et al. (2011) and others. For A.U.~0, oxygen values were taken from a cluster of ureilites and a chondrule from the Kakangari chondrite (Prinz et al., 1989). For R chondrites, the average excludes hot desert finds. *Gamma ray spectrometer results for Vega 1 and 2 (Surkov et al., 1987).

5.2 Chemical trends in the Solar System

The trends in Table 2 are coherent and regular. Similarities in the K/U ratio among disparate rock types on Earth, and the large difference of this ratio from that in chondrites, were established long ago by Wasserburg et al. (1964).

Note that the K/U ratio of HED meteorites is much more similar to rocky materials from the inner Solar System than it is to materials presumed to have originated outside Earth's orbit.

The Rb/Ba ratio is also consistent and characteristic for each of the rock suites under consideration (Fig. 10). This and the Ga/Al ratio increase markedly in the order that the K/U ratio increases (Table 2). The remarkable coherency of Rb/Ba ratios in lunar materials, which are distinct from the chondritic ratio, has been demonstrated (BVSP, 1981, p. 251).

5.2.1 Attribution of the HEDs to protoMercury: implications for elemental compositions

Not all element ratios will vary in the same manner as those in Table 2, nor would such coherency be expected given the profound differences in planetary masses, timescales and extents of differentiation, and other factors. What is remarkable, however, is that so many compelling geochemical signatures vary in the same order, with the HED family near one extreme. These regular variations involving both major and minor elements, as well as the stable isotopic variations of the most abundant elements, are consistent with an origin from the inner Solar System, most logically, from early Mercury. The inventory of volatile and large ion lithophile elements in protoMercury would have been greatly reduced by spalling, leaving a remnant planet highly depleted in these elements, including K, U, and Th. Chemical data provided by Nittler et al (2011) and Peplowski et al. (2011) confirm our deduction.

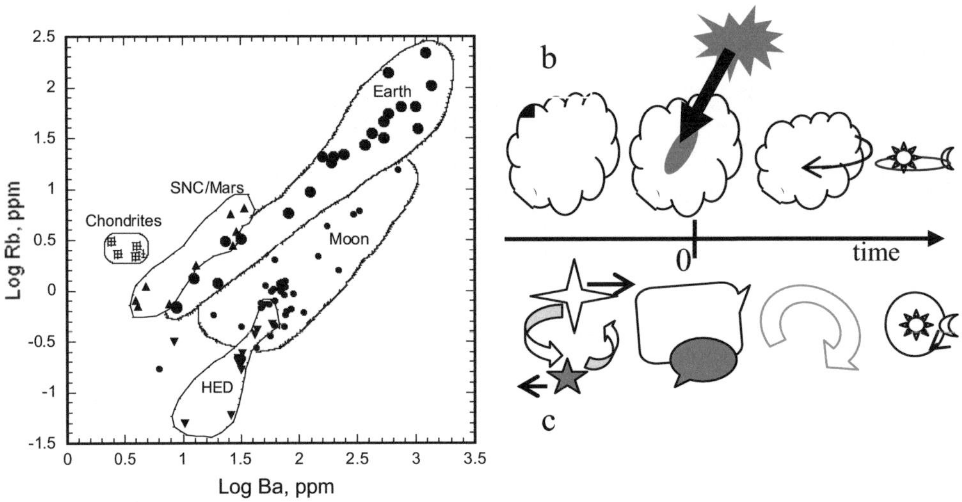

Fig. 10. Zonation in the pre-solar nebula. (a) Rb vs. Ba concentrations for Solar System materials, suggesting HEDs are refractory inner Solar System material. Sources as in Table 2. (b) One hypothesis for origination of the zonation. From left to right: the pre-solar nebula of dust and gas (white cloud) is stable, but a nova injects material (grey) including refractory CAIs, while producing or inducing short-lived isotopes such as ^{26}Al (grey). The injection diffuses, but the center of its mass provides a gravitational instability, precipitating nebula contraction, rotation around this locus, and planetary accretion. (c) Alternatively, two passing and interacting stars provide distinct reservoirs of dusty gas.

5.3 Oxygen isotope variations in the Solar System

Graphs of $\delta^{17}O$ vs. $\delta^{18}O$ values of achondrites and other meteorite types (Fig. 11) display a lattice-work of 1:1 and 1:2 lines, which respectively represent a mass-independent fractionation trend and an ordinary mass-dependent fractionation trend. Practically all Solar system materials lie between the 1:1 lines known as EC (Equilibrated chondrites) and CCAM (carbonaceous chondrite anhydrous minerals). The CCAM line includes CAI inclusions from Allende which are very old, refractory, were rich in short lived isotopes and extend to incredibly low values. CAIs from enstatite and ordinary chondrites have similarly low values (McKeagen et al., 1998; Guan et al., 2000). Several lines of evidence point to CAI's being near the center of the Solar System and representing a distant reservoir (MacPherson et al., 2005).

The 1:2 trends are superimposed on the 1:1 trends and are due to ordinary fractionation processes that continue today on Earth (the TFL). For any given planet, surface materials will lie to the upper right along these 1:2 lines (Fig. 11), while deep interior materials will lie to the lower left. Mars' meteorites lie above the TFL, whereas the HEDs lie below by a similar amount. These 1:2 trends develop during late stage heating of formed bodies by various processes and may onset early from impact heating

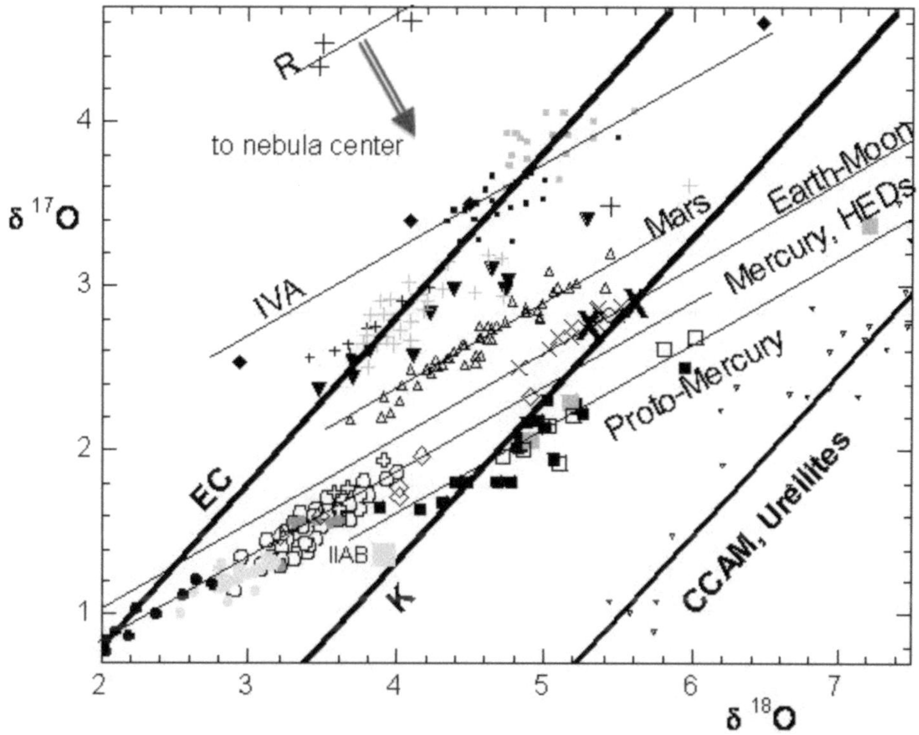

Fig. 11. Oxygen isotopes. Tiny triangles = ureilites. Open squares = winonaites. Grey square = IIICD irons. Black squares = IAB irons. Circles = HEDS. Dark grey dots = mesosiderites. Light grey dots = main group pallasites. Black dots = IIIAB irons. Open cross = angrites. Open diamond = brachinites. Light X = aubrites. Heavy X = enstatite chondrites. Open triangles = Mars meteorites. Filled triangles =IIE irons. Filled diamonds = IVA irons. Small plus = H chondrites. Tiny black squares = L chondrites. Tiny grey squares = LL chondrites. Large plus = R-chondrites. Light lines = fractionation trends for various bodies, as labeled. Heavy lines = solid-gas interaction, as labeled. Double arrow indicates the Solar System gradient. Data from Clayton and Mayeda (1996); Franchi et al. (2008); and other sources.

6. Geophysical implications

To obtain bulk compositions of the silicate Earth and its core requires coherently explaining Fig. 11 in general terms that are consistent with fundamental principles and the data, and understanding the assembly, spalling, and resultant layers of Mercury.

6.1 Attribution of the HEDs to evolving Mercury: implications for layer compositions

The large size of Mercury's core shows that it was deeply spalled both during and after core formation. Spalling was caused by late-stage impacts from the outer Solar System (Section 1.2), which added material from distance because impactors bury themselves while ejecting target material (French, 1988). Thus, the HED family trend lies closer to the TF line than proto-Mercury did. The upper layers of early Mercury must be represented in the meteorite collection. The trend of winonaites, IAB and IIICD irons lies below the HEDs (Fig. 11). This material is more pristine, consisting of non-magmatic irons and is similar to the enstatite chondrites associated with the Earth (e.g., Javoy, 1995) corroborating proto-Mercury as the source.

The HEDs represent early Mercury's present mantle, the mesosiderites are a combination of poorly sorted core-mantle boundary material and late-stage additions, and the IIIAB irons represent core material. Main-group pallasites represent material ejected from the zone of gravitational settling during core formation. Approximate compositions are in Table 3.

Object or layer	**Meteorite analogy or composition**
Proto-Mercury	Winonaites + IAB + IIICD + ices
Bulk silicate early Mercury	HEDs
Core-mantle sorting	Pallasites, main group
Mercury's evolving core	IIIAB
Proto-Earth	enstatite chondrites + ices
Bulk silicate Earth	½ enstatite chondrites + ½ ordinary chondrites + ices
Earth's core	~85%Fe + ~5%Ni + ~10%C with minor S, N, P

Table 3. Zones in Mercury and Earth and their connections with meteorite types.

6.2 Explanation of solar system gradients and trends

We dismiss the hypothesis that the material which formed the planets post-dates the Sun. The "condensation gradient" is invalid in view of recent data on comets (Zolensky et al., 2006). Simultaneous formation of the Sun and planets by 3-d nebular collapse is supported by the axial spins and orbital characteristics of the planets (Hofmeister & Criss, 2012) and indicates rapid accretion over ~1-3 Ma. What did the Solar System form from? Pre-solar grains indicate diverse sources (Bernatowicz & Zinner, 1997). For simplicity, given Table 2 and Fig. 11, we consider that two reservoirs of dust existed, which were embedded in an immense reservoir of gas. Gas constituents are mostly H_2 with He and CO. Its mass is ~100 times larger than the dust mass, based on solar composition (Basu & Antia, 2008).

Our analysis is also based on the mineralogy of dust in astronomical environments. Infrared dust emissions of the exploded circumstellar dust of NGC 6302 contain many peaks, consistent with the presence of pure, magnesium endmember silicates (forsterite, enstatite, and diopside), possibly calcium carbonate (Molster et al., 2001; Kemper et al., 2002), along with CAI minerals (Hofmeister et al., 2004). Hydrosilicates are possible (Hofmeister & Bowey, 2006). Dense dust clouds around young stellar objects contain ices (H_2O, CO, CO_2, CH_4, hydrocarbons) thought to nucleate on the dust, summarized by Bowey and Hofmeister (2005) who identified melilite (a CAI phase) and Mg-rich amphiboles (hydrated pyroxenes).

Both dust reservoirs condensed from hot gas expelled from pre-existing stars (Fig. 10bc). The large, and possibly pre-existing, dust reservoir involves O, Mg, Si, Fe, consistent with major elements in the Earth and with stellar nucleosynthesis. This reservoir is exemplified by the ordinary and R chondrites and dominates the outer reaches (Fig. 12). The large dust reservoir also contained lesser amounts of high-temperature phases, and lower temperature phases with volatile elements. The small, probably injected, dust reservoir provided the CAI inclusions, short-lived isotopes, refractory elements, and lesser amounts of iron and other materials which dominate the large dust reservoir. For a simple explanation of the 1:1 trends, we assume that the gas phase is greatly enriched in ^{16}O.

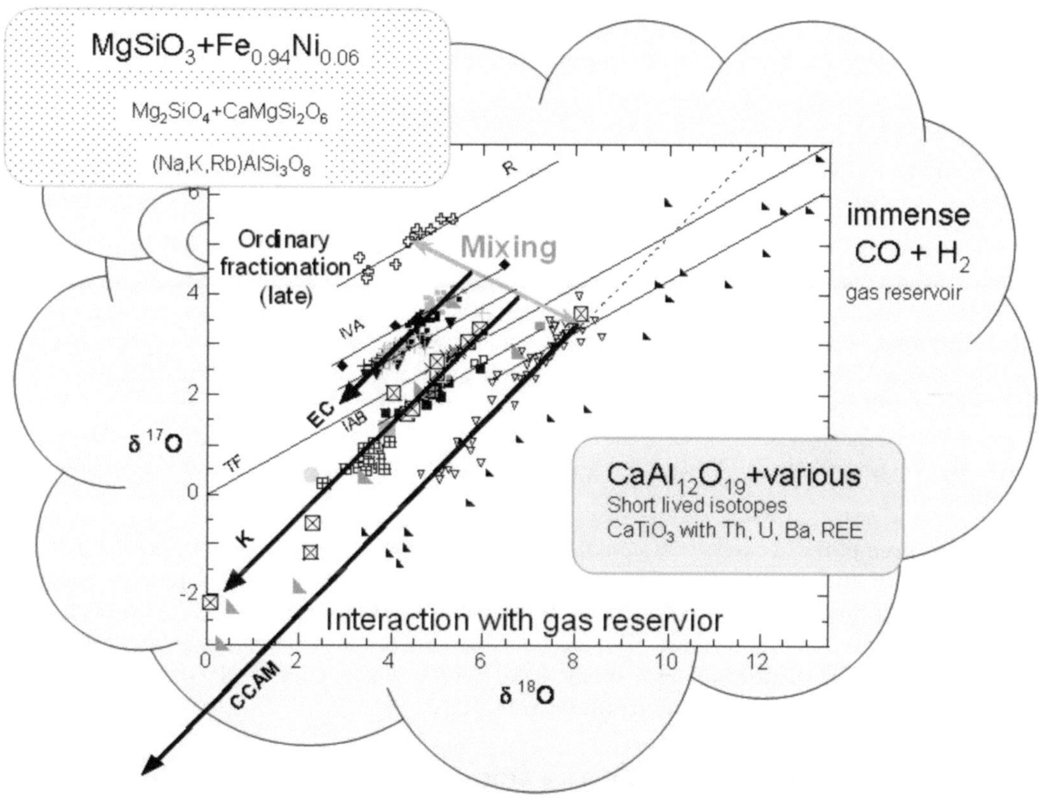

Fig. 12. Reservoirs and processes altering oxygen isotopes and mineral phases and compositions. Symbols for meteorites as in Fig. 11 with the exception of open cross = R-chondrites, and a few additions: Square with cross = acapulcoites. Square with X = K-chondrites. Right triangles = dark inclusions in CAIs. Light lines = late, ordinary fractionation within the various bodies, as labeled. Heavy lines = mixing or gas fractionation, as labeled. Grey double arrow indicates the Solar System gradient.

Upon assembly of the planets, and heating via impacts or radioactivity, dust in the main reservoir reacted with trapped ices, exemplified by CO:

$$2Fe^0 + CO + MgSiO_3 \text{ (enstatite)} \rightarrow FeC + (Mg,Fe)SiO_4 \text{ (olivine)}. \quad (4)$$

This reaction converted end-member pyroxene to solid-solution olivine. The isotopic drive was towards the CO reservoir, which was rich in ^{16}O. This reaction is germane to core formation, suggesting that carbon is the core's light element, consistent with solar abundances. Earth's core has ~10% light elements, mostly C, consistent with ubiquitous occurrence of olivine with Fe/(Mg+Fe) ~ 0.1 in meteorites, mantle samples, and terrestrial basalts. Equation 4 and similar reactions provide for enrichment of Ni accompanying incorporation of C, S, N and P in the core while lithophiles (e.g., Ge and Ga) are transferred to the silicates. Comparison of the composition of IAB to IIIAB irons associated with evolving Mercury (see Mittlefehldt et al., 1998) supports our proposal.

The small dust reservoir (CAIs, ureilites) has many phases and many reactions are possible. Because metals that condense at high *T* are rare (e.g., V, Mo, Nb, Pt group), reactions similar to Eq. 4 in the small reservoir could go to completion. Melilite could be converted to forsterite and fassitic pyroxene would be formed, for example. Without ample Fe^0, the following reaction is suggested to be important with dust serving as nuclei:

$$CO + H_2 \rightarrow C + H_2O. \quad (5)$$

Metals such as V were incorporated in hibonite, ample graphite was produced, material was hydrated and/or carbonated, and the CAI phases were mass-independently fractionated.

Reactions like Eq. 4 and 5 proceed with time and this sequence is recorded in meteorities which were ejected from evolving Mercury and other planets as the late stages of accretion progressed. Nearly full-formed planets provided the ejecta, in accord with early assembly involving dust, with large impacts being associated with the late influx from great distance (Hofmeister & Criss, 2012). The earliest ages deduced from isotopic studies correspond to condensation of the dust reservoirs from their stellar sources. Early chondrule ages (Amelin et al., 2002) of 2.5 Ma later record the onset of large impacts, and the end of proto-planet assembly, consistent with Hofmeister and Criss' (2012) analysis using the time-dependent virial theorem.

Mixing of these two dust reservoirs created a graded solar system (Figs. 11 and 12). Because of contraction and events like the LHB, the proportion of the main (outer) reservoir increased with time. The EC line pertains to the asteroid belt, whereas the K line pertains to the dust in the inner Solar System. The current planets lie about half way between these 1:1 lines. To this, CO and H_2O and other ices, which were frozen on dust drawn into protoplanets, were added in indeterminate amounts.

6.3 Fractionation trends and evolutionary processes

After assembly of the bodies and heating, mass-dependent fractionation occurs, superimposing lines parallel to TF for individual planets and large masses. All 1:2 lines for meteorite types are about the same length, consistent with similar thermal histories of the body after ejection to the asteroid belt. The longer line for the HEDs and related meteorites is consistent with origination in Mercury, given that the large Earth has an even longer 1:2 line. The degree of ordinary fractionation directly depends on the mass of the object (Fig. 13).

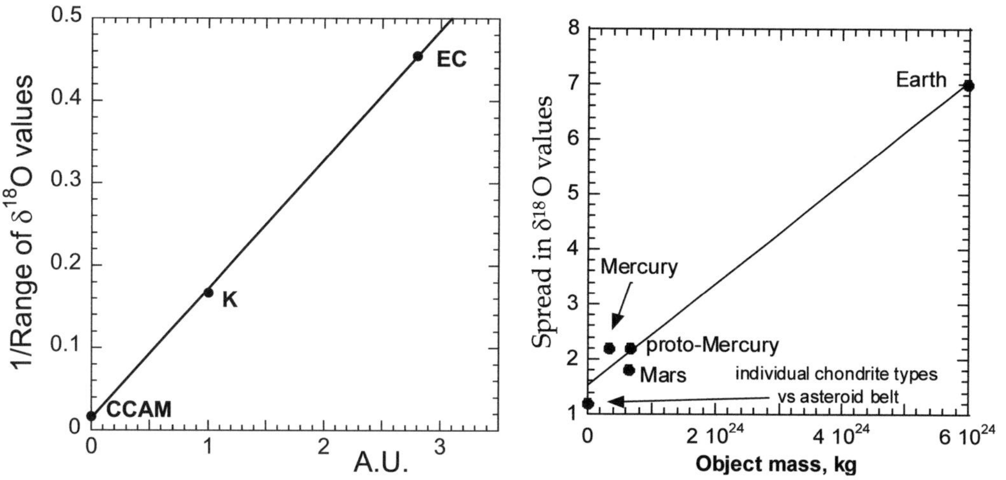

Fig. 13. Dependence of mass dependent fractionation on mass and of the inverse mass-independent fractionation on distance from the center. Objects and fractionation trends as labeled.

This trend is consistent with larger bodies being hotter and staying active longer. This trend may aid in deciphering parent bodies.

In contrast, the inverse of the length of the 1:1 lines depends on distance from the nebula center (Fig. 13). This important new finding corroborates that the small reservoir with short-lived isotopes was indeed at the center of the nebula, and suggests that our Solar System formed after a nova. Involvement of a second star is mandated by the existence of two dust reservoirs. The 1:1 line lengths could be derived from light-energy from the nova.

7. Conclusions and future research

We provide an alternative view of the origins of meteorites and the asteroid belt, based on fundamental principles. The connection of HEDs and related meteorites with early Mercury provides a coherent view of the ~40,000 samples of meteorites, suggests new relationships with the planets, and provides a consistent picture of their formation. Our analysis removes guesswork in estimating chemical compositions for the planets and provides many testable predictions to guide future work:

- The light elements in the core are C, S, N and P.
- Core formation involved redox reactions dominated by C, and released divalent iron into mantle minerals. This process increased the proportion of olivine to pyroxene.
- The $\delta^{18}O$ values of Earth are vertically zoned, with the lower mantle being lower in ^{18}O than the upper mantle. Earth's bulk $\delta^{18}O$ value is close to +4.
- Carbonaceous chondrites are too refractory to describe the Earth's composition.
- The smaller the object's core, the more Fe-rich the basalts.
- The oxygen isotope composition of Mercury will be shown to be similar to that of the HEDs.
- Asteroid 4Vesta is a mesosiderite that has undergone minor surface melting and rapid quenching due to impacts.

- Studies of Vesta's moment of inertia will disclose that it is internally homogeneous and lacks a core.
- Many Earth-crossing orbits are not elliptical but aperiodic bounded orbits.
- Properly analyzed emission spectra will help correct the asteroid classification system.
- All differentiated meteorites will be eventually linked to planetary bodies and other large objects such as the Moon that have a minimum radius of 1200 km.
- All meteorite materials are processed.
- The preSolar nebula was grossly homogeneous, but chemically and isotopically zoned in detail, with refractory elements more concentrated in the center, and volatile elements concentrated in outer zones.
- Mineral dust in the preSolar nebula was dominated by pure enstatite and iron metal.

8. Acknowledgment

Special thanks to Janet E. Bowey (University College London) for help with ISO data. We thank our Washington U. colleagues (R. F. Dymek, B. Fegley, B. L. Jolliff, K. Lodders and R. Korotev) and W. Hamilton (Colorado School of Mines) for helpful discussions. Partial support was provided by NSF EAR-0757841 and AST-0908309 funds.

9. References

Amelin, Y., Krot, A.N., Hutcheon, I.D., Ulyanov, A.A. (2002) Lead Isotopic Ages of Chondrules and Calcium-Aluminum-Rich Inclusions. *Science,* Vol. 297 pp. 1678-1683, ISSN: 0036-8075

Armytage, R.M.G., Georg, R.B., Savage, P.S., Williams, H.M., Halliday, A.N. (2011) Silicon isotopes in meteorites and planetary core formation. Geochim. Cosmochim. Acta, vol. 75, 3662-3676

Armitage, P.J. (2011). Dynamics of Protoplanetary Disks. *Annual Review of Astronomy and Astrophysics,* Vol. 49, pp. 195-236, ISSN: 0066-4146

Basu, S., Antia, H. M. (2008). Helioseismology and Solar Abundances. Physics Reports, Vol. 457, No. 5–6, pp. 217-283, ISSN: 0370-1573

Bates, J.B. (1978) Infrared emission spectroscopy. *Fourier Transform IR Spectr.* Vol. 1, pp. 99-142

Benz, W., Slattery W.L., and Cameron A.G.W. (1988) Collisional stripping of Mercury's mantle. *Icarus,* Vol. 74, pp. 516-528, ISSN: 0019-1035

Bernatowicz, T.J., Zinner, E.K. (1997). *Astrophysical implications of the laboratory study of presolar materials.* AIP, New York.

Boss, A.P. (1998). Temperatures in protoplanetary disks. *Ann. Rev. Earth Planet Sci.,* Vol. 26, pp. 53-80, ISSN: 0084-6597

Bowey, J.E. and Hofmeister, A.M. (2005). Overtones and the 5-8 μm spectra of deeply embedded objects. *Monthly Notices of the Royal Astronomical Soc.* Vol. 358, pp. 1383-1393, ISSN: 0035-8711

Brearly, A.J., Jones, R.H., 1998. Chondritic meteorites. *Rev. Mineral.,* Vol. 36, pp. 1-398, ISSN: 1529-6466

Brewster, M.Q. (1992) *Thermal Radiative Transfer and Properties.* John Wiley and Sons, Inc.

Burbine, T.H., Buchanan P.C., Binzel, R.P., Bus S.J., Hiroi T., Hinrichs J.L., Meibom A., and McCoy T.J. (2001) Vesta, vestoids, and the howardite, eucrite, diogenite group: Relationships and the origin of spectral differences. *Meteorites Planet. Sci.,* Vol. 36, pp. 761-781, ISSN: 1086-9379

Burbine, T.H.; Greenwood, R.C.; Buchanan, P.C.; Franchi, I.A., Smith, C.L. (2007). Reflectance spectra of Mesosiderites: Implications for asteroid 4 Vesta. *38th Lunar and Planetary Science Conference*, League City, Texas, USA, 12-16 March 2007

Bus, S.J., Binzel, R.P. (2002). Phase II of the Small Main-Belt Asteroid Spectroscopic Survey. *Icarus*, Vol. 158, pp. 146-177, ISSN: 0019-1035

Basaltic Volcanism Study Project (1981). Basaltic Volcanism on the Terrestrial Planets. Pergamon, New York, ISBN: 0-08-028086-2

Cameron, A.G.W., Fegley, B., Benz, W., & Slattery, W. L. (1988). The strange density of Mercury: theoretical considerations, In: *Mercury*, C. Chapman, C. & Vilas, F. (Eds), pp. 692-708. University of Arizona Press, ISBN 0-8165-1085-7, Tucson, AZ.

Chamberlin, M.A., Sykes, M.V., Tedesco, E.F. (2011). Mid-Infared lightcurve of Vesta. *Icarus*, Vol. 215, (06-2011), pp. 57-61, ISSN: 0019-1035

Christensen, P.R., Bandfield J. L., Hamilton V.E., Howard D.A., Lane M.D., Piatek J.L., Ruff S.W., and Stefanov W.L. (2000) A thermal emission spectral library of rock-forming minerals. *J. Geophys. Res.*, Vol. 105, pp. 9735-9739, ISSN: 0885-3401

Clayton, R.N. and Mayeda T.K. (1996) Oxygen isotope studies of achondrites. *Geochim. Cosmochim. Acta*, Vol. 60, pp. 1999-2017, ISSN: 0016-7037

Consolmagno, G.J., Brit, D.T. (2003). Stony meteorite porosities and densities: A review of the data through 2001. *Meteoritics and Planetary Science*, Vol. 38, No. 8, pp. 1161-1180, ISSN: 1086-9379

Criss, R.E., Farquhar, J. (2008). Abundance, Notation, and Fractionation of Light Stable Isotopes. *Reviews in Mineralogy & Geochemistry; Oxygen in the Solar System*, Vol. 68, pp. 15-30, ISSN: 1529-6466

Dotto, E., Muller. T.G., Barucci M.A., Encrenaz Th. Knacke R.F. Lellouch E, Doressoundiram A., Crovisier J., Brucato J.R., Colangeli L., and Mennalla V. (2000). ISO results on bright Main Belt asteroids: PHT-S observations. *Astron. Astrophys.*, Vol. 358, pp. 1133-1141, ISSN: 0004-6361

Drake, M.J. (2001). The eucrite/Vesta story. *Meteor. Planet. Sci.*, Vol. 36, pp. 501-513, ISSN: 1086-9379

Duke, M.B., Silver L.T. (1967). Petrology of eucrites, howardites and mesosiderites. *Geochim. Cosmochim. Acta*, Vol. 58, pp. 3921-3929, ISSN: 0016-7037

Dymek, R.F., Albee A.L., Chodos A.A., and Wasserburg G.J. (1976). Petrography of isotopically-dated clasts in the Kapoeta howardite and petrologic constraints on the evolution of its parent body. *Geochim. Cosmochim. Acta*, Vol. 40, pp. 115-1130, ISSN: 0016-7037

French, B.M. (1998) *Traces of Catastrophe: A Handbook of Shock-Metamorphic Effects in Terrestrial Meteorite Impact Structures*. LPI Contribution No. 954, Lunar and Planetary Institute, Houston.

Franchi, I.A. (2008). Oxygen Isotopes in Asteroidal Materials. *Reviews in Mineralogy & Geochemistry; Oxygen in the Solar System*, Vol. 68, pp. 345-397, ISSN: 1529-6466

Gaffey, M.J. (1976). Spectral reflectance characteristics of the meteorite classes. *J. Geophys. Res.*, Vol. 81, pp. 905-920, ISSN: 0885-3401

Gaffey, M.J. (1997) Surface lithologic heterogeneity of asteroid 4 Vesta. *Icarus*, Vol. 127, pp. 130-157, ISSN: 0019-1035

Gando, A., Gando, Y., Ichimura, K. et al. (2011). Partial radiogenic heat model for Earth revealed by geoneutrino measurements. *Nature Geoscience*, Vol. 4, (09-2011), pp. 647-651, ISSN: 1752-0894

Goldstein, H. (1950). *Classical Mechanics.* Addison-Wesley Publishing Co., Inc., Reading, Massachusetts, ISBN: 0-201-02510-8

Gomes, R., Levinson, H.F., Tsiganis, K., & Morbidelli, A., (2005). Origin of the cataclysmic late heavy bombardment period of the terrestrial planets. *Nature,* Vol. 435, pp. 466-469, ISSN: 0028-0836

Goodrich, C.A.and Bird J.M. (1985). Formation of iron-carbon alloys in basaltic magma at Uivfaq, Disko Island: the role of carbon in mafic magmas. *J. Geology,* Vol. 93, pp. 475-492, ISSN: 0022-1376

Gounelle, M., Spurny, P., Bland, P.A. (2006). The orbit and atmospheric trajectory of the Orgueil meteorite from historical records. *Meteoritics & Planetary Science,* Vol. 41, No. 1, pp. 135-150, ISSN: 1086-9379

Grady, M.M. (2000). *Catalogue of Meteorites.* Cambridge University Press.

Grady, M.M. (2006). The history of research on meteorites from Mars. In: McCall, G.J.H., Bowden, A.J., Howarth, R.J. (Eds.). (2006). *The History of Meteoritics and Key Meteorite Collections: Fireballs, Falls and Finds.* Geological Society, Special Publication 256. 13: 978-1-86239-194-9.

Guan, Y., McKeegan, K.D., MacPherson, G.J. (2000). Oxygen isotopes in calcium-aluminum-rich inclusions from enstatite chondrites: new evidence for a single CAI source in the solar nebula. *Earth and Planetary Science Letters,* Vol. 181, pp. 271-277, ISSN 0012-821X

Heras A.M., Morris P.W., Vandenbussche B., Müller T.G. (2000). Asteroid 4 Vesta as seen with the ISO short wavelength spectrometer. In: *Thermal Emission Spectroscopy and Analysis of Dust, Disks, and Regoliths.* M.L. Sitko, A.L. Sprague, and D.K. Lynch, pp. 205-212, ISBN: 978-1-58381-532-8

Hiroi T., Binzel R.P., Shunshine, J.M., Pieters, C.M., Takeda, H. (1995). Grain sizes and mineral composition of Vesta-like asteroids. *Icarus,* Vol. 115, pp. 374-386, ISSN: 0019-1035

Hofmeister, A.M., Bowey, J.E. (2006). Quantitative IR spectra of hydrosilicates and related minerals. *Monthly Notices Royal Astronomical Society,* Vol. 367, pp. 577-591, ISSN: 0035-8711

Hofmeister, A.M and Criss, R.E. (2005) Earth's heat flux revisited and linked to chemistry. *Tectonophysics,* Vol. 395, pp. 159-177, ISSN: 0040-1951

Hofmeister, A.M and Criss, R.E. (2012). A thermodynamic model for formation of the Solar System via 3-dimensional collapse of the dusty nebula. *Planetary and Space Science,* ISSN 0032-0633 Vol. 62, pp. 111-131

Hofmeister, A.M., Rossman, G.R. (1984). Determination of Fe^{3+} and Fe^{2+} concentrations in feldspar by optical and EPR spectroscopy. *Phys. Chem. Minerals,* Vol. 11, pp. 213-224, ISSN: 0342-1791

Hofmeister, A.M., Keppel E., Speck A.K. (2003). Absorption and reflection spectra of MgO and other diatomic compounds. *Mon. Not. R. Astron. Soc.,* Vol. 345, No. 1, (03-2003), pp. 16-38, ISSN: 0035-8711

Hofmeister, A.M., Wopenka, B, Locock, A., (2004). Spectroscopy and structure of hibonite, grossite, and $CaAl_2O_4$: implications for astronomical environments. *Geochim. Cosmochim. Acta,* Vol. 68, pp. 4485-4503, ISSN: 0016-7037

Javoy, M., 1995. The integral enstatite chondrite model of the earth. *Geophys. Res. Lett.,* Vol. 22, pp. 2219-2222, ISSN: 0094-8276

Kemper, F., Jaeger, C., & Waters L.B.F.M. (2002). Detection of carbonates in dust shells around evolved stars. *Nature,* Vol. 415, pp. 295–297, ISSN: 0028-0836

Kitts K., Lodders, K. (1998). Survey and evaluation of eucrite bulk compositions. *Meteoritics Planet. Sci.,* Vol. 33, pp. 197-213, ISSN: 1086-9379

Lewis, J.S., (1974). The temperature gradient in the solar nebula. *Science,* Vol. 186, pp. 440-43, ISSN: 0036-8075

Lim, L.F., McConnochie, T.H., Bell, J.F., Hayward, T.L. (2005) "Thermal infrared (8-13 μm) spectra of 29 asteroids: the Cornell Mid-Infrared Asteroid Spectroscopy (MIDAS) Survey." *Icarus,* Vol. 173, pp. 385-408, ISSN: 0019-1035

Lodders, K. (1998). A survey of shergottite, nakhlite and chassigny meteorites whole-rock compositions. *Meteoritics Planet. Sci.,* Vol. 33, pp. 183-190, ISSN: 1086-9379.

Lodders, K. (2000). An oxygen isotope mixing model for the accretion and composition of rocky planets. *Space Sci. Rev.,* Vol. 92, pp. 341-354, ISSN: 0038-6308

Lodders, K. (2003). Solar system abundances and condensation temperatures of the elements. *The Astrophysical Journal,* Vol. 591, (06-2003), pp. 1220-1247, ISSN: 0004-637X.

Lodders, K., Fegley, B. (1998). *The Planetary Scientist's Companion.* Oxford, New York.

Love, S.G., Kiel, K. (1995). Recognizing Mercurina meteorites. Meteoritics, Vol. 30, pp269-278, ISSN: 1086-9379.

Low, M.J.D., Coleman, I. (1966). Measurement of the spectral emission of infrared radiation of minerals and rocks using multiple-scan interferometry. *Appl. Opt.* Vol. 5, pp. 1453-1455, ISSN: 1559-128X

Lucey, P.G., Keil K., Whitely, R. (1998). The influence of temperature on the spectra of the A-asteroids and implications for their silicate chemistry. *J. Geophys. Res.,* Vol. 103, pp. 5865-5871, ISSN: 0885-3401

MacPherson, G.J., Simon, S.B., Davis, A.M., Grossman, L., Krot, A.N. (2005). Calcium-Aluminum-rich Inclusions: Major Unanswered Questions. In: Krot, A.N., Scott, E.R.D., Reipurth, B. (2005). *Chondrites and the Protoplanetary Disk ASP Conference Series,* Vol. 341, pp. 225-250, ISSN: 1-58381-208-3

McKeegan, K.D., Leshin, L.A., Russell, S.S., MacPherson, G.J. (1998). Oxygen Isotopic Abundances in Calcium-Aluminium-Rich Inclusions from Ordinary Chondrites: Implications for Nebular Heterogeneity. *Science,* Vol. 280, pp. 414-418 ISSN: 0036-8075

Melosh, H.J., Tonks, W.B. (1993). Swapping Rocks: Ejection and Exchange of Surface Material Among the Terrestrial Planets. *Meteoritics,* Vol. 28, No. 3, vol. 28, (1993), pp. 398, ISSN: 1086-9379

Mittlefehldt, D.W., McCoy, T.J., Goodrich, C.A., Kracher, A. (1998). Non-chondritic meteorites from asteroidal bodies. *Rev. Mineral,* 36, pp. 1-195, 1529-6466

Molini-Velsko, C., Mayeda T.K. and Clayton R.N. (1986) Isotopic composition of silicon in meteorites. *Geochim. Cosmochim. Acta,* Vol. 50, pp. 2719-2726, ISSN: 0016-7037

Molster, F.J., Lim, T.L., Sylvester, R.J., Waters, L.B.F.M. Barlow, M.J., Beintema, D.A., Cohen, M., Cox, P., Schmitt, B., (2001). The complete ISO spectrum of NGC 6302. *Astron. Astrophys.,* Vol. 372, pp. 165–172, ISSN: 0004-6361

Nittler, L.R. and 14 others (2011) The major-element composition of Mercury's surface from MESSENGER X-ray spectrometry. Science, Vol. 333, pp. 1847-1850. ISSN: 0036-8075.

Pepowski, P.N. and 16 others (2011) Radioactive elements on Mercury's surface from MESSENGER: implications for the Planet's formation and evolution. Science, Vol. 333, pp. 1850-1852. ISSN: 0036-8075.

Prinz, M., Weisberg, M.K., Nehru, C.E., MacPherson, G.J., Clayton, R.N., Mayeda, T. (1989). K. Petrologic and Stable Isotope Study of the Kakangari (K-Group) Chondrite: Chondrules, Matrix, CAI's. *Abstracts of the Lunar and Planetary Science Conference*, Vol. 20, p. 870

Ruzicka, A., Snyder G.A., Taylor L.A. (1997). Vesta as the howardite, eucrite and diogenite parent body: implications for the size of a core and for large-scale differentiation. *Meteoritics Planet. Sci.* Vol. 32, pp. 825-840, ISSN: 1086-9379

Shearer, C.K., Fowler G.W., Papike, J.J. (1997). Petrogenic models for magmatism on the eucrite parent body: evidence from orthopyroxene in diogenites. *Meteoritics Planet. Sci.* Vol. 32, p. 877, 1086-9379

Spudis, P.D., Guest, J.E. (1988). Stratigraphy and Geologic History of Mercury. In: *Mercury* The University of Arizona Press, Tuscon, ISBN: 0-8165-1085-7

Surkov, Yu. A., Kirnozov, F.F., Glazov, V.N., Dunchenko, A.G., Tatsy, L.P., Sobornov, O.P. (1987). Uranium, thorium, and potassium in the Venusian rocks at the landing sites of Vega 1 and 2. *J. Geophys. Res.*, Vol. 92, pp. 537-540, ISSN: 0885-3401

Symon, K.R. (1971). *Mechanics.* Addison-Wesley. Reading, Massachusetts.

Taylor, H.P., Duke, M.B., Silver, L.T., Epstein, S. (1965). Oxygen isotope studies of minerals in stony meteorites. *Geochi. Cosmochim. Acta* , Vol. 29, pp. 489-512, ISSN: 0016-7037

Tholen, D.J. (1989). "Asteroid taxonomic classifications". *Asteroids II.* Tucson: University of Arizona Press, ISBN:08165-11233

van Breemen, J.M., Min, M., Chiar, J.E., Waters, L.B.F.M., Kemper, F., Boogert, A.C.A., Cami, J., Decin, L., Knez, C., Sloan, G.C., and Tielens, A.G.G.M., (2011). The 9.7 and 18 mm silicate absoption profiles towards diffuse and molecular cloud lines-of-sight. *Astron. Astrophys.* Vol. 526, in press, ISSN: 0004-6361

Vickery, A.M., Melosh, H.J. (1983). The Origin of SNC Meteorites: An Alternative to Mars. *Icarus*, Vol. 56, pp. 299-318, ISSN: 0019-1035

Wasserburg, G.J., MacDonald G., Hoyle F., and Fowler W.A. (1964). Relative contributions of uranium, thorium and potassium to heat production in the Earth. *Science,* Vol. 143, pp. 465-467, ISSN: 0026-8075

Wetherill, G.W. (1988). Accumulation of Mercury from planetesimals, In: *Mercury*, C. Chapman, C. & Vilas, F. (Eds), pp. 670-691. University of Arizona Press, ISBN 0-8165-1085-7, Tucson, AZ.

Zolensky, M.E. et al., (2006). Mineralogy and petrology of Comet 81P/Wild 2 Nucleus samples. *Science*, Vol. 314, pp. 1735-1739, ISSN: 0036-8075

8

The Eruptions of Sarychev Peak Volcano, Kurile Arc: Particularities of Activity and Influence on the Environment

Alexander Rybin[1], Nadezhda Razjigaeva[2], Artem Degterev[1], Kirill Ganzey[2] and Marina Chibisova[1]
[1]Institute of Marine Geology and Geophysics of Far East Branch, of Russian Academy of Sciences, Yuzhno-Sakhalinsk,
[2]Pacific Institute of Geography of Far East Branch, of Russian Academy of Sciences, Vladivostok,
Russia

1. Introduction

More then 68 quaternary volcanic edifices are picked out in the Kurile Islands, among them 36 are active and potentially dangerous. During 45 thousand years not less then 12 large explosive eruptions, connected with the formation of the calderas, have occurred. During these eruptions a great amount of ash emitted and large pumice-pyroclastic covers formed, that caused the climate and landscape changes (Melekestsev et al., 1988).

During the historical time (about 250 years for the Kurile Islands) about 29 great and catastrophic eruptions were fixed (Gorshkov, 1967). In XXth century the volcanoes of the Central and Northern Kurils were the most active and productive by the volume of the erupted material. These tendencies have continued in current century. Explosive eruptions occurred many times on Chikurachki volcano (Girina et al., 2008). Phreatic and phreato-magmatic explosions were detected on the Ebeko, Berga, Chirinkotan, Severgin volcanoes.

Because of rare population of the Kurile Islands the damage at the eruptions was not so considerable as it usually was in densely populated island countries such as Indonesia, Japan and others. However, some cases were written in the historical documents, when the eruptions in the Kurile Islands caused the material damage and victims. In 1778 during the strong eruption of Raikoke volcano 15 Russian manufacturers died. There are the historical materials about the destruction of ainu settlements in Shiashkotan Island after the eruption of Sinarka volcano (Gorshkov, 1967). In 1933 at the eruption of Severgin volcano in Harimkotan Island the Japanese settlement was destroyed, tsunami waves formed by the eruption, caused the death of several people in the neighboring islands Onekotan and Paramushir (Miyakate, 1934).

During the eruptions of Sarychev Peak volcano in 1946 and 1973 the evacuation of military unit was conducted in Matua Island. At Tyatya volcano eruption in 1973 the military camp was destroyed on the Lovtsov peninsular (the north of Kunashir Island). In settlement Yuzhno-Kurilsk located at the distance of 60 km from the volcano, the ash fell and great panic among the population was. Great panic was also fixed at small phreatic eruption of Ivan Grozny volcano in Iturup Island in 1989 (Abdurakhmanov et al., 1990).

At present time the constant population in the Kurile Islands lives only in the southern (Kunashir, Iturup, Shikotan) and northern (Paramushir) islands, and practically all the settlements, excepting those located in Shikotan Island (the Small Kurile Arc) are in the zone of different volcanic danger.

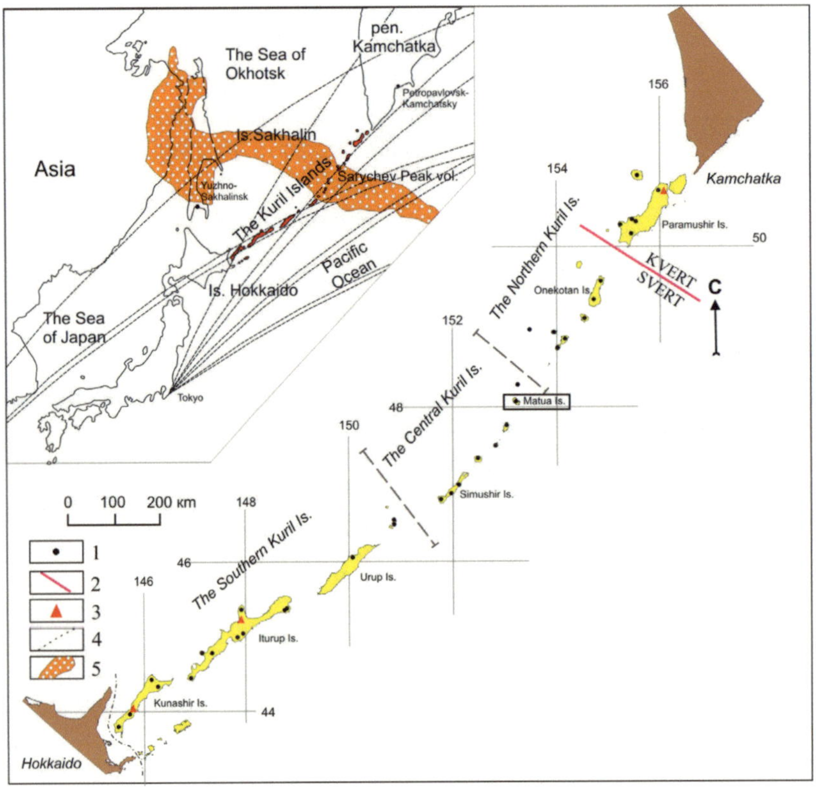

Fig. 1. Map of the Kuril Islands and the zone of the responsibility of the monitoring of the volcanic activity of Sakhalin Volcanic Eruption Response Team (SVERT) and Kamchatka Volcanic Eruption Response Team (KVERT). KVERT issues information for the northernmost Kuriles (Paramushir and Atlasova Islands). The remaining Kuriles are monitored by SVERT. Sarychev Peak is located on Matua Island in the central Kuriles. Inset shows a schematic version of the primary air routes in the vicinity of the northwest Pacific.

During last decades the cases when airplanes fall into the clouds of ash, become more frequent because of increasing of volume and geography of airtraffic. The last information is represented very important because the most part of air routs connecting the North America and the East-Asia region goes along the Kurile Islands (fig. 1). During passed 40 years 4 serious cases, when airplanes fell into the clouds of ash, occurred only on Alaska, three last incidents were during last 15 years.

After 33 years of break in eruptive activity, Sarychev Peak volcano, located in Matua Island in the Central part of the Kurile arc and being one of the most active volcanoes, began its work (fig. 1, 2). The eruption in 2009 several days impeded the work of airtransport passed along the Kurile arc (Salinas, 2009). Sulfur aerosols emitted as a result of explosive activity of the volcano could possibly influence to the climate of region (Haywood et al., 2010). Besides that a large emission of pyroclastic material such as pyroclastic flows and tephra produced considerable geological and ecological effects (fig. 2). The landscape structure of Matua Island underwent the cardinal changes: in the bounds of Sarychev Peak volcano edifice the full reconstruction of the landscapes occurred, the view of the adjacent surroundings was considerably changed (Ganzey et al., 2010).

Fig. 2. Sarychev Peak volcano: A - explosive activity of Sarychev Peak volcano at the eruption in 2009. On the image made by the astronauts of ISS the eruptive column, reached the height of several kilometers, and pyroclastic flows, descended along the slopes of the volcano are seen, 12 June 2009. (Published by Earth Sciences and Image Analysis Laboratory, NASA Johnson Space Center, http://earthobservatory.nasa.gov/NaturalHazards/view.php?id=38985);
B - Sarychev Peak after the eruption on June 27, 2009. Robust gas and water vapor cloud rises from the summit vent. The green vegetated terrain in the near ground is the southeastern sector of the island that was only minimally impacted by the eruption (fig. 2). Photo A.V. Rybin, IMGG FEB RAS; C - Sarychev Peak volcano in 2010, the north-western view. Photo A.V. Rybin, IMGG FEB RAS.

In given work we represent the results of the researches of Sarychev Peak volcano eruption in 2009: (1) a common characteristic of Sarychev Peak volcano is shown - geological structure, morphology, data about its historical eruptions; (2) the chronology of the eruptive activity at the eruption 2009 is described; (3) data of study of the eruption products, their facial and chemical composition are presented; (4) the influence of Sarychev Peak volcano eruption to the nature of Matua Island is described.

2. Common characteristic of Sarychev Peak volcano

Sarychev Peak volcano is one of the most active volcanoes of the Kurile Island arc (fig. 1). The edifice of Sarychev Peak Volcano occupies the north-western part of Matua Island that is located in the central part of the Kurile Islands. The island has a form of sublongitude ellipse with the sizes 6×12 km, the area of it before the eruption of Sarychev Peak volcano in June 2009 was 52.5 km^2 (fig. 2).

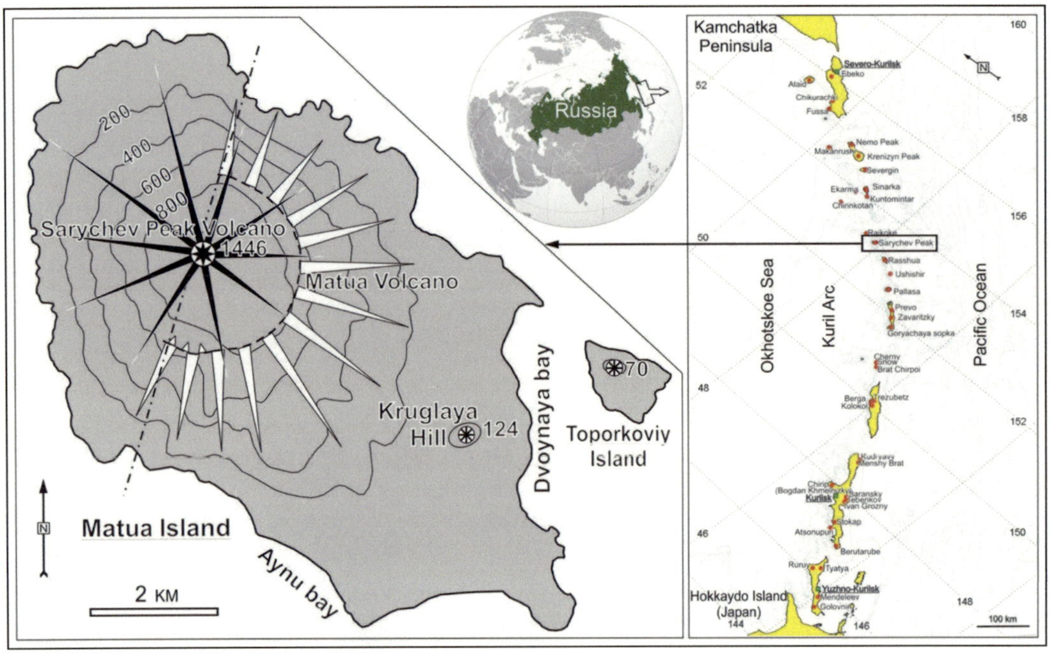

Fig. 3. Geological scheme of Sarychev Peak volcano, stroke shows the line of fault picked out by G.S. Gorshkov (1967). The map of the Kurile Islands is on the additional map.

Sarychev Peak volcano is intracalderal stratovolcano (elevation 1446 m; coordinates: 48.92° N., 153.20° E), formed mainly by pyroclastic material (fig. 3). The volcano is located in the caldera of Pleistocene volcano Matua, the ruins of which form the north-western part of Matua Isl. In spite of the volcano is formed by "Somma-Vesuvius" type, the characteristic elements of its structure is closed to a large measure. The north-western part of the island with a half of somma of Matua volcano was sank along the fault plane (fig. 3), detected by G.S. Gorshkov (1967) and overlapped by the formations of young cone - pyroclastic and lava flows. The basement of the volcanic edifices consists of volcanogenic rocks with absolute age according to data of K-Ar-dating Ishizuka et al. (2011) 1.61 mln. years.

3. Historical eruptions of Sarychev Peak volcano

Since 1760th not less than 10 eruptions of Sarychev Peak volcano were fixed in the historical chronics and described with different degree of details (Andreev et al., 1978; Glavatsky & Efremov, 1948; Gorshkov, 1948, 1967; Grishin & Melekestsev, 2010; Levin et al., 2009; Markhinin, 1964; Rybin et al., 2010; Shilov, 1962). Among them the most studied events occurred in the XXth century: 1946, 1960, 1976 и 2009. The behavior of these eruptions in comparison with previous events is reconstructed in details, and their sequences were studied soon by specialists-volcanologists. Lower the brief description of historical eruption of Sarychev Peak volcano is given.

3.1 The eruptions of the volcano in XVIII-XXI centuries

XVIII century

The eruption in 1760th: strong explosive eruption, the data about it are contained in the descriptions of Cossack sotnik I. Chyorny, visited the Kurile Islands in 1766-1769 and known about it from inhabitants (Polonsky, 1994). According to this information, as a result of this eruption in considerable degree in Matua Isl. the vegetation was destroyed, and Toporkovy Isl. was fully burnt.

XIX century

The eruptions in 1878-1879: weak effusive-explosive eruption, a brief mention about it is in the work of English trader G. Snow (1992). It was reported about lava flows slowly descended to the sea.

XX century

The eruptions on the 17-22 of January 1923: weak explosive eruption when "the explosions with emission of ash and scoria" were marked (Kamio, 1931; Gorshkov, 1967).

The eruptions on the 14th of February 1928: moderate explosive eruption, accompanied with "storm of bombs and lappili in the surroundings of the crater" (Tanakadate, 1931; Gorshkov, 1948).

The eruptions on the 13th of February 1930: explosive eruption is of moderate power. There are no the concrete estimations of volume of erupted products; it was reported only, that it was erupted "colossal" amount of volcanic material. The accumulation of pyroclastic in the southern part of the island caused the increasing of the shore line up to 30 m (Tanakadate, 1931; Gorshkov, 1948)

The eruptions on the 9-17th of November 1946: strong explosive-effusive eruption, the process and the sequences of it were reconstructed on the base of polling data, collected by S. Glavatsky in Kamchatka and G. Efremov in Sakhalin (Glavatsky & Efremov, 1948).

The eruption was accompanied by emission of large amount of heated clastic material, mainly pyroclastic flows and tephra. The accumulation of pyroclastic material in the shore zone caused to local increasing of the shore line, the sites of new formed surface were found by the eye-witnesses in the north-eastern, north-western and south-western parts of the island. Lava flows formed several new capes. The height of ash clouds according to data of the eye-witnesses reached 7 km. Ash-falls caused by this eruption were fixed on the territory of neighboring islands and even in Kamchatka (Glavatsky & Efremov, 1948).

The eruptions in 1954: weak explosive activity of the volcano at the end of summer and autumn 1954, expressed in weak ash emissions. Luminescence above the crater was fixed, however lava eruption did not occur, after lifting it solidified in the form of lava plug (Gorshkov, 1967; Shilov, 1962).

The eruptions from the 30th of August till the 3d of September 1960: moderate explosive eruption was accompanied by emission of ash and formation of pyroclastic flows. By the information of inhabitants of the island the first explosion elevated the eruptive column up to 4.5 km, the height of next emissions did not excess 0.5 km (Shilov, 1962).

The eruptions from the 23d of September till 2d of October 1976: moderate explosive-effusive eruption, the only historical eruption of Sarychev Peak volcano that was observed by the volcanologists (Andreev et al., 1978). The eruption consisted of series of explosions up to the height from 0.5 to 2.5 km. Along the north-western and western slopes of the cone the pyroclastic flows descended, they burnt the soil-vegetation cover on their way. Lava flows are fixed on the western, south-western and north-western slopes of the edifice; two of them (western and south-western) reached the sea and formed two new capes.

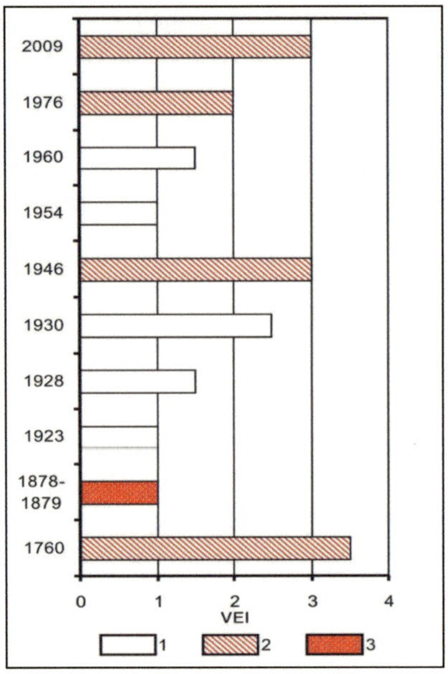

Fig. 4. The chronology of the eruptions of modern eruptive stage of Sarychev Peak volcano. Legend: 1 - explosive eruption; 2 - explosive-effusive eruption; 3 - effusive - explosive eruption. The evaluation of eruptions Volcanic Explosive Index (VEI) was done on the base of characteristics of historical eruptions found in the literature.

After the eruption in 1976 the information about the state of the volcano was received from local inhabitants and rare air observations. By data of SEAN Bulletin (Scientific Event Alert Network Bulletin), published by Smithson's Institute (USA) in 1986 and 1989 two episodes of volcano activity - weak phreatic eruptions were detected. (http://www.volcano.si.edu/world/volcano.cfm?vnum=0900-24=&volpage=var). It is probably that it was small activity expressed in increasing of steam-gas emission or in the form of weak phreatic outbursts.

In August 2007 A.K. Klitin (Yuzhno-Sakhalinsk) and S.A. Chirkov (Institute of volcanology and seismology FEB RAS, Petropavlovsk-Kamchatsky) visited the crater of the volcano. In 2008 the crater was researched by the scientists of the Institute of Marine Geology and Geophysics FEB RAS R.V. Zharkov, D.N. Kozlov and A.V. Degterev. Great fumarolic activity was observed in the crater of the volcano, but because of large gas-laden the fumarolic areas were not investigated. The temperature on the crater rim did not exceed 100°C.

As it is seen from the given data, the character of the eruptive activity of Sarychev Peak volcano is mainly explosive and explosive-effusive – from weak phreatic emissions with VEI 0-1 to strong explosive eruptions of sup-plinian type with VEI 3-4 (fig. 4). The characteristic features of the eruptions of Sarychev Peak volcano are their high explosiveness at which emitted fragmental material forms pyroclastic flows and long ash plumes. Sarychev Peak volcano is one of few volcanoes of the Kurile Island Arc at the historical eruptions of which the ash fell not only in the limits of the island arc but also to the north-east in Kamchatka (Glavatsky & Efremov 1948) and to the west on the territory of Khabarovsky Kray and Sakhalin (Levin et al., 2009). The products of all the historical eruptions of Sarychev Peak volcano are represented by andesite-basalts with 53.78 wt. %, SiO_2, 3.07 wt.% Na_2O, 0.96 wt.% K_2O (Table 1).

Date	1930		1946		1960		1976				2009	
№	1	2	3	4	5	6	7	8	9	10	11	12
SiO_2	53.4	50.85	53.84	52.95	53.6	54.82	54.72	54.76	53.22	54.04	54.41	54.85
TiO_2	1.09	1.07	0.96	0.8	0.93	0.88	0.93	0.93	0.85	0.87	0.88	0.93
Al_2O_3	19.14	18.88	18.58	19.14	16.3	18.02	18.1	18.16	17.95	18.27	18.23	17.78
Fe_2O_3	2.64	4.83	4.43	3.93	3.71	3.76	4.31	4.2	4.75	9.61	9.74	9.3
FeO	5.76	5.06	4.26	5.76	4.47	5.14	4.68	4.74	5.08			
MnO	0.28	0.42	0.22	0.18	0.11	0.2	0.2	0.2	0.2	0.2	0.2	0.2
MgO	4.42	4.38	3.96	3.65	3.83	3.74	3.7	3.75	4.13	4.16	4.12	3.86
CaO	8.7	9.3	8.91	8.8	7.89	8.76	8.82	8.72	9.24	9.17	9.1	8.91
Na_2O	3.22	2.88	3.24	3.23	2.95	3.06	3.18	3.13	2.98	2.97	2.84	3.25
K_2O	1.08	0.99	1.06	0.93	0.22	1.06	1.1	1.15	1.06	0.91	0.93	1.02
P_2O_5	0.21	0.1	0.08	0.07	0.17	0.36	0.3	0.36	0.48	0.21	0.21	0.23
Sum	100.11	100.35	99.61	99.57	100.38	100.07	100.29	100.34	100.15	100.52	100.76	100.33

Table 1. Summary composition of lavas and pyroclastic material of some historical eruptions of Sarychev Peak volcano (mas. %).

Note. All the analyses are given by reference data excepting № 12: 1. Volcanic bomb 1930 , analyst V.P. Enman, Institute of Volcanology (IV) (Gorshkov, 1967); 2. Matrix of pyroclastic flow 1946, analyst N.S.Klassova, IV (Gorshkov, 1967); 3. Volcanic bomb 1946 (average of two analyses), (Gorshkov, 1967); 4. The material of pyroclastic flows 1946(?),Sakhalin Complex Scientific Research Institute (SakhCSRI) (Fedorchenko et al., 1989); 5. Volcanic sand 1960, SakhCSRI (Shilov, 1962); 6. Lava 1976 (Andreev et al., 1978); 7. Lava 1976 (Andreev et al., 1978); 8. Volcanic bomb 1976 (Andreev et al., 1978); 9. Volcanic ash 1976, analyst T.S. Osetrova, IV (Andreev et al., 1978); 10. Volcanic bomb 2009, Alaska University (Fairbanks, USA) (Rybin et al., 2011); 11. Volcanic bomb 2009, Alaska University (Fairbanks, USA) (Rybin et al., 2011); Lava 2009, analysts Gorbach G.A., Tkalina E.A., Hurkalo N.V., Far East Geological Institute.

4. The eruption of Sarychev Peak volcano on the 11-17th of June 2009

Strong explosive-effusive eruption on the 11-17 of June 2009 came to the list of the strongest volcanic events occurred in the Kurile Islands during last 300 years and became the first eruption in the Central Kurile Islands in XXI century. Eruptive clouds by data of Tokyo VAAC (Volcanic Ash Advisory Center) raised the height 8-16 km, the plume of the volcanic ash stretched to the western and north-western direction at the distance of 1.5 thousand km, to the east and south-eastern direction at the distance more then 3.0 th. km, this corresponds with the sector of covering from Amur Region to Alaska Peninsula. For the first time during the historical period the ashfalls were fixed on the territory of Sakhalin Island and Khabarovsky Kray (Grishin et al., 2010; Levin et al., 2010).

4.1 The chronology of the eruptive activity of Sarychev Peak volcano at the eruption in 2009

The first signs of volcanic activity of Sarychev Peak volcano were fixed by Sakhalin Volcanic Eruption Response Team (SVERT): on satellite images of NOAA (spectroradiometer AVHRR) and Terra (spectroradiometer MODIS) on 11th of June, a thermal anomaly testified to the increasing of fumarolic activity (Rybin et al., 2010). Further monitoring of the eruptive process was also done on the base of data of distance sounding.

On 12th of June the increase of the activity of the volcano began, eight volcanic explosions occurred; their height above the rim of the crater was from 5 to 12 km (fig. 5). The largest explosion during this day occurred at 07:57, as a result of this a dense ash cloud with diameter 35 km formed on the height 12 km. 3.5 hours later the cloud without changing of form moved eastward 30 km from the volcano, further ash plumes stretched in two directions: to the south-east at the distance 200 km and to the south-west 185 km. Beginning from 14:57 the series of the explosions occurred and ash plumes from them increased in sizes and stretched to the south-eastern direction more than 500 km and to the south-western direction more 150 km. The eruption accompanied by the descending of heat pyroclastic flows along the slopes of the volcano.

On 13th of June between 01:30 and 04:50 a series of explosions occurred, the maximal height of ash columns above crater rim reached 10 km. At 01:30 on the satellite images the ash cloud with diameter more than 50 km was observed, in 2 hours it moved 80 km from Matua Island to the south-eastern direction.

Next two explosions had the same characteristics. Ash cloud of isometric form with diameter 18 km originated at the explosion which had occurred at 04:50. Over an hour the cloud had increased up to 60 km, them the dense part of it began to move eastward formed a plume more than 500 km in the south-eastern direction. 5 hours later the explosion with ash emission up to 10 km occurred, it formed ash cloud 60 km in diameter during 5 hours grew up to 120 km and also moved in the south-eastern direction, the width of originated plume reached 200 km. At 21:30 the greatest explosion for all the period of eruption was fixed, the diameter of dense part of ash cloud reached 65 km. During four hours it increased and reached 140 km in diameter. Ash plumes moved in two directions, the denser part stretched in the south-eastern direction, and the part with lower content of ash traveled in the north-western direction. After this explosion the considerable pause of volcanic activity was observed, it lasted 14 hours.

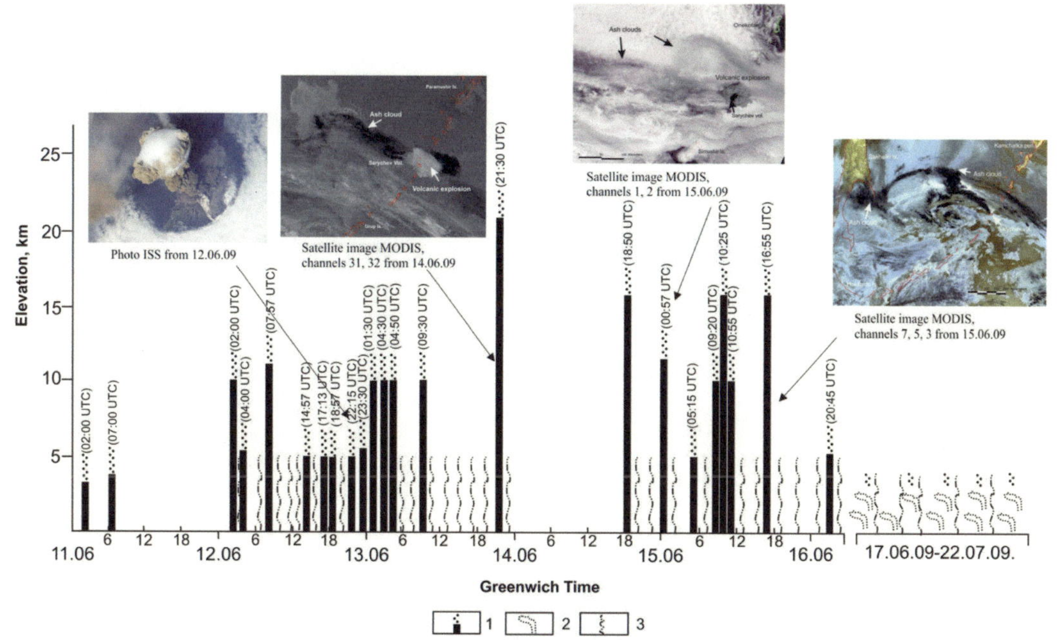

Fig. 5. Summary of satellite observations, inferred explosive events and representative imagery through the eruption, adapted from Levin et al. (2010). Heights of individual ash clouds were determined by using the altitude temperature (Kienle & Shaw, 1979; Sparks et al., 1997) and parallax (referred to in Oppenheimer 1998) methods from the MTSAT satellite data and those reported by the Tokyo VAAC. Symbols are as follows: 1 - gas-steam emissions, 2 - explosions, 3 - continuous ash emission, 4 - times are reported in Greenwich Time or UTC.

The volcanic explosion in June, 14 at 18:57 began the next series of the events. Ash cloud of the first explosion had the diameter about 20 km, then in 3 hours its size increased to 120 km and at 21:30 ash cloud lost its round form and stretched in the western and eastern directions.

On 15th of June seven explosions were detected, ash clouds had diameters from 62 to 170 km (fig. 6). Ash plumes move in the north-western and south-eastern directions. Continuous emissions of ash occurred between the explosions. Last large explosion was at 16:55.

On the 16th of June the eruption lasted. A continuous supply of volcanic material and weak volcanic explosions were fixed during this day. Such situation remained until June, 19.

Since June, 20 the volcano went to the stage of large steam-gas activity, which was accompanied by rare explosions with small amount of ash. During the period from June, 20 till October the thermal anomaly, connected with heated pyroclastic flows on the slopes of the volcano, was often detected in the satellite images.

Ash falls during the period of eruption (collections of the material and the eye-witness accounts) were detected in Raikoke, Rasshua, Ushishir, Ketoi, Simushir, in the northern part of Urup and on all the territory of Sakhalin and in Khabarovsky Kray.

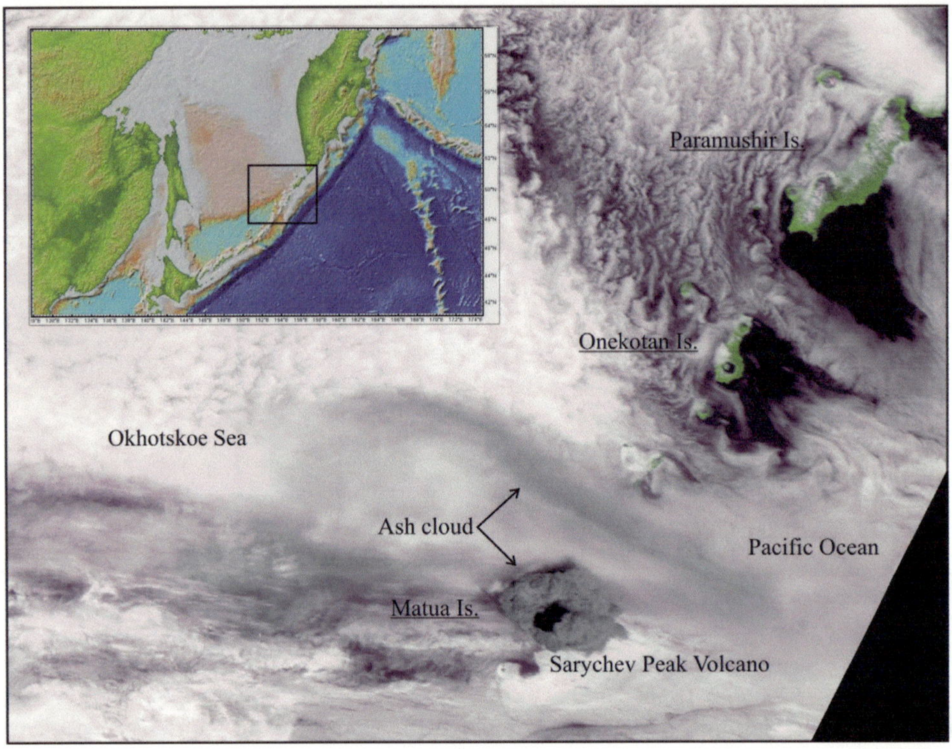

Fig. 6. Explosive activity of Sarychev Peak volcano on 15th of June 2009. Ash cloud formed above the volcano and its diameter is more then 100 km (satellite image MODIS).

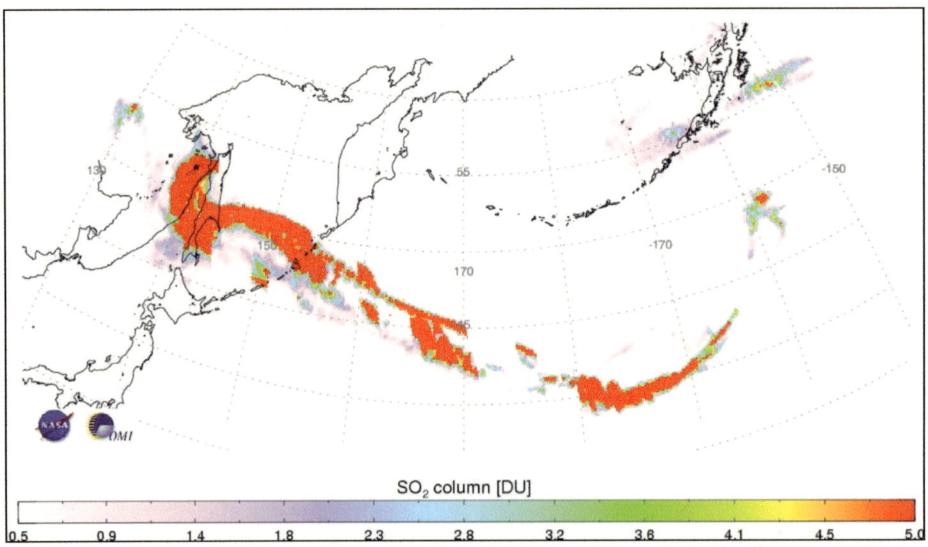

Fig. 7. Summary concentration of sulfur dioxide from 11th till 17th of June 2009, emitted during the eruption of Sarychev Peak volcano (by data Aura – OMI, published by Earth Sciences and Image Analysis Laboratory, NASA Johnson Space Center, http://earthobservatory.nasa.gov/IOTD/view.php?id=38975).

According to data of the group of ozone monitoring of NASA, the total concentration of sulfur dioxide in aerosol from 11th till 17th of June was the maximal for the Pacific region in 2009, aerosol clouds by the instrumental methods (satellite AURA) were fixed up to the western shore of Alaska (fig. 7). Data of satellite CALIPSO corroborate the emission of ash material on the height up to 15 km and possibly up to 21 km. The ions of sulfur on such heights can live for a long time and, reflecting the sun light cause the fall of temperature.

Eruptive clouds, connected fine-dispersed pyroclastic material and lifted during the explosive activity on the considerable height, represented a serious danger for airplanes. By data (Salinas, 2009) at the eruption of Sarychev Peak volcano in 2009 65 routs, passed along the Kurile Islands, were changed; 6 changed the course; 2 planes returned to airports of departure; besides 12 non-planed landings were made for re-fuelling. Additional expenses of air-companies because of eruption were evaluated $1.8 mln (Salinas, 2009). Due to operative acts of Sakhalin Volcanic Eruption Response Team (SVERT) in collaboration with Alaska Volcano Observatory (AVO) the negative sequences were avoided.

4.2 Pyroclastic deposits of the Sarychev Peak volcano eruption in 2009

Occurred eruption also as previous events of Sarychev Peak volcano was characterized by high explosiveness, at which the material, erupted by the volcano, formed numerous pyroclastic flows. The process of their formation was fixed in photographs, made by ISS astronauts (fig. 1 a). Ten days later after the ending of active phase of the eruption the products of the explosive activity of the volcano were studied by us during the field works (fig. 8). On the base of the complex of stratigraphical and lithological data the next types of fascias were detected (according to the classification in the work (Fisher & Schminke, 1984): the deposits of pyroclastic flows; the deposits of ground surge; the deposits of ash cloud surge; the deposits of ash cloud of pyroclastic flow; tephra fall. Tephra of the eruption 2009 are represented by two benches of ashes: brown and grey, the bound between them is rather clear. The ash of brown color was probably connected with the initial stage of the eruption, when the voiding of volcanic vent had occurred after considerable pause in its eruptive activity. The ash of brown color is covered with the ash of grey color. Its fall possibly occurred during the final stage of the eruption.

Total volume of volcanites erupted during the active phase of the eruption (11-17 June 2009) was, according to different estimations from 0.1-0.2 (Grishin et al., 2010) to 0.2-0.4 km^3 (Levin et al., 2010b; Rybin et al., 2010). The products of the eruption were presented by andesite-basalts that are typical rocks of its modern eruptions with the content 54.04-54.41 wt.% SiO_2, 18.23-18.27 wt.% Al_2O_3, and 0.91-0.93 wt.% K_2O (Rybin et al., 2011).

The deposits of pyroclastic flows strip mainly in the bounds of near-shore part of the island, forming new-formed surface in the small harbors of shore line. On one of the sites of such surface we found a small lake (~10×25 m), its formation occurred as a result of supply of storm water or, that is more probably, by local tsunami of volcanogenic origin (fig. 8). Tsunami of such type, but in considerably larger quantity, was fixed at the eruption of Severgin Peak volcano (Harimkotan Island) in January 1933, when the supply of pyroclastic material of directed explosion to near-sea water caused the formation of tsunami waves with height ~20 m (Gorshkov, 1967).

Fig. 8. Pyroclastic deposits: (a) the site of land formed as a result of accumulation of pyroclastic material in near-shore area and the lake appeared on its surface; (b) the frontal part of pyroclastic flow; (c) the surface of pyroclastic flow covered by the ridges of fragments; (d) fumaroles activity of new pyroclastic deposits. Photos (a, c, d) were made 10 days later after the end of the eruption.

The analysis of satellite images allow to mark out, at least 8 pyroclastic flows, the most of them occupy the northern and western slopes of the volcano. This estimation in definite degree is conventional, because the flows cover each other and in the lower part of the cone form practically interflowing cover, setting off the foot of the volcano. The total area of pyroclastic flows is about 25 km^2 (Ganzey et al., 2010). During the period of conducting of the field works in 2009 the intensive solfataric activity connected with residual heat of pyroclastic material was seen on the flows (fig. 8). By measuring of R.V. Zharkov the maximal temperature of output of gases was 476°C. Visible thickness of the flows in the near-shore part of the island varies from <1 to 4 m, depending on their distance from eruptive center. They are formed by porous lightly rounded fragments of andesite-basalts with size 15-30 cm and account for about 50-80% of total volume of visible thickness. The separate blocks >1 m are met. The maximal concentration of the fragments is characteristic for the surface of the flow, where they form lengthwise stretched ridges (fig. 8).

4.3 The changes of the landscapes of Matua Island after the eruption of Sarychev Peak volcano in 2009

As a result of Sarychev Peak volcano eruption the landscapes of Matua Isl. underwent the considerable changes. The full reconstruction of the landscapes occurred in the bounds of the Sarychev Peak volcano cone that is connected with moving of pyroclastic flows, which effected thermal (in June, 28 in the northern part of the island on the pyroclastic flow on the depth 30 cm the temperature was 420°C) and mechanical influence. All the vegetation and soil cover were destroyed (fig. 9). The summit part of the volcano was occupied by the landscapes of stratovolcanic cone with thick layer of pyroclastic deposits. The landscapes of steep slopes and slopes of average steepness are dominant on the slopes of the volcanic cone; they are covered with loose and weakly lithified pyroclastic deposits (fig. 9). In separate sites the landscapes of steep slopes and average steepness slopes of lava flows and disjoint scarps remain. (Levin et al., 2010).

Near the south-eastern foot of volcanic cone the belt stretched, which represented the ecoton between the zones of full ("dead zone") and the least reconstruction of the landscapes. The boundary of distribution of the brush-woods of alder descended again and now it was located on the mark about 450 m. In transition zone we observed shrinkage of practically all the brushes of alder. Low plants suffered larger from the ash-falls especially such small bushes as Rhododendron *aureum, Empetrum sibiricum, Cassiope lycopodioides, Phyllodoce aleutica* et al. Some of them near "dead zone" were covered with the layer of ash, but continue to blossom. Projective vegetation on the sites, where ash cover reached 10-12 cm, was only 10-15%. Low bushes such as cowberries suffered mostly; part of them on the sites of intense ash-fall was buried fully (Levin et al., 2010).

The landscapes of terraces surfaces in the south-eastern part of the island subjected to the least volcanic transformation. They remained in previous boundaries, a fall of volcanic ash did not considerably influence to vegetative and soil covers. The bushes of alder *(Duschekia fruticosa)* suffered least of all in the southern part of the island; they on the same sites in contrast to mountain ash *(Sorbus sambucifolia)* have not the signs of negative influence (fig. 10 a, b, c). The leaves of mountain ash have yellow edging, though such changes do not influence greatly to the vegetation of the bushes.

The brush-woods of alder, growing in the zone of fall of rudaceous ash and scoria, have on the leaves the signs of shrinkage and spots. Thin layer of ash of silt size remains on some of them. High-grasses, located in the south-eastern part of the island did not suffer from ash falls. In given zone the elementary plants (green moss and lichens) were subjected to great influence, they were fully covered with ash (Levin et al., 2010).

Before the eruption along all the shore of the island, the landscapes of abrasion- denudation cliffs with boulder-pebble beaches and storm ramparts with meadow associations with high-grasses on the meadow-sod soils were located (8,31 % of island area). Passing of pyroclastic flows to the shore destroyed them in the northern part of the island (0,89 % of island area).

Before the eruption practically all central cone was covered with numerous firn. It is probably that during the initial stage of the eruption (sooner on 12th of June) as a result of temperature rising on the surface after the influence of erupted volcanic material, several lahars descended along the slopes of the volcano, the signs of their movement were retained in the eastern and south-eastern parts of the island. The analogues process was marked by E.K. Markhinin (1964) after the eruption in November 1960. The thickest and longest flow was observed on the southern slope of the volcano up to old runways at the distance 2,4 km. The width of the zone of lahars influence was not more than 10-15 m, the vegetative and soil cover in the zone of lahars movement was fully destroyed (fig. 11 a, b). Intense melting of the firns during the eruption caused the formation of numerous temporary water flows on the surface of small bog in Ainu bay; this was not observed in 2007-2008.

During the year and a half after the volcanic event, the view of the island considerably changed. The pioneer landscapes began to form, because of loose composition of volcanic material the erosion processes develop intensely on the different hypsometrical levels, the change of shore line is observed. The formation of system of ravines goes on the slopes of the volcano. Their depth can reach 5-6 m. Weak lithification of pyroclastic deposits and great amount of atmospheric precipitations cause the formation of small mud flows, descending along the slopes and penetrating to the zones with remained vegetation and soil cover. The signs of mud flows are seen on the slopes of southern and south-eastern expositions (Ganzey et al., 2011).

The eruption of lava flows did not cause the considerable changes in geomorphologic view of the island. The flows of lava are clear seen on the landscape maps. There are no vegetation and soil cover on their surface. The trunks of alder with the sings of charring on the side close to the flow are found near the foot of the north-eastern flow. As S. Yu. Grishin with co-authors (2010) noticed, the restocking of vegetative-soil cover can lasted many hundreds years, but "it is not real in the conditions of very intense activity of the volcano".

Main changes of geomorphic structure of the island are expressed in smoothing of shore line. It is connected with that the frontal parts of pyroclastic flows, penetrated to near-shore zone (the area of the island was increased at the cost of this), began to destroyed because of loose character of forming rocks and active development of abrasion processes. The material began to redeposit by alongshore currents in nearest bays. As a result of this the sharp increasing of beach zone occurred. In some bays these widths reached more than 150 m. In spite of increasing of beach zone, in total the decreasing of Matua Island area is observed from 53,67 km2 in 2009 to 53,48 km2 in 2010.

Vegetative cover on the slopes of Sarychev Peak volcano began to restock, but this process has a local character. The appearance of grass vegetation occurs in the zones, where the deposits of pyroclastic material have small thickness. These zones are the sites of volcano edifice with projections of the relief, characterized by the presence of rather steep slopes, which are not able to keep great amount of loose material. And also the scarps such as frontal parts of old lava flows are turned to opposite side from the volcano (Ganzey et al., 2011).

From remained root system the new shoots began to grow through the deposits of pyroclastic flows, waves and volcanic ash. In the near-shore zone in the northern, north-western and north-eastern parts of the island, for example on Lisiy Cape, the vegetation was burnt by pyroclastic waves; the trunks of alder with the sings of charring were found. Here greatly thinned grass vegetation without soil cover or on the primitive-turf soils began to form in the conditions of the landscapes of steep and average steep slopes of weakly lithofying pyroclastic deposits and sub-volcanic bodies

The analogous tendency of restocking of the vegetation along the shore line was noticed by E.K. Markhinin (1964), who had visited the island in the autumn 1960. After the eruptions in 1930, 1946 and 1960 the low parts of volcano slopes were covered with thick grass vegetation. In separate sites the branches of the alder had the signs of burning and charring by volcanic material. Grass vegetation was represented in the area of Lisiy Cape and Sivuch Cape.

On the south-eastern slopes of Sarychev Peak volcano, where in 2009 ecoton zone had been marked, the vegetation began to penetrate on the slopes of the volcano. The restocking of vegetation is seen along the boundaries of pyroclastic flows, where the thickness of deposits is lesser and the vegetation is destroyed by pyroclastic waves. At present time the thin grass cover with separate greatly depressed bushes of alder are met on the height up to 650 m.

Fig. 9. The landscapes of the volcanic cone (the southern slope) before (upper photo, August 2008) and after (low photo, June 2009) the eruption 2009.

Fig. 10. The state of vegetative cover buried under the layer of ash: (a) rhododendron goldish (Rhododendron aureum Georgi); (b) Cassiope lycopodioides (Pall.) D. Don; (c) blueberry (Empetrum).

Fig. 11. The largest lahar of the eruption 2009: (a) the frontal part. The edifice of Sarychev Peak volcano is seen on the background; (b) the valley along which the mud flow moved.

The analogous processes as on the slopes of the volcano are found in the zones of lahars movement. Washing out of the floor of the valleys along which the lahars descended is of small-scale manifestation, the restocking of the vegetation has dotted character. The representatives of cereals prevail in the vegetation; the bushes of alder and mountain ash are met in the single cases. The restocking goes more actively owing to ferns *(Drypteris expana)* (fig. 12 a).

The different species of mushrooms intensely grow on the separate sites (fig. 12 b, c). The analogous vegetative associations form on the boards of the valleys, but the processes go considerably quicker. Soil cover was either destroyed fully both on the board and floors of the valleys, or buried under the great thickness of pyroclastic material. Only in the places with conserved soil cover the pioneer vegetation forms. In the valleys dried bushes of alder are met everywhere, the trunks of them are polished. The view of damage of the trees shows that pyroclastic waves were the main damage factor; the surface of the trunks is smoothly polished, the ends of the branches are sharpened, but the integrity is not damaged. Such character of damage is caused by high speed of the waves and small size of the particles of solid components.

So, after the analysis of the landscape structure of Matua Isl. from 1964 till 2009 the definite evolutional tendencies of the landscapes were revealed. The landscapes in the south-eastern part of the island were subjected the least change by the eruption. Even at the strong eruptions in 1946 and 2009 the thickness of fallen volcanic ash did not exceed 1 sm, this did not considerably effect to vegetative and soil covers. At the influence of volcanic activity the changes of the landscapes occurred in the bounds of Sarychev Peak volcano edifice, first of all they expressed in destroying and burying of vegetative and soil cover. The boundaries of the distribution of alder bushes on the southern slopes of Sarychev Peak volcano changed from 600 m in 1964 to 840 m in 2008. Catastrophic eruption in June 2009 sank this boundary to the height 450 m (Ganzey et al., 2010).

By 2010 the greatest transformations are connected with the change of shore line after the destruction of the frontal parts of pyroclastic flows, penetrated to near-shore zone and redeposition of the material in small bays. It is observed the development of active erosion processes on the slopes of the volcano that connected with weak lithification of deposits.

Restocking of vegetative and soil components of the landscapes on the slopes of Sarychev Peak volcano has dotted character and goes in the zones where the thickness of volcanic deposits is not large. There is no doubt that vegetative groups penetrate to new hypsometrical levels. However, given scheme is possible only in the condition of absence of volcanic activity in the nearest time.

Interesting data were received after the quantitative analysis of built landscape maps. In the table the indexes for 2009 were obtained two weeks later the eruption of Sarychev Peak volcano. As it was noticed above, a year later the eruption the decrease of the area of Matua Island was by 0,19 km^2, this was connected with washing out of the frontal parts of pyroclastic flows and redeposition of the material along shore line.

Number of landscape contours increased twice. During a year after the eruption the complexity of landscape picture of Matua Isl. increased more than five times. Given index shows the interaction between quantity and average area of landscape contours. Analogous situation relates to the coefficient of breaking up, which since the moment of eruption increased two times (index has the reciprocal dependence).

The process of increase of given values complex is connected with that after the eruption the restocking of vegetative and soil cover goes on the small sites, where the thickness of pyroclastic deposits is not large. However the increase of amount of natural-territorial complex (NTC) is not so considerable. This is connected with that the formation of pioneer landscapes goes on identical scheme all over the island. It is probable that given index in future will also vary in the analogous limits.

The eruption of Sarychev Peak volcano caused the increase of index of entropic degree of complexity of landscape picture. It reflects the probability of change of one landscapes site by another, we can indirectly judge about the balance and stability of landscape structure. So, the tendency of increase of entropic degree of complexity shows the increase of instability of formed system NTC at present time.

The estimation of landscape variety for Matua Island was done during finishing stage of the works. As it is seen from the table the tendency of its increase remains, this is caused by the increase of quantity of landscape contours, NTC, dotted character of restocking of the landscapes. It is probable that in future the tendency of increase of landscape variety will remain up to reaching of climax of the system, when landscape complexes will reach stable state and will be in equilibrium with the conditions of environment. It is very difficult to say about time periods for this process. Will it last decades or centuries? However, reaching of such state is possible only under the conditions of absence of strong volcanic events in the island, this is hardly probable, data on the activity of the volcano corroborate about this not only during the historic time but also in Holocene.

It is necessary to notice that the change of landscape contour of Matua Island went under the influence of inner forces during the volcanic silence. For characteristic of the process of landscape change of islands-volcanoes it is expedient to use term "relaxation of environment" of theory of island biogeography (MacArthur, Wilson, 1967; Diamond, 1972), which the nature is subjected after their separation from the continent. Only the process of relaxation of the landscapes during volcanically quiet periods favors the decrease of landscape areas, breaking of landscapes, their complexity and variety; and the eruptions of the volcanoes cause a sharp sometimes one-moment increase of given indexes.

Fig. 12. Restocking of vegetative cover and penetration of vegetation to the territory of volcanic desert: (a) Drypteris expana; (b, c) mushrooms.

5. Conclusions

Occurred eruption is a characteristic episode of the newest stage of the eruptive history of Sarychev Peak volcano, which is the most active volcano. Sarychev Peak volcano intensely worked, during last 250 years in average its eruptions occurred every 25 years; in connection with this the possibility of its eruptions is high in the nearest future.

The peculiarities of the eruption 2009 is analogues to previous events, such as strong eruptions in 1765±5, 1930 and 1945, that also characterized by emissions of great amount of pyroclastic rocks and the considerable influence to nature of the island.

Considering a high activity of Sarychev Peak volcano during historical period, it is unlikely that in nearest time it will change cardinally its regime of the eruptive activity and become less dangerous. Taking into account a high explosiveness of the volcano, nearly all of its eruptions will represent a hazard in different degree for the aviation. Also people in the island will be subjected danger at the strongest eruptions (periodically scientific and tourist groups visit Matua). The main striking factors near the edifice will be pyroclastic flows and waves, also lahars, at the distance - volcanic ash with its highly expressed abrasive features and eruptive gases. So, at monitoring of the volcanic activity, much attention must be paid to this object.

6. Acknowledgement

The investigations are made under the support of Russian Found of Fundamental Investigations 10-05-10032к, 10-05-00797а, 10-05-10052а, 09-05-00003, 09-05-00364; Far East Branch of Russian Academy of Sciences Yu-S-D-08-043, 11-III-B-08-015; National Scientific Found № ARC-0508109. The authors would like to thank T.Yu. Novikova for English translation.

7. References

Abdurakhmanov, A.; Zlobin, T.; Markhinin, E. & Tarakanov, R. (1990). The eruption of Ivan Grozny volcano in 1989, *Volcanology and seismology*, № 4. pp. 3-9.

Andreev, V.; Shantser, A.; Khrenov A.; Okrugin, V. & Nechaev, V. (1978). The eruption of Sarychev Peak volcano in 1976, *Bulletin of volcanology stations*, № 55. pp. 35-40.

Diamond, J. (1972). Biogeographic kinetics: estimation of relaxation times for aviafaunas of south-west Pacific islands, *Proc. nat. Acad. Sci.*, V. 69, 1972. pp. 209-235.

Fedorchenko, V.; Abdurakhmanov, A. & Rodionova, R. (1989). Volcanism of the Kurile Island arc, Nauka, Moscow, Russia, 239 P.

Fisher, R. & Schminke, H. (1984). Pyroclastic rocks, Springer-Verlag, Berlin, Heidelberg, N.Y., Tokyo, 472 P.

Ganzey, K.; Razzhigaeva, N. & Rybin, A. (2010). The change of landscape structure of Matua Island from the second half of XXth till the beginning of XXIth centuries (The Kurile Archipelago), *Geography and natural resources*, № 3. pp. 87-93.

Girina, O.; Ushakov, S.; Malik, N.; Manevich, A.; Melnikov, D.; Nuzhdaev, A.; Demyanchuk, Yu. & Kotenko, L. (2008). The active volcanoes of Kamchatka and Paramushir Island, North Kurils in 2007, *Volcanology and seismology*, № 6. pp. 1-18.

Glavatsky, P. & Efremov, G. (1948). The Eruption of Sarychev Peak volcano in November 1946, *Bulletin of volcanology stations*, № 15. pp. 8-12.

Gorshkov, G. (1967). Volcanism of the Kurile Island Arc, Nauka, Moscow, Russia, 287 P.

Grishin, P.; Girina, O.; Vereshchaga, E. & Viter, I. (2010). A strong eruption of Sarychev Peak volcano (the Kurile Islands, 2009) and its influence on vegetative cover, *Vestnik FEB RAS*, № 3. pp. 40-50.

Haywood, J.; Jones, A.; Clarisse, L.; Bourassa, A.; Barnes, J.; Telford, P.; Bellouin, N., Boucher, O.; Agnew, P.; Clerbaux, C.; Coheur, P.; Degenstein, D. & Braesicke, P. (2010). Observations of the eruption of the Sarychev volcano and simulations using the HadGEM2 climate model, *Journal of Geophysical Resechers*, Vol. 115. D21212.

Ishizuka, Y.; Nakagawa, M.; Baba, A.; Hasegawa, T.; Kosugi, A.; Uesawa, S.; Matsumoto, A. & Rybin, A. (2011). Along-arc variations of K-Ar ages for the submarine volcanic rocks in the Kurile Islands, *7th Biennial Workshop on Japan-Kamchatka-Alaska Subduction Processes (JKSP-2011)*, pp. 279-280, Petropavlovsk-Kamchatskiy, Russia, august 25-30, 2011.

Kamio, X. (1931). The earthquake in Moroton Bay in Simushir island in June 1920 and the eruptions in Matua Island in January 1923, *Geological Journal*, V. 38, № 1. (In Japanese).

Kienle, J. & Shaw, G. (1979). Plume dynamics, thermal energy and long distance transport of Vulcanian eruption clouds from Augustine Volcano, Alaska, *Journal of Geophysical Resechers*, 6 (1-2):139-164.

Levin, B.; Razzhigaeva, N.; Ganzey, K.; Rybin, A. & Degterev, A. (2010). Change of landscape structure of Matua Is. after the eruption in 2009, *Doklady Akademii Nauk*, V. 431, № 5. pp. 692-695.

Levin, B.; Rybin, A.; Razzhigaeva, N.; Frolov, D.; Salyuk, P.; Mayor, A.; Vasilenko, N.; Zharkov, R.; Prytkov, A.; Kozlov, D.; Chernov, A.; Chibisova, M.; Guryanov, V.; Koroteev, I. & Degterev, A. (2009). Complex expedition «Sarychev Peak volcano - 2009» (The Kurile Islands), *Vestnik FEB PAS*, № 6. pp. 98-104.

MacArthur, R. & Wilson E. (1967). The theory of island biogeography, Prinston Univ. Press, Prinston, 1967. P. 203.

Melekestsev, I.; Braitseva, O. & Sulerzhitsky, L. (1988). The catastrophic explosive eruptions of the volcanoes of the Kurile-Kamchatka region at the end of Pleistocene - beginning of Holocene, *Doklady Akademii Nauk USSR*, V. 300, № 1. pp. 175-181.

Miyatake, K. (1934). About the eruption of ythe volcano in Harumkotan Island (The Central Kurile Islands) on 8th of January 1933, *Bulletin of volcanology Society of Japan*, V. 3. № 1. (In Japanese).

Oppenheimer, C. (1998). Volcanological applications of meteorological satellites, *Int. J. Rem. Sens.*, 19:2829-286.

Polonsky, A. (1994). The Kuriles, *Local Studies Bulletin*, № 3. pp. 3-86.

Rybin, A.; Chibisova, M.; Webley, P.; Steensen, T.; Izbekov, P.; Neal, C & Realmuto, V. (2010). Satellite and ground observations of the June 2009 eruption of Sarychev Peak volcano, Matua Island, Central Kuriles, *Bulletin Volcanology*, V. 73. № 4. pp. 40-56.

Salinas, L. (2010). United Airlines Flight Dispatch, *Congressional Hazards Caucus*, http://www.agiweb.org.

Shilov, B. (1962). The eruptions of Sarychev Peak volcano in 1960, *Transactions SakhNII*, Iss. 12. 1962. pp. 143-149.

Snow, G. (1992). The reports about the Kurile Islands, *Local Studies Bulletin*, № 1. pp. 89-127.

Sparks, R.; Bursik, M.; Carey, S.; Gilbert, J.; Glaze, L.; Sigurdsson, H. & Woods, A. (1997). Volcanic plumes, Wiley, Chichester, P. 589.

Tanakadate, H. (1931). Volcanic activity in Japan and vicinity during the period between 1924 and 1931, *Japanese Journal of Astronomy and Geophysics*, V. 9. № 1. pp. 47-64.

9

Asymmetric Continuum Theories – Fracture Processes in Seismology and Extreme Fluid Dynamics

Teisseyre Roman
Institute of Geophysics, Polish Academy of Sciences, Warszawa, Poland

1. Introduction

Our consideration has its origin in a well known fact that the classical continuum mechanics cannot well describe the seismic processes, seismic wave emission and even wave propagation. Classical elasticity describes perfectly the small deformations, but in its frame we cannot imagine the recording of the very long seismic displacement waves. In fact we record the deformation waves which, in case of a constant pressure load, directly relate to shear and rotation waves. Only in a more general theories, like the very complicated micromorphic theory, the rotation waves may exist: some recent observation data, obtained by very sensitive sensors, confirm existence of the rotational waves. Our aim is to present a relatively simple theoretical frame, based on the Newton law (exactly its first order derivatives) applied for the deformation strains (first order derivatives of displacements); in this way we arrive directly to the fundament of the Asymmetric Continuum Theory. Following this way we like also to present an approach to Asymmetric Molecular Strain Theory, such way may lead us to some explanations related to the vortex problems in fluids and even mechanism of turbulence.

The Asymmetric Continuum Theory (Teisseyre, 2008, 2009, 2011) has been developed due to many insufficiencies of the classic symmetric theory and some phenomena found by seismological observations, e.g., the trials to record the rotation motions has started already at beginning of last century, while only at end of that century such reliable records have been discovered by means of the Sagnac measurement system (cf., Lee et al., 2009).

The similar asymmetric approach can be introduced for the fluids; the molecular strains may present the symmetric and anti-symmetric molecular deformations (Teisseyre, 2009). However, we will meet there the additional complications when considering the non-laminar and turbulent motions; moreover, the density variation in gases will have a great importance. We will try to present here some related trials.

However, firstly we may shortly explain the difficulties found with an use of the classic elasticity:

- Classic elasticity include only displacements as basic motions; rotations and strains do not appear as independent deformations. The independently released rotation and strain fields should have the own independent motion relations, e.g., a balance of angular momentum shall be independently introduced, while in the classical elasticity this balance holds automatically due to the symmetry of stresses;

- Fault slip solutions used in seismology could be included only by the additional friction constitutive laws;
- Fracture pattern reveals usually an asymmetric pattern with a main slip plane; fracture deformations with the granulation and fragmentation processes include the rotation motions;
- Edge dislocations present the asymmetric strains; relation between a density of the edge dislocations and stresses cannot be found in the frame of the classical theory.

Of course the classical elasticity describes almost perfectly the small deformations, but in observations in a near field earthquake zone we meet with a number of the well-recognized insufficiencies, moreover the recoding of the very long seismic waves means that in fact we record the deformation fields, $D_{ik} = \partial u_k / \partial x_i$, which becomes integrated during the adequate time by a seismometer to reveal in abstraction the long displacement motion ($u_k = \int \sum (\partial u_k / \partial x_i) \Delta x_i \mathrm{d}t$; where Δx means a rigid element of seismometer platform).

We might mention some numerous attempts used to improve the classical elasticity; however we will confine ourselves to the following:

- The Cosserat brothers' theory of elasticity (Cosserat, E. and F. 1909) included the displacements and rotations;
- The micropolar and micromorphic elastic theories (cf., Eringen and Suhubi, 1964; Mindlin, 1965; Nowacki, 1986; for an advanced review, see Eringen 1999, 2001).

These micropolar and micromorphic theories present a very powerful tool to describe many complicated problems (see, e.g, Teisseyre, 1973; 1974). These theories seem to be too much complicated for an use in common seismological studies and require a knowledge of many additional material constants.

The similar approach can be introduced for the fluids; the molecular strains may present the symmetric and anti-symmetric molecular deformations (Teisseyre, 2009). We will meet there the additional complications when considering the non-laminar and turbulent motions; moreover, the density variation in gases will have a great importance. We will present here the first trials in these directions; however, we should underline that such efforts have been already undertaken in a frame of the micropolar and micromorphic theories (Eringen, 2001).

From the presented reason this contribution is divided into two parts: **Fracture earthquake processes** and **Extreme fluid dynamics**;

2. Fracture earthquake processes

2.1 Motion relations for fields released in independent fractures

Following the Asymmetric Theory (cf., Teisseyre, R. 2009, Teisseyre, 2011) we present the relations for the basic motions and deformations. For the basic fields, displacements, rotations, axial and strains treated as the independent motions we present the related motion equations. For the symmetric strain and for rotation strain we write:

$$E_{(ik)} = \frac{1}{2}\left(\frac{\partial u_k}{\partial x_i} + \frac{\partial u_i}{\partial x_k}\right) \; ; \; \breve{E}_{[ik]} = \frac{1}{2}\left(\frac{\partial \breve{u}_k}{\partial x_i} - \frac{\partial \breve{u}_i}{\partial x_k}\right)$$
$$D_{ik} = \frac{\partial u_k}{\partial x_i} \; , \; \breve{D}_{ik} = \frac{\partial \breve{u}_k}{\partial x_i} \; ; \; E_{ik} = E_{(ik)} + \breve{E}_{[ik]} \tag{1}$$

where their sum present the asymmetric strain field, E_{ik}, and the related displacement might be different.

The strains relate to the physical deformations, $D_{ik} = \frac{\partial u_k}{\partial x_i}$ or $\breve{D}_{ik} = \frac{\partial \breve{u}_k}{\partial x_i}$; The symmetric strains, $E_{(ik)}$, can be expressed by the derivatives of displacements, u_k, which may originally relate to the fracture slip motion at seismic source, while the rotation strain represents a rotation on a molecular level, which is given by the expression, $\breve{E}_{[ik]} = \frac{1}{2}\left(\breve{D}_{ik} - \breve{D}_{ki}\right)$, related to the displacements, $\breve{u}_k$, (having only the mathematical sense):

$$\omega_n = \frac{1}{2}\varepsilon_{nik}\omega_{ik} = \frac{1}{2}\varepsilon_{nik}\frac{\partial \breve{u}_k}{\partial x_i} \; , \; \breve{E}_{[ik]} \equiv \omega_{ik} = \frac{1}{2}\left(\frac{\partial \breve{u}_k}{\partial x_i} - \frac{\partial \breve{u}_k}{\partial x_i}\right) \tag{2}$$

where $\varepsilon_{nik} = \{1,-1,0\}$ is the fully antisymmetric tensor.

The total asymmetric strains, E_{ik}, may be presented as follows:

$$E_{kl} = E_{(kl)} + \breve{E}_{[kl]} = \delta_{kl}\overline{E} + \hat{E}_{(kl)} + \breve{E}_{[kl]} \; ;$$
$$\overline{E} = \frac{1}{3}\sum_{s=1}^{3} E_{(ss)} \; ; \; \hat{E}_{(ik)} = E_{(ik)} - \delta_{ik}\frac{1}{3}\sum_{s=1}^{3} E_{(ss)} \tag{3}$$

where the total axial strain, $\overline{E} = -p$, may relate to pressure and the deviatoric strains, $\hat{E}_{(ik)}$, relate to the shear field.

All these strain fields, total axial, deviatoric shear and rotational, can be released at a seismic source quite independently or some of these strains might be mutually related through their relations to the reference displacements, e.g., $\overline{u}_s$, $\hat{u}_s$, $\breve{u}_s$:

$$\overline{E} = \frac{1}{3}\sum_{s} E_{ss} = \frac{1}{3}\sum_{s}\frac{\partial \overline{u}_s}{\partial x_s} \quad \text{- total axial strain} \tag{4a}$$

$$\hat{E}_{(ik)} = E_{(ik)} - \delta_{ik}\frac{1}{3}\sum_{s=1}^{3} E_{(ss)} = \frac{1}{2}\left(\frac{\partial \hat{u}_k}{\partial x_i} + \frac{\partial \hat{u}_i}{\partial x_k}\right) - \delta_{ik}\frac{1}{3}\sum_{s=1}^{3}\frac{\partial \hat{u}_s}{\partial x_s} \quad \text{- deviatoric shears} \tag{4b}$$

$$\breve{E}_{[ik]} = \frac{1}{2}\left(\frac{\partial \breve{u}_k}{\partial x_i} - \frac{\partial \breve{u}_i}{\partial x_k}\right) \quad \text{- rotation strain} \tag{4c}$$

where some inter-relations between the written reference displacements may be related to the join source processes; in general the relation to common, phase shifted, or independent displacements may be expressed as:

$$\bar{u}_s=\xi^0 u_s\ ,\ \hat{u}_s=e^0 u_s\ ,\ \breve{u}_s=\chi^0 u_s\ ;\qquad \left\{\xi^0, e^0, \chi^0\right\}=\{0,\pm 1,\pm \mathrm{i}\} \tag{5}$$

Strain rotation has a different sense that a simple rotation motion . However, in relation to the displacement derivatives both might be expressed by the similar mathematical formulae.

We assume that any independent motion or deformation may be represented by the independent equations of motion (cf., Teiseyre and Gorski, 2009, Teisseyre, 2009, Teisseyre, 2011).

The motion equation for deformation, $D_{ni}=\frac{\partial u_i}{\partial x_n}$, follows from the derivatives of the classic Newton formula as follows:

$$\mu\sum_s \frac{\partial^2 u_i}{\partial x_s \partial x_s}-\rho\,\frac{\partial^2 u_i}{\partial t^2}+(\lambda+\mu)\frac{\partial}{\partial x_i}\sum_s \frac{\partial u_s}{\partial x_s}=0 \tag{6a}$$

Accordingly for symmetric and anti-symmetric strains we obtain:

$$\mu\sum_s \frac{\partial^2}{\partial x_s \partial x_s}E_{(nl)}-\rho\,\frac{\partial^2}{\partial t^2}E_{(nl)}=-(\lambda+\mu)\frac{\partial^2}{\partial x_n \partial x_l}\sum_s E_{(ss)}\ ;\qquad E_{(nl)}=\frac{1}{2}\left(\frac{\partial u_l}{\partial x_n}+\frac{\partial u_n}{\partial x_l}\right) \tag{6b}$$

$$\mu\sum_s \frac{\partial^2}{\partial x_s \partial x_s}\breve{E}_{[nl]}-\rho\,\frac{\partial^2}{\partial t^2}\breve{E}_{[nl]}=0\ ;\qquad \breve{E}_{[nl]}=\frac{1}{2}\left(\frac{\partial u_l}{\partial x_n}-\frac{\partial u_n}{\partial x_l}\right) \tag{6c}$$

where in all these formulae we may use the different reference displacement motions as explained above (see, eqs. 4a-4c); we have also omitted the external forces.

For the constitutional relations, joining stresses and strains, we choice the most simple relations:

$$S_{(ik)}=2\mu E_{(ik)}+\lambda\delta_{ik}E_{(ss)}\qquad ;\quad S_{[ik]}=2\mu E_{[ik]} \tag{7a}$$

$$\bar{S}=\left(2\mu+3\lambda\delta_{ik}\right)\bar{E}\ ;\quad \hat{S}_{(ik)}=2\mu\hat{E}_{(ik)}\ ;\quad \breve{S}_{[ik]}=2\mu\breve{E}_{[ik]} \tag{7b}$$

and the stresses may be separated for the axial, deviatoric and rotation stresses:

$$\begin{gathered} S_{kl}=S_{(kl)}+S_{[kl]}=\delta_{kl}\bar{S}+\hat{S}_{(kl)}+\breve{S}_{[kl]}\ ; \\ \bar{S}=\frac{1}{3}\sum_{s=1}^{3}S_{(ss)}\ ,\ \hat{S}_{(ik)}=S_{(ik)}-\delta_{ik}\frac{1}{3}\sum_{s=1}^{3}S_{(ss)} \end{gathered} \tag{7c}$$

2.2 Release-rebound fractures and propagation of waves

We have already mentioned that the fracture source processes could run according to the release - rebound related processes; this can have its influence on a propagation pattern. We follow some result presented by Teisseyre, (1985, 2009, 2011).

When a total axial stress is constant, the release-rebound system may be described by the linear relations between the time and space derivatives; the unique relations might remind the Maxwell-like ones. To this end we should choice the special coordinate system in which the deviatoric strains can be represented by the off- diagonal tensor in the specially chosen coordinate system:

$$\hat{E}_{(ik)} = E_{(ik)} - \delta_{ik}\frac{1}{3}\sum_{s=1}^{3} E_{(ss)} \rightarrow \hat{E}_{(ik)} = \begin{bmatrix} 0 & \hat{E}_{(12)} & \hat{E}_{(13)} \\ \hat{E}_{(12)} & 0 & \hat{E}_{(23)} \\ \hat{E}_{(13)} & \hat{E}_{(23)} & 0 \end{bmatrix} \tag{8}$$

In this system we can define the shear vector as $\hat{E}_i = \left\{\hat{E}_{(23)}, \hat{E}_{(31)}, \hat{E}_{(12)}\right\}$, and the rotation vector $\breve{E}_i$ as $\breve{E}_i = \left\{\breve{E}_{[23]}, \breve{E}_{[31]}, \breve{E}_{[12]}\right\}$. We may mention that when using the 4D approach, the vector $\hat{E}_i$ can be defined invariantly (cf., Teisseyre, 2009)

The release-rebound process may mean that a break of molecular bonds on molecular level releases a rotation field, $\partial\breve{E}/\partial t$, and than in a rebound motion will appear, $\text{rot}\,\hat{E}$, and reversely a release of shear field, $\partial\hat{E}/\partial t$, causes the rebound change: a rotation of the angular strains, rot $\breve{E}$; a shear strain may cause a slip motion at a source.

Such a release-rebound processes are adequately described by the Maxwell-like relations (Teisseyre, 2009, 2011):

$$\text{rot}\,\breve{E} - \frac{\partial\hat{E}}{V\partial t} = 0\ , \quad \text{rot}\,\hat{E} + \frac{\partial\breve{E}}{V\partial t} = 0\ ; \quad V = V^S = \sqrt{\frac{\mu}{\rho}} \tag{9}$$

From these relations we obtain the wave equations which coincide with the earlier derived formulae (6a) and (6c). When $\sum_s \frac{\partial u_s}{\partial x_s}$ remains constant we obtain the wave relations for rotation and shear fields, which, de facto, are mutually correlated as shown in the previous equation:

$$\Delta\breve{E} - \frac{\partial^2\breve{E}}{V^2\partial t^2} = 0 \quad \text{and} \quad \Delta\hat{E} - \frac{\partial^2\hat{E}}{V^2\partial t} = 0 \tag{10}$$

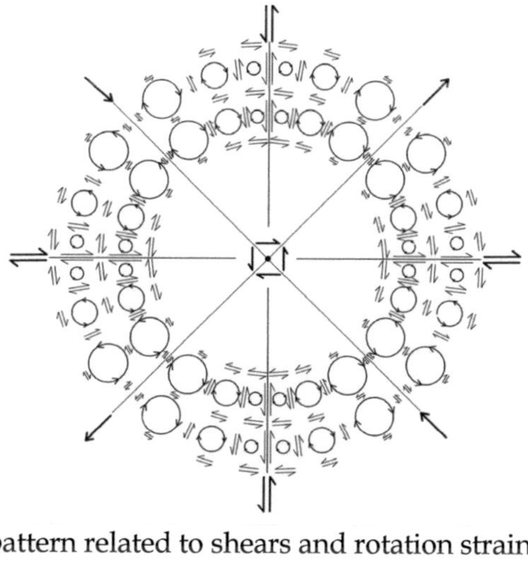

Fig. 1. Wave interaction pattern related to shears and rotation strains.

The related wave mosaic (Teisseyre and Gorski 2011) explains the interrelated propagation pattern of the shear and rotation motions as presented on Fig. 1.

Such pattern visualizes the wave interactions corresponding to the release-rebound processes at a fracture source.

2.3 Induced strains

In a seismic active zones we have a very high concentration of the defects; the classic paper by Eshelby et al. (1951) states that an array of the n linear defects leads to a concentration of the applied external stress field, S, to a multiple value of the applied field at a top of an array:

$$S^{Top} = nS \tag{11}$$

Such stress concentration can form a crack. The Peach - Koehler forces exerted on defect lines (Peach and Koehler, 1950) are related to the dislocation slip vector, b , dislocation line versor, v , and the applied stress field , S_{sk} :

$$F_n = \varepsilon_{nsq} \sum_k S_{sk} b_k \nu_q \tag{12}$$

where the fully asymmetric tensor, ε_{nsq}, expresses the vector product between the stress field multiplied by a slip vector and a dislocation line versors.

Two opposite cracks could mutually join to form a bigger crack; in such a process the stresses concentrated at the opposite edges of these cracks (or arrays) become mutually annihilated; that means their energies become released. This simple fracture model (cf., Droste , Teisseyre, 1959) could be applied to a continuum with a high defect content. To this end we should generalize the original Peach - Koehler expression (12) assuming, first, that the stresses, S_{sk}, represent an asymmetric field and, second, that we can describe the defect density by means of dislocation slip vector and dislocation line versor:

$$\alpha_{qk} = \nu_q b_k \tag{13}$$

Then using relations (12) and (13) we can define the induced stresses as product of the Peach-Koehler force and normal to a defect plane, $F_n n_p$:

$$\begin{aligned}
S_{np}^{Ind} &= S_{(np)}^{Ind} + S_{[np]}^{Ind} = F_n n_p = \varepsilon_{nsq} \sum_k \alpha_{qk} S_{sk} n_p \\
S_{(np)}^{Ind} &= \frac{1}{2} \varepsilon_{nsq} \sum_k \alpha_{qk} S_{sk} n_p + \frac{1}{2} \varepsilon_{psq} \sum_k \alpha_{qk} S_{sk} n_n \\
S_{[np]}^{Ind} &= \frac{1}{2} \varepsilon_{nsq} \sum_k \alpha_{qk} S_{sk} n_p - \frac{1}{2} \varepsilon_{psq} \sum_k \alpha_{qk} S_{sk} n_n
\end{aligned} \tag{14}$$

The total stresses would means a reorganized stress system due to a defect influence:

$$S_{np}^{T} = S_{np} + S_{np}^{Ind} = S_{(np)} + S_{[np]} + S_{(np)}^{Ind} + S_{[np]}^{Ind} \tag{15}$$

Considering the applied field as given by the constant total axial stresses only (that is applied stresses equivalent to an applied constant pressure, $S^T_{np} = -p\delta_{np} + S^{Ind}_{np}$) we obtain from eqs. 14 and 15 that the non diagonal component of asymmetric total stresses will relate to the expressions with a defect contain:

$$S^T_{12} = p(\alpha_{23} - \alpha_{32})n_2 \,,\; S^T_{23} = p(\alpha_{31} - \alpha_{13})n_3 \,,\; S^T_{31} = p(\alpha_{12} - \alpha_{21})n_1 \qquad (16a)$$

while the axial fields will be given as follows

$$S^T_{11} = (\alpha_{23} - \alpha_{32} - 1)pn_1,\; S^T_{22} = (\alpha_{31} - \alpha_{13} - 1)pn_2,\; S^T_{33} = p(\alpha_{12} - \alpha_{21} - 1)pn_3 \qquad (16b)$$

Thus, we come to an important expression that a fracture under pressure will run as the shear and strain rotation mutually related processes; this result confirms a known experimental fact that under applied pressure a fragmentation has a shear and rotational character.

2.4 Seismological networks

The global seismological network is based only on recording of displacements by the system of seismometers; only in some separate seismically active regions we have the local system of strain-meters and in some places also the devises to record a rotation motion.

However, a desired global system could watch any strain changes and related waves in a global scale.

At a given time moment the recorded displacement data presents a total sum of all deformation released and appears only due to the integration effect of a sum of deformations, $\frac{\partial u_q}{\partial x_p}$, over an interval, Δx, which means a length of a seismometer platform much more rigid than soil layers. At a given time moment we obtain:

$$\Delta u_k = \int \frac{\partial u_k}{\partial x_i} \Delta x_i \qquad (17a)$$

and the total displacement would be given as:

$$u_k = \int \sum \frac{\partial u_k}{\partial x_i} \Delta x_i dt \qquad (17b)$$

Only in such a way we might be inform on a displacement amplitude of the long strain waves.

Thus, it is extremely important to have the new word-wide seismic network based on the strain and rotation recording systems. We should note that only in the few seismic active regions such recording systems already exist, this is not enough as the strain waves can propagate over the whole word.

3. Extreme fluid dynamics

3.1 Asymmetric relations for momentum flux

We follow the asymmetric continuum theory developed in our former papers (Teisseyre, 2007; 2008; 2009; 2011); we assume that beside the symmetric molecular strains (strain rates) there appear also the antisymmetric molecular strain rates (spins); for the related molecular stresses we write:

$$\tilde{S}_{kl} = \tilde{S}_{(kl)} + \tilde{S}_{[kl]} \tag{18}$$

These fields, stress and strain rates, can be related by means of the following constitutive relations

$$\begin{gathered}\frac{1}{3}\sum_s \tilde{S}_{ss} = -p = \frac{1}{3}k\sum_s \tilde{E}_{ss}\ ; \qquad \tilde{S}_{kl} = \tilde{S}_{(kl)} + \breve{\tilde{S}}_{[kl]}\,, \quad \tilde{E}_{kl} = \tilde{E}_{(kl)} + \breve{\tilde{E}}_{[kl]} \\ \tilde{S}^D_{(kl)} = \eta \tilde{E}^D_{(kl)}\,, \quad \tilde{S}_{[kl]} = \eta\, \breve{\tilde{E}}_{[kl]}; \quad \tilde{S}^D_{(kl)} = \tilde{S}_{(kl)} - \frac{1}{3}\delta_{kl}\tilde{S}_{ss}\,, \quad \tilde{E}^D_{(kl)} = \tilde{E}_{(kl)} - \frac{1}{3}\delta_{kl}\tilde{E}_{ss}\end{gathered} \tag{19}$$

where $\tilde{S}^D_{kl}$ and $\tilde{E}^D_{kl}$ mean the deviatoric parts of tensors and η is the viscosity; for a simplicity we have assumed the same constant, η, for symmetric and antisymmetric parts; the points in the considered fluid continuum will have six degrees of freedom: displacement velocity and rotation motion - spin.

The molecular strain rates could be related the derivatives of some reference displacement velocities, υ_l and υ'_l :

$$\tilde{E}_{kl} = \frac{1}{2}\left(\frac{\partial \upsilon_l}{\partial x_k} + \frac{\partial \upsilon_k}{\partial x_l}\right), \quad \breve{\tilde{E}}_{[kl]} = \frac{1}{2}\left[\frac{\partial \upsilon'_l}{\partial x_k} - \frac{\partial \upsilon'_k}{\partial x_l}\right] \tag{20}$$

where the field υ' may have only a mathematical sense.

For the equal reference velocities, $\upsilon' = \upsilon$, we can obtain the relation between the molecular stresses and displacement velocity as a sum of molecular strain rate and molecular spin (having another sense that a simple point rotation velocity (cf., eqs. 1, 2 and Fig.1):

$$\tilde{S}_{kl} = \tilde{S}_{(kl)} + \breve{\tilde{S}}_{[kl]} = \eta\frac{\partial \upsilon_l}{\partial x_k} = \eta(\tilde{E}_{kl} + \breve{\tilde{E}}_{[kl]}) \tag{21}$$

A macroscopic transport field, υ, we will treat as an imposed external field, while the internal transport related to the rotation velocity could be treated as the additional motion introduced to the system of equations. Thus, in motions with advanced vorticity dynamics we assume that a transport can be related both to displacement velocity and the vortex motions (cf., Teisseyre, 2009).

First we may remember the Navier-Stokes relations:

$$\frac{\mathrm{d}(\rho\upsilon_i)}{\mathrm{d}t} = \upsilon_i\frac{\partial \rho}{\partial t} + \rho\frac{\partial \upsilon_i}{\partial t} + \rho\upsilon_s\frac{\partial \upsilon_i}{\partial x_s} = \eta\frac{\partial^2 \upsilon_i}{\partial x_k \partial x_k} - \tilde{F}_i \tag{22a}$$

where $\tilde{\mathrm{F}}$ are the body forces, υ is the displacement velocity, η is the dynamic viscosity.

We may present this relation in the different forms, with the independent fields $\tilde{E}_{ki}$ and $\breve{\tilde{E}}_{[ki]}$, and valid both for $\tilde{E}_{ki}(\upsilon)$ and $\breve{\tilde{E}}_{[ki]}(\upsilon')$ with $\upsilon' = \upsilon$, or with the phase shifted reference displacements, $\upsilon' = \pm i\upsilon$:

$$\frac{\mathrm{d}(\rho\upsilon_i)}{\mathrm{d}t} \rightarrow \upsilon_i \frac{\partial \rho}{\partial t} + \rho\frac{\partial \upsilon_i}{\partial t} + \rho\upsilon_s \frac{\partial \upsilon_i}{\partial x_s} = \eta \frac{\partial}{\partial x_k}\left(\tilde{E}_{(ki)} + \breve{\tilde{E}}_{[ki]}\right) - \tilde{F}_i \tag{22b}$$

This is an equivalent form expressing a possible relation between the molecular strain and spin fields.

For the molecular symmetric stresses we can write after Landau and Lipshitz (1959; new edition in Russian , 2001, cited as its Polish translantion, 2009):

$$\begin{aligned} \tilde{S}_{(ij)} &= 2\eta\left(\tilde{E}_{(ij)} - \frac{\delta_{ij}}{3}\sum_s \frac{\partial \upsilon_s}{\partial x_s}\right) + \varepsilon\delta_{ij}\sum_s \frac{\partial \upsilon_s}{\partial x_s}, \quad \text{or} \\ \tilde{S}_{(ij)} &= \eta\left(\frac{\partial \upsilon_j}{\partial x_i} + \frac{\partial \upsilon_i}{\partial x_j} - \frac{2\delta_{ij}}{3}\sum_s \frac{\partial \upsilon_s}{\partial x_s}\right) + \varepsilon\delta_{ij}\sum_s \frac{\partial \upsilon_s}{\partial x_s} \end{aligned} \tag{25a}$$

In our new approach we should add an influence of rotation molecular processes; therefore we introduce the asymmetric fluid viscous stress tensor adding an influence of the antisymmetric molecular stresses (Teisseyre, 2009):

$$\breve{\tilde{S}}_{[ij]} = 2\eta\breve{\tilde{E}}_{[ij]} = \eta\left(\frac{\partial \upsilon_j}{\partial x_i} - \frac{\partial \upsilon_i}{\partial x_j}\right) \tag{25b}$$

The total stresses become give as :

$$\tilde{S}_{ij} = \tilde{S}_{(ij)} + \tilde{S}_{[ij]}\ ; \quad \tilde{S}_{ij} = 2\eta\left(\frac{\partial \upsilon_j}{\partial x_i} - \frac{\delta_{ij}}{3}\sum_s \frac{\partial \upsilon_s}{\partial x_s}\right) + \varepsilon\delta_{ij}\sum_s \frac{\partial \upsilon_s}{\partial x_s} \tag{25c}$$

and the total molecular axial strain as:

$$\sum_s \tilde{S}_{ss} = -3p \tag{25d}$$

Now for the momentum flux we write the modified asymmetric tensor:

$$\begin{aligned} T_{kn} &= -S_{kn} + \rho\upsilon_k\upsilon_n + \rho\tilde{\upsilon}_k\tilde{\upsilon}_n\ ; \ \tilde{\upsilon} = \left\{\dot{\tilde{r}}, \tilde{r}\tilde{\omega}, \dot{\tilde{z}}\right\} \\ T_{(kn)} &= -S_{(kn)} + \rho\upsilon_k\upsilon_n + \rho\tilde{\upsilon}_k\tilde{\upsilon}_n\ , \quad T_{[kn]} = -S_{[kn]} \end{aligned} \tag{26a}$$

or we could write

$$T_{kn} = -2\eta\left(\frac{\partial \upsilon_n}{\partial x_k} - \frac{\delta_{kn}}{3}\sum_s \frac{\partial \upsilon_s}{\partial x_s}\right) - \varepsilon\delta_{kn}\sum_s \frac{\partial \upsilon_s}{\partial x_s} + \rho\upsilon_k\upsilon_n + \rho\tilde{\upsilon}_k\tilde{\upsilon}_n \tag{26b}$$

We arrive to the general expression for the motion of fluids given as follows

$$\frac{\partial(\rho\upsilon_n + \rho\tilde{\upsilon}_n)}{\partial t} + \sum_k \frac{\partial T_{kn}}{\partial x_k} = \rho g\delta_{zn} + F_n \tag{27}$$

and for a mass conservation we will have

$$\frac{d\rho}{dt}=\frac{\partial\rho}{\partial t}+\sum_s\frac{\partial(\rho\upsilon_s)}{\partial x_s}+\sum_s\frac{\partial(\rho\tilde{\upsilon}_s)}{\partial x_s}=0 \tag{28}$$

3.2 Vortex transport processes

In some problems, a macroscopic transport field, υ, can be treated as an imposed external field, while the internal transport rotation shall enter into the additional terms introduced to the system of equations. Thus, in motions with advanced vorticity dynamics we assume that a transport can be related both to displacement velocity and the vortex motions; we follow the approach applied to fragmentation and slip in the fracture processes (Teisseyre, 2009).

For great Reynolds number the laminar motions become unstable and the double transport process with displacement velocities and spin motions may generate the micro-vortices; a kind of dynamic vortex structure can be formed, simultaneously undergoing a displacement transport process.

However, under some special conditions, an isolated vortex center can be formed; inside it, the displacement velocity transport might be negligible. However, to consider formally the transport related to such extreme conditions we should return to the basic problem: how to incorporate the spin motion into the system of basic relations.

The independent rotational motions related to angular moment and the variable arm and spin could well describe the vortex transport phenomena (Teisseyre, 2009); we should underline that our approach differs from the classical one defining a vorticity as rotation of transport velocity:

$$\varsigma=\text{rot }\tilde{\upsilon}\quad\left(\text{or}\qquad \varsigma_n=\varepsilon_{npi}\partial\tilde{\upsilon}_i/\partial x_p\right).$$

Following that paper (Teisseyre, 2009) the rotation transport, $\tilde{\upsilon}_k$, can be given as radial velocity, $\tilde{r}$, angular velocity, $\tilde{\omega}$ and axial velocity, $\tilde{z}$:

$$\tilde{\upsilon}_k\rightarrow\{\tilde{r},\,\tilde{\omega},\,\tilde{z}\} \tag{29}$$

We obtain for a rotation transport in the vortices along the z-axis:

$$\begin{aligned}
&\frac{\mathrm{d}}{\mathrm{d}t}=\frac{\partial}{\partial t}+\tilde{\upsilon}_k\frac{\partial}{\partial x_k}\rightarrow\frac{\mathrm{d}}{\mathrm{d}t}=\frac{\partial}{\partial t}+\tilde{r}\frac{\partial}{\partial r}+\tilde{\omega}\frac{\partial}{\partial\varphi}+\tilde{z}\frac{\partial}{\partial z}\rightarrow\\
&\frac{\mathrm{d}(\rho\tilde{r})}{\mathrm{d}t}=\tilde{r}\frac{\partial\rho}{\partial t}+\rho\frac{\partial\tilde{r}}{\partial t}+\rho\tilde{r}\frac{\partial\tilde{r}}{\partial r}+\rho\tilde{\omega}\frac{\partial\tilde{r}}{\partial\varphi}+\rho\tilde{z}\frac{\partial\tilde{r}}{\partial z}\\
&\frac{\mathrm{d}(\rho\tilde{\omega})}{\mathrm{d}t}=\tilde{\omega}\frac{\partial\rho}{\partial t}+\rho\frac{\partial\tilde{\omega}}{\partial t}++\rho\tilde{r}\frac{\partial\tilde{\omega}}{\partial r}+\rho\tilde{\omega}\frac{\partial\tilde{\omega}}{\partial\varphi}+\rho\tilde{z}\frac{\partial\tilde{\omega}}{\partial z}\\
&\frac{\mathrm{d}(\rho\tilde{z})}{\mathrm{d}t}=\tilde{z}\frac{\partial\rho}{\partial t}+\rho\frac{\partial\tilde{z}}{\partial t}++\rho\tilde{r}\frac{\partial\tilde{z}}{\partial r}+\rho\tilde{\omega}\frac{\partial\tilde{z}}{\partial\varphi}+\rho\tilde{z}\frac{\partial\tilde{z}}{\partial z}
\end{aligned} \tag{30}$$

At such the vortex motion we obtain from the equations (26-28) for the momentum flux only a simple expression:

$$T_{kn} = \rho\tilde{\upsilon}_k\tilde{\upsilon}_n \ ; \ \tilde{\upsilon} = \{\tilde{r}, \tilde{\omega}, \tilde{z}\} :$$

$$\frac{\partial(\rho\tilde{\upsilon}_n)}{\partial t} + \sum_k \frac{\partial T_{kn}}{\partial x_k} = \rho g \delta_{zn} \tag{31}$$

that is

$$\begin{aligned}
&\frac{\partial(\rho\tilde{r})}{\partial t} + \frac{\partial(\rho\tilde{r}^2)}{\partial r} + \frac{\partial(\rho\tilde{\omega}\tilde{r})}{\partial\varphi} + \frac{\partial(\rho\tilde{r}\tilde{z})}{\partial z} = 0 \\
&\frac{\partial(\rho\tilde{\omega})}{\partial t} + \frac{\partial(\rho\tilde{r}\tilde{\omega})}{\partial r} + \frac{\partial(\rho\tilde{\omega}^2)}{\partial\varphi} + \frac{\partial(\rho\tilde{\omega}\tilde{z})}{\partial z} = 0 \\
&\frac{\partial(\rho\tilde{z})}{\partial t} + \frac{\partial(\rho\tilde{r}\tilde{z})}{\partial r} + \frac{\partial(\rho\tilde{\omega}\tilde{z})}{\partial\varphi} + \frac{\partial(\rho\tilde{z}^2)}{\partial z} = \rho g \delta_{zn}
\end{aligned} \tag{32a}$$

and for a mass conservation we will have

$$\frac{d\rho}{dt} = \frac{\partial\rho}{\partial t} + \tilde{r}\frac{\partial\rho}{\partial r} + \tilde{\omega}\frac{\partial\rho}{\partial\varphi} + \tilde{z}\frac{\partial\rho}{\partial z} = 0 \tag{32b}$$

These relation present the dynamic relation under a given thermodynamical condition. At constant thermodynamical condition, $\frac{\partial\rho}{\partial t} = 0$, we obtain:

$$\begin{aligned}
&\rho\frac{\partial\tilde{r}}{\partial t} + \frac{\partial(\rho\tilde{r}^2)}{\partial r} + \frac{\partial(\rho\tilde{\omega}\tilde{r})}{\partial\varphi} + \frac{\partial(\rho\tilde{r}\tilde{z})}{\partial z} = 0 \\
&\rho\frac{\partial\tilde{\omega}}{\partial t} + \frac{\partial(\rho\tilde{r}\tilde{\omega})}{\partial r} + \frac{\partial(\rho\tilde{\omega}^2)}{\partial\varphi} + \frac{\partial(\rho\tilde{\omega}\tilde{z})}{\partial z} = 0 \\
&\rho\frac{\partial\tilde{z}}{\partial t} + \frac{\partial(\rho\tilde{r}\tilde{z})}{\partial r} + \frac{\partial(\rho\tilde{\omega}\tilde{z})}{\partial\varphi} + \frac{\partial(\rho\tilde{z}^2)}{\partial z} = \rho g \delta_{zn}
\end{aligned} \tag{33a}$$

and for a mass conservation we will have

$$\frac{d\rho}{dt} = \tilde{r}\frac{\partial\rho}{\partial r} + \tilde{\omega}\frac{\partial\rho}{\partial\varphi} + \tilde{z}\frac{\partial\rho}{\partial z} = 0 \tag{33b}$$

A complex system presents relations given by the equations 25 and 33. We might solve these equations for two cases; first with constant density $\rho = \rho_0$ outside the vortices, (e.g., for liquids), given by the relations (25) and with a variable density inside the vortices, $\rho(r,\varphi,z)$; still this complex system with relations (25 and 33) will remain quite complicated. Outsite the vortices we could have the curvilinear system of flow around and between the vortices.

3.3 Stationary vortex relations

We will confine our consideration to a single vortex space, the related stationary equations at constant thermodynamical condition with $\frac{\partial\rho}{\partial t} = 0$ become:

$$\frac{\partial(\rho\tilde{r}^2)}{\partial r}+\frac{\partial(\rho\tilde{\omega}\tilde{r})}{\partial\varphi}+\frac{\partial(\rho\tilde{r}\tilde{z})}{\partial z}=0$$
$$\frac{\partial(\rho\tilde{r}\tilde{\omega})}{\partial r}+\frac{\partial(\rho\tilde{\omega}^2)}{\partial\varphi}+\frac{\partial(\rho\tilde{\omega}\tilde{z})}{\partial z}=0 \tag{34a}$$
$$\frac{\partial(\rho\tilde{r}\tilde{z})}{\partial r}+\frac{\partial(\rho\tilde{\omega}\tilde{z})}{\partial\varphi}+\frac{\partial(\rho\tilde{z}^2)}{\partial z}=\rho g\delta_{zn}$$

and

$$\tilde{r}\frac{\partial\rho}{\partial r}+\tilde{\omega}\frac{\partial\rho}{\partial\varphi}+\tilde{z}\frac{\partial\rho}{\partial z}=0 \tag{34b}$$

Note that such the flows can occur effectively for the Reynolds numbers above the critical value. It would be possible to solve these stationary cases with a help of the following types of relations:

$$\tilde{r}=\tilde{r}_0\left[1-\varepsilon\frac{\varphi}{2\pi}\right]\exp(-\alpha z)\exp(\gamma r),\quad \tilde{\omega}=\tilde{\omega}_0\left[1-\varsigma\frac{\varphi}{2\pi}\right]\exp(\beta z)\exp(\delta r),$$
$$\tilde{z}=\tilde{z}_0\exp(\gamma z)\exp(\kappa r) \tag{35}$$

where $\alpha>0$, $\beta>0$.

The profiles of related vortices, $\tilde{r}(z)$, may differ from almost linear for very small, α (that is for $\exp[-\alpha z]\approx 1-\alpha z$) up to very rapidly decreasing, $\tilde{r}_0(z)$, for the great α values. The spin motion should adequately increase and vertical vortex component, $\tilde{z}$, might increase or decrease.

We may simplify these relations assuming that density can change only along the z- axis:

$$\rho=\rho_0\exp[\vartheta z] \tag{36a}$$

and further more we might put instead of eq. 35:

$$\tilde{r}=\tilde{r}_0\exp(-\alpha z)\exp(\chi r),\quad \tilde{\omega}=\tilde{\omega}_0\exp(\beta z)\exp(\delta r),\quad \tilde{z}=\tilde{z}_0\exp(\gamma z)\exp(\kappa r) \tag{36b}$$

Here to solve the equations

$$\rho\frac{\partial(\tilde{r}^2)}{\partial r}+\rho\frac{\partial(\tilde{r}\tilde{z})}{\partial z}+\tilde{r}\tilde{z}\frac{\partial\rho}{\partial z}=0$$
$$\rho\frac{\partial(\tilde{r}\tilde{\omega})}{\partial r}+\rho\frac{\partial(\tilde{\omega}\tilde{z})}{\partial z}+\tilde{\omega}\tilde{z}\frac{\partial\rho}{\partial z}=0 \tag{34a}$$
$$\rho\frac{\partial(\tilde{r}\tilde{z})}{\partial r}+\rho\frac{\partial(\tilde{z}^2)}{\partial z}+\tilde{z}^2\frac{\partial\rho}{\partial z}=\rho g\delta_{zn}$$

and

$$\tilde{r}\frac{\partial\rho}{\partial r}+\tilde{\omega}\frac{\partial\rho}{\partial\varphi}+\tilde{z}\frac{\partial\rho}{\partial z}=0 \quad (34b)$$

we will have enough parameters, ϑ, α, β, γ, χ, δ, κ, to solve the same number of linear equations as the three equations (34a) will split into six relations: three additional equtions should express the equality of the exponnetial parameters appearing there.

Note that such the flows can occur effectively for the Reynolds numbers above the critical value.

4. Conclusions

Our consideration on the continuum theories used in seismology clearly indicate that the global seismic recording network based on the 3-components seismometer stations is quite insufficient to trace the strain waves, especially the long period waves spreading inside a whole Earth interior. To confirm this important need of a global strain and rotation seismic network we should mention that in some seismically very active regions the strain-meter system and rotation sensor system already exist.

Following this new approach to solid state continuum theory, we presented a new asymmetric molecular theory for fluids with a hope that this new approach opens a way to consider a mechanical model of vortex and turbulence motions; here, we have confined our considerations to processes under a constant thermodynamical condition.

5. References

Cosserat, E., and F. Cosserat (1909), *Théorie des Corps Déformables*, A. Hermann et Fils, Paris.

Droste, Z., and R. Teisseyre (1959), The mechanism of earthquakes according to dislocation theory, *Sci. Rep. Tohoku Univ., Ser. 5, Geophys.* 11, 1, 55-71.

Eringen, A.C. (1999), *Microcontinuum Field Theories*, Springer Verlag, Berlin.

Eshelby, J.D., F.C. Frank, and F.R.N. Nabarro (1951), The equilibrium of linear arrays of dislocations, *Philos. Mag.* 42, 351-364.

Eringen, A.C. (2001), *Microcontinuum Field Theories*, Springer Verlag, Berlin.

Eshelby, J.D., F.C. Frank, and F.R.N. Nabarro (1951), The equilibrium of linear arrays of dislocations, *Philos. Mag.* 42, 351-364.

Landau, L.D. and Lifshitz, J.M.. 1959, Fluid mechanics (Theoretical Physics, v. 6), (translated from Russian by J.B. Sykes and W.H. Reid), London, Pergamon Press, pp 536.

Landau, L.D. and Lifszyc, J.M.. 2009, Hydrodynamika, (in Polish) Wydawnictwo Naukowe PWN Warszawa, pp 671.

Lee W.H.K., Celebi, M., Todorovska M.I. and Igel, H., 2009, Introduction to Special issue on Rotational Seismology and Engineering Applications, Bull. Seismol. Soc. Am. 62, no. 2B 945-957.

Mindlin R.D,1965, On the equations of elastic materials with microstructure, Int. J. Solid Struct., vol 1(1), p. 73

Nowacki, W. 1986, *Theory of Asymmetric Elasticity*, PWN, Warszawa and Pergamon Press.

Peach, M., and J.S. Koehler (1950), The forces exerted on dislocations and the stress fields produced by them, *Phys. Rev.* 80, 436-439.

Teisseyre, R. (1973), Earthquake processes in a micromorphic continuum, *Pure Appl. Geophys.* 102, 15-28.

Teisseyre, R. (1974), Symmetric micromorphic continuum: wave propagation, point source solutions and some applications to earthquake processes. In: P. Thoft-Christensen (ed.), *Continuum Mechanics Aspects of Geodynamics and Rock Fracture Mechanics*, 201-244.

Teisseyre, R. (1985), New earthquake rebound theory, *Phys. Earth Planet. Inter.* 39, 1, 1-4.

Teisseyre, R. (2009), Tutorial on new development in physics of rotation motions, *Bull. Seismol. Soc. Am.* 99, 2B, 1028-1039.

Teisseyre R, 2010, Fluid Theory with Asymmetric Molecular Stresses: Difference between Vorticity and Spin Equations, Acta Geophys. vol 58, 6, 1056-1071T

Teisseyre, R. (2011), Why rotation seismology: Confrontation between classic and asymmetric theories, *Bull. Seismol. Soc. Am.* 101, 4 , 1683-1691.

Teisseyre, R., and M. Górski (2009), Fundamental deformations in asymmetric continuum: Motions and fracturing, *Bull. Seismol. Soc. Am.* 99, 2B, 1028-1039.